# International Series in Operations Research & Management Science

Volume 273

More information about this series at http://www.springer.com/series/6161

Turan Paksoy • Gerhard-Wilhelm Weber •
Sandra Huber
Editors

# Lean and Green Supply Chain Management

## Optimization Models and Algorithms

Springer

*Editors*
Turan Paksoy
Department of Industrial Engineering
Konya Technical University
Konya, Turkey

Gerhard-Wilhelm Weber
Poznan University of Technology
Poznan, Poland

Sandra Huber
Department of Logistics Management
Helmut Schmidt University
Hamburg, Germany

ISSN 0884-8289 ISSN 2214-7934 (electronic)
International Series in Operations Research & Management Science
ISBN 978-3-319-97510-8 ISBN 978-3-319-97511-5 (eBook)
https://doi.org/10.1007/978-3-319-97511-5

Library of Congress Control Number: 2018959616



This Springer imprint is published by the registered company Springer Nature Switzerland AG
The registered company address is: Gewerbestrasse 11, 6330 Cham, Switzerland

# Foreword

Recent studies demonstrate that (Garza-Reyes, J. A., et al. 2018, IJPE, 200, 170–180) there is a strong relationship between lean and green, and the integrated use of lean and green tools and methods creates a synergy in organizations and increases both operational and environmental performance simultaneously. This is not surprising! Because waste reduction (materials, water, energy etc.), which is one of the main objectives of lean philosophy, is also directly related to green philosophy.

Lean and green manufacturing philosophy and techniques facilitate companies to improve their business performance in terms of reducing manufacturing lead times and costs and improving on product quality and on-time delivery performance while becoming greener by minimizing material usage and waste and applying recycling and reuse.

Thus, while aiming at elimination of wastes, both lean manufacturing and green manufacturing work for a common purpose: sustainability. Lean approach proposes to eliminate the traditional "seven deadly wastes" (*transportation, inventory, motion, waits, overproduction, overprocess, and defects*) for a more efficient workplace and achieving operational excellence. Meanwhile, green approach suggests elimination of environmental wastes, minimization of usage of earth sources, and reuse and recycle of products and materials.

In other words, lean and green SCM supports sustainability in accordance with the triple bottom line approach:

1. Economic impact; by increasing productivity
2. Social impact; by generating human-centric workplaces
3. Environmental impact; by eliminating wastes and reducing material usage.

In this book, *Lean and Green Supply Chain Management: Optimization Models and Algorithms*, I and my colleagues try to draw a comprehensive picture of lean and green supply chain management. The purpose of this book is to deeply inquire the synergy between lean and green issues and comprehensively expose the emerging-cutting edge issues related to original, innovative, and novel real-world applications of the emerging area of the lean and green supply chain management. The book chapters focus on new development and contribution of current research to the body of the knowledge in the areas of lean and green supply chain management, supported by modeling, operational research, and optimization. All chapters in the book can be grouped under four main parts: *Lean and Green Supply Chain Optimization*, *Lean and Green Supplier Selection*, *Lean and Green Vehicle/Fleet Management*, and *Lean and Green Operations Management*.

The contributing authors of this book are experienced scientists and practitioners from different countries; they use and further refine the model-based methods of operations research, algorithms, and heuristics.

To all the authors of these valuable chapters, I extend my appreciation and gratitude for having shared their devotion, knowledge, and vision with the academic community and mankind. I also would like to thank all the referees for their valuable time and great effort in reviewing all the book chapters here.

I am very thankful to *Springer* and its publishing team. I cordially very much appreciate the support of Christian Rauscher, *Senior Editor, Business, Operations Research & Information Systems/Springer*, for having provided the chance for this wonderful work. Also, I want to express my special thank to my precious colleague and coeditor Professor Gerhard Wilhelm Weber—*Willi*—for his marvelous support and encouragement in making this book project very successful. Last but not least, I would like to thank my outstanding colleague and my coeditor Sandra Huber for her rigorous and disciplined work here. I express my gratitude to them for having ensured a premium book of a high standard academic, applied, and social importance.

Now, I wish all of you a lot of joy in reading this interesting book, and I hope that a great benefit is gained from it both personally and globally!

Konya Technical University, Turkey — Turan Paksoy
June 2018

# Preface

The dynamics of *supply chain management* (*SCM*) has changed and accelerated over the years; new paradigms have been added into SCM to respond to changes based on the increasing economic, environmental, and social concerns and pressures. In today's global and competitive environment, *lean* and *green* practices have become most important factors to enhance the companies' performance. *Lean* and *green* manufacturing philosophy and techniques facilitate companies to improve their business performance in terms of reducing manufacturing lead times and costs and to improve on-time delivery and product quality, while becoming greener by minimizing material usage and waste and applying recycling and reuse. All these *lean* and *green* activities increase productivity, reduce carbon footprint, and enhance customer satisfaction, which in turn make companies sustainable and competitive.

In this context, our book *Lean and Green Supply Chain Management: Optimization Models and Algorithms* aims to give the reader some of the most recent research dealing with *lean and green supply chain management*. Furthermore, the book presents a comprehensive reference resource, including novel optimization algorithms to achieve lean and green supply chains that enable and enhance business performance as well as environmental performance simultaneously.

Our book includes contributions and reviews of the latest developments in optimization and mathematical models, exact, approximate, and hybrid methods, and their applications for lean and green supply chains, including:

- *A detailed literature surveys on models and algorithms developed for lean and green supply chains*
- *Case studies and experimental applications on lean and green supply chain network design and modeling*
- *Supply chain network design and optimization considering customer behaviors and environment*
- *Sustainable supply chain management considering lean, green, agile/responsive and resilient networks*
- *Applications of fuzzy sets theory and extensions in lean and green supply chain management*

- *Multi-criteria decision making for lean and green supply chain management*
- *Risk management in lean and green supply chains*
- *Supplier selection with lean and green considerations*
- *Lean and green vehicle/fleet management*
- *Lean and green operations management*

The book gathers 9 chapters, which are prepared by experts from various countries of the world on the lean and green supply chain management and on its theory, methods, and application. Short descriptions of all chapters are given subsequently, within some grouping by subjects.

## Lean and Green Supply Chain Optimization

**Chapter 1,** named as *"Lean and Green Supply Chain Management: A Comprehensive Review,"* by *Batuhan Eren Engin, Maren Martens, and Turan Paksoy*, focuses on a comprehensive literature survey on lean and green supply chain management. In total, 41 articles, published between 2000 and 2017, are analyzed by the authors. One of the main goals is to classify the existing literature and analyze future directions in this research area. First, lean and green practices are described for important contributions. Second, measures and indicators are established to evaluate the impact of lean and green methods on the overall supply chain performance.

**Chapter 2,** entitled *"A New Model for Lean and Green Closed-Loop Supply Chain Optimization,"* by *Turan Paksoy, Ahmet Çalik, Alexander Kumpf, and Gerhard-Wilhelm Weber*, is dedicated to a new model for a lean and green closed loop automotive supply chain in order to provide an optimal distribution plan for several periods as well as for several echelons in the network. To solve such problems, six different objectives are included, such as the minimization of transportation cost, late deliveries, etc. One of the main contributions is the usage of a fuzzy weighted additive method, which employs a fuzzy analytic hierarchy process for the determination of the weights. In addition, the investigated framework can be utilized to understand the drivers in an integrated model.

## Lean and Green Supplier Selection

**Chapter 3**, called as *"Risk Management in Lean & Green Supply Chain: A Novel Fuzzy Linguistic Risk Assessment Approach"* by *Turan Paksoy, Ahmet Çalik, Abdullah Yildizbaşı, and Sandra Huber*, proposes a novel fuzzy linguistic risk assessment approach to evaluate risk management in lean and green supply chains. A solution approach with four stages is presented that, e.g., includes the determination of weights by the decision maker, the assignment of suppliers to risk groups, etc. Furthermore, a numerical investigation is carried out to show the applicability of the proposed method. In order to present managerial insights, the authors provide action plans which correspond to the different risk levels.

**Chapter 4**, named *"A New Multi Objective Linear Programming Model for Lean and Green Supplier Selection with Fuzzy TOPSIS,"* by *Belkız Torğul and Turan Paksoy*, examines the assessment of a multi-objective linear programming model for a supplier selection problem. Apart from financial factors, quantitative as well as qualitative factors that incorporate lean and green supply chain management are included in the model. The authors use a real-world case study involving the delivery of plastic parts in the automotive industry in Turkey to assess the performance and effectiveness of the proposed model.

## Lean and Green Vehicle/Fleet Management

**Chapter 5**, entitled *"The Impact of Routing on $CO_2$ Emissions at a Retail Grocery Store Chain: A GIS-based Solution Approach,"* by *ÇağrıKoç, Mehmet Erbaş, and Eren Özceylan*, introduces a variant of a pollution-routing problem in order to analyze a retail grocery store chain in Turkey. Thereby, special attention is paid to fuel consumption and $CO_2$ emissions of the vehicles. The presented tabu search heuristic is able to provide managerial insights by investigating the trade-off between the number of vehicles and the shop demands. Furthermore, it is beneficial to gain further insights by visualizing the data and the solutions on a graphical user interface.

**Chapter 6**, named as *"A Simulated Annealing Algorithm Based Solution Method for a Green Vehicle Routing Problem with Fuel Consumption,"* by *Kenan Karagul, Yusuf Sahin, Erdal Aydemir, and Aykut Oral*, presents the development of a simulated annealing algorithm that provides decision support for a green vehicle routing problem with fuel consumption. Thereby, the total vehicle weight, the changes of the weight, and the distances have to be taken into account in a heuristic. The methodology is illustrated for selected test instances, and statistical analysis techniques are applied to compare the different methods.

**Chapter 7**, called *"Development of a Web-Based Decision Support System for Strategic and Tactical Sustainable Fleet Management Problems in Intermodal Transportation Networks,"* by *Adil Baykasoğlu, Kemal Subulan, A. Serdar Taşan, Nurhan Dudaklı, Murat Turan, Erdin Çelik, and Özgür Ülker*, is devoted to a web-based decision support system for sustainable fleet management in intermodal transportation networks. The promotion of interactive decision support is of major importance in sustainable fleet planning. A decision maker seeks to minimize the overall transportation costs, the total transit times, and the $CO_2$ emissions through the introduction of strategic as well as tactical fleet management. In this context, a decision support system is introduced to foster an international logistics company of Turkey.

## Lean and Green Operations Management

**Chapter 8**, given the name *"Integrated Production Scheduling and Distribution Planning with Time Windows,"* by *Saadettin Erhan Kesen and Tolga Bektaş*, tackles an integrated production scheduling and distribution planning problem with time windows. Mainly, decisions must be made on two decision levels: the orders are processed through a set of identical parallel machines, and then, they must be delivered with a fleet of vehicles taking into consideration the time windows of the customers. After presenting a mathematical model, an in-depth analysis is given on the computational results.

**Chapter 9**, entitled *"Achieving Shojinka by Integrated Balancing of Multiple Straight Lines with Resource Dependent Task Times,"* by *Yakup Atasagun, Yakup Kara, and Gözde Can Atasagun*, addresses a mathematical model for a problem called multiple straight lines with resource-dependent task times. The goal is to incorporate the concept of Shojinka in order to attain flexibility on the number of workers. An illustrative example is given to validate the introduced model. Furthermore, the experimental study shows the benefits of an integrated approach of multiple assembly lines, such as, e.g., an average total cost improvement of around 23%.

With this book, we editors pursue various **aims**, stated as follows:

- *To provide a **compendium*** for all researchers who could become able to familiarize with those emerging research subjects that are of common practical and methodological interest to representatives of management, engineering, economics, and environmental protection;
- *To offer an **encyclopedia** and a **dictionary*** that will enable researchers and practitioners to access in a fast way the key notions of their domains and, via suitable methods and references, to proceed and to branch further on the way to treat their own, their institutes', or institutions' problems and challenges;
- *To create an **atmosphere*** in which the new generation of young scientists can advance, grow, and mature when reading this book and develop on their way of becoming recognized scholars;
- *To initiate a **momentum*** of encouragement and excitement, which could strongly foster interdisciplinary research, the ***solution*** of striking real-world problems, and a creative and fruitful cooperation and synergy among experts from all over the globe;
- *To further introduce* ***Springer Verlag*** as a *premium publishing house* on newest and most timely fields of scientific investigation and application and as a *center of excellence*.

Our work provides an excellent reference to graduate, postgraduate students, decision makers, and researchers in private domains, universities, industries, and governmental and non-governmental organizations, in the areas of various operational research, engineering, management, business, and finance, wherever one has to represent and solve uncertainty-affected practical and real-world problems.

In the future, mathematicians, statisticians, game theorist, physicist, chemists, and computer scientists, but also biologists, neuroscientists, social scientists, and representatives of the humanities, are warmly welcome to enter into this discourse and join the collaboration for reaching even more advanced and sustainable solutions. It is well accepted that predictability in uncertain environments is a core request and an issue in any field of engineering, science, and management. In this respect, this book opens a very new perspective; what is more, it has the promise to become very significant in academia and practice and very important for humankind!

Konya, Turkey — Turan Paksoy
Poznan, Poland — Gerhard-Wilhelm Weber
Hamburg, Germany — Sandra Huber

# Acknowledgments

We, as the editors, express our cordial gratitude to all the authors for their important contributions and to the reviewers for their valuable service, care, and helpful comments. Eventually, we are very thankful to prestigious *Springer-Verlag* for the precious opportunity of preparing this compendium, especially to *Christian Rauscher* and *Barbara Bethke*, who were with us, guided us and supported us from the earliest stage of preparation of this book until today.

# Organization

## Program Chairs

| | |
|---|---|
| Turan Paksoy | Konya Technical University, Turkey |

## Program Committee

| | |
|---|---|
| Gerhard-Wilhelm Weber | Poznan University of Technology, Poland |
| Sandra Huber | Helmut-Schmidt University, Germany |

# Contents

# About the Editors

**Turan Paksoy** received his B.S. degree in industrial engineering from Gazi University, M.S. degree in industrial engineering, and PhD degree in production management from Selçuk University in 2004. He received the associate's degree in production management in 2008 from National Academic Council of Turkey. In 2015, he was promoted to full professor at Selçuk University. Dr. Paksoy currently is working as a faculty member at Konya Technical University, Department of Industrial Engineering. He attended London Kingston University as visiting professor in 2015. Dr. Paksoy also has the affiliation of founding director of KTO Karatay University Lean & Green SCM Lab. He has been honored with being chosen for "Global Sustainable Leaders" by U.S. Green Chamber of Commerce in 2018. Dr. Paksoy has over 200 published papers (about 35 papers in SCI indexed journals). His papers appear frequently in reputable and internationally recognized journals such as *Computers and Industrial Engineering, European Journal of Operational Research, Transportation Research, and International Journal of Production Research*. His academic studies have been cited about 900 times in SCI journals. Dr. Paksoy current academic interest includes lean manufacturing, green logistics, closed-loop supply chain management, and fuzzy decision making.

**Gerhard-Wilhelm Weber** is a Professor at Poznan University of Technology, Poznan, Poland, at the Faculty of Engineering Management, in the Chair of Marketing and Economic Engineering. His research is on mathematics, OR (operational research), finance, economics, optimization and control, neuro- and biosciences, data mining, analytics, AI, and education and development. He received diploma and doctorate in mathematics and economics/business administration at RWTH Aachen and habilitation at TU Darmstadt (Germany). He held Professorships at the University of Cologne and TU Chemnitz, Germany. At the Institute of Applied Mathematics, METU, Ankara, Turkey, he was a Professor in the programs of financial mathematics and scientific computing and has been a member of further METU graduate schools, institutes, and departments. He has different affiliations at Universities of Siegen (Germany), Federation University (Ballarat, Australia), University of Aveiro (Portugal), University of North Sumatra (Medan, Indonesia),

Malaysia University of Technology, Chinese University of Hong Kong, KTO Karatay University (Konya, Turkey),; at EURO (Association of European OR Societies) where he is "Advisor to EURO Conferences," at IFORS (International Federation of OR Societies), where he is member in many national OR societies and working groups, at POP (Pacific Optimization Research Activity Group), etc. He has publications in a variety of fields, he is a member in numerous editorial boards, and he has supervised many young people.

**Sandra Huber** is a member of the Logistics-Management Department at the Helmut-Schmidt-University, University of the Federal Armed Forces in Hamburg, Germany, since 2013. Prior to this position, she received her M.Sc. of Science at the Helmut-Schmidt-University. In 2016, she was awarded with her Ph.D. degree with a thesis on "Interactive multiobjective optimization in supply chain management—local search strategies for rich vehicle routing problems." Her research area is in decision support and supply chain management. She is particularly interested in multi-objective combinatorial optimization, interactive approaches, heuristics for vehicle routing problems, and the simulation of preference information. She visited the research institute CIRRELT in Montréal, Canada, as well as the Université Bretagne Sud.

# Lean and Green Supply Chain Management: A Comprehensive Review

**Batuhan Eren Engin, Maren Martens, and Turan Paksoy**

## 1 Introduction

There is a challenging competition between the companies in the business market in terms of cost, price, quality, efficiency and even in social responsibility for brand perception. One way to gain an advantage is to decrease any cost incurred in manufacturing and supply chain (SC) without detriment to the quality of the product. By integrating lean and green thinking into SC management (SCM), companies can take advantages of significant cost savings, improved quality, reduced costs and risk, and improved services across SC by identifying, reducing and eliminating unnecessary usage of resources or processes that generate waste (Mollenkopf et al. 2010; Wiese et al. 2015).

On the other hand, while eliminating and minimizing waste across the SC, companies are required to monitor their environmental impacts and comply with the regulations regarding the environment. This being sad, Green SC practices help companies to increase environmental efficiency. A report by United States Environmental Protection Agency (2000) investigated firms in which lean and green principles were being implemented and reported that these firms have seen significant savings. Besides these benefits, going green (i.e. green thinking) can also improve the reputation of a business, as more and more people are becoming environmentally aware, and thus prefer to purchase from corporations with green thinking.

B. E. Engin (✉) · T. Paksoy
Department of Industrial Engineering, Konya Technical University, Konya, Turkey

M. Martens
Faculty of Business Management, University of Applied Sciences, Landshut, Germany
e-mail: maren.martens@haw-landshut.de



T. Paksoy et al. (eds.), *Lean and Green Supply Chain Management*, International Series in Operations Research & Management Science 273,
https://doi.org/10.1007/978-3-319-97511-5_1

The growing importance of Lean and Green SC is driven mainly by the escalating deterioration of the environment, e.g. increasing levels of pollution and waste, and diminishing raw material resources (Srivastava 2007). Lean and green activities are gaining increasing interest among researchers and practitioners of operations and supply chains. Based on this growing attention, it is necessary to have a comprehensive review of the current state of the research in this emerging field to help researchers and practitioners focus on future directions. Recently, no review papers could be found in this field that had undertaken a systematic classified analysis of research papers. In order to fill this gap, this paper aims at systematically collecting (reviewing) and analyzing the existing works that added lean and green practices to supply chain management (SCM). Considering this, the main research questions addressed in this review are:

- *What are the most frequent issues in studies dealing with lean and green paradigms in the context of supply chain?*
- *What could be future directions for researchers and practitioners on the lean-green SCM area?*

The remainder of the paper is structured as follows: Section 2 digs into the details of lean and green paradigms. Section 3 covers some earlier review/partial-review papers and clarifies the distinction of our paper and our research methodology are discussed. Section 4 includes the detailed analyses and classifications of reviewed papers are discussed. The current gaps analysis results, future research directions and conclusion remarks are presented and discussed in Sect. 5.

## 2 Lean and Green Paradigms in the Context of Supply Chain

### *2.1 Lean Paradigm*

The concept of lean refers to the elimination of waste and is the basic philosophy that originated as part of Toyota Production Systems (Garza-Reyes 2015). The key in lean manufacturing is to identify the activities that do not directly add any value to the system or product and eliminate these unnecessary activities to reduce cost (Thanki et al. 2016). These non-value activities are the ones for which the customer would not be willing to pay. Ohno (1988) defined seven determinants that cause waste, i.e. overproduction, waiting, transportation, inappropriate processing, unnecessary inventory, unnecessary motion and defects in manufacturing. The definitions of these parameters, or mudas as referred to in lean philosophy, in the context of manufacturing and their equivalent activities in the context of logistics are given in Table 1.

Lean supply chain (LSC) is defined as "as minimizing waste in the downstream SC, while making the right product available to the end customer at the right time and location" by Reichhart and Holweg (2007). LSC represents a strategy based

**Table 1** Definition of mudas in SC context

| Muda[a] | Definition in the context of manufacturing[b] | Equivalent activity for corresponding muda, in the context of SCM |
|---|---|---|
| Overproduction | Manufacturing an item before it is actually required | Shipping an item before it is actually needed |
| Waiting | Goods are not moving, or waiting to be processed | Waiting before or in loading/unloading, waiting for drivers/operators |
| Transportation | Transporting the products between places/processes | Transporting the products between places/processes |
| Inappropriate processing | Using high precision equipment when simpler tools would be sufficient | Empty backhauling, shipping half-empty large size trucks instead of smaller truck |
| Unnecessary inventory | Keeping excess inventory than what is needed | Keeping excess inventory than what is needed |
| Unnecessary motion | Any movement related to ergonomics, bending, stretching, walking, lifting, reaching, rotating a piece | Re-picking, re-packaging/unpacking, unoptimized routes for trucks |
| Defects | Item resulting in control, rework or scrap due to low quality, which is a tremendous cost to organizations | Deterioration of goods, miscommunication/lack of communication with suppliers and customers resulting in shipping of goods to a wrong place |

Sources: [a]Ohno (1988), [b]Rachid et al. (2017)

on cost reduction and flexibility, focused on process improvements, through the reduction or elimination of non-value-adding operations related to excess time, labor, equipment, space, and inventories across the SC (H. Carvalho et al. 2010; Corbett and Klassen 2006).

By a LSC that is free from waste and non-value adding activities, companies can save a considerable amount of capital for valuable investments, and thereby increase their efficiency and competitiveness. Several lean practices used in the literature are given in Table 6.

## 2.2 *Green Paradigm*

As to environmental degradation that raises concerns among the governments, legislations and protective regulations have been put into action to make companies reduce their environmental footprint. This is where green paradigm emerged from as a philosophy. Since then, green practices in supply chains received great interest from practitioners and academicians due to the pressure of various stakeholders, i.e. customers, government and law enforcement bodies to reduce the environmental impact of supply chains. Green supply chain (GSC) is formally defined as "the coordination of the supply chain in a form that integrates environmental concerns and considers inter-organizational activities" by Machado and Davim (2017) and "integrating environmental thinking into supply-chain management, including product design, material sourcing and selection, manufacturing processes, delivery of the final product to the consumers as well as end-life management of the product after its useful life" by Srivastava (2007). Cost-effective planning of distribution routes in coordination with the suppliers and customers, capacity and efficiency maximization throughout the chain, the elimination of expedited and extra shipping costs and sustainable initiatives, along with the reduction of the impact of logistics processes on the environment are the components of a green supply chain (Machado and Davim 2017). Green supply chain management practices can be listed as energy efficiency, reduction of greenhouse gases emissions, water conservation or processing, waste reduction, reduced packaging/increased use of biodegradable packaging, product and packaging re-cycling/re-use and other green procurement practices (Hervani et al. 2005; Machado and Davim 2017). In addition to those practices, green operations involve operational aspects related to network design (collection; inspection/sorting; pre-processing), green manufacturing and remanufacturing (production planning and scheduling; inventory management, product and material recovery) and waste management (source reduction; pollution prevention and disposal) (Srivastava 2007). Some of these topics are also related to lean paradigms and therefore we allocate a subsection to unfold this subject reviewing the studies on the relationship between lean and green paradigms.

## 2.3 Lean and Green Supply Chain

While there are separate streams of research on green and lean SC, authors have started to address the intersection of these strategic initiatives in the supply chain context. Corporations have started to realize that the implementation of lean, green and social practices have positive impacts on companies' triple bottom line performance (economic, environmental and social), and even stronger impacts with the integration of these practices (Wu et al. 2015). This surely motivates firms to adopt some combination of lean and green strategies into SC. Toyota is an example of a firm that has successfully integrated green and lean strategies into its SC and many other firms, such as Wal-Mart, General Motors, Andersen Corporation, Intel, etc., have seen economic and environmental benefits by adopting lean and green SCs (United States Environmental Protection Agency 2000). Besides these benefits, it is recognized that Wal-Mart has grown respect within the market (Friedman 2008). The initial purpose of adopting lean and green initiatives across the SC is to increase the effectiveness of the SC by reducing numerous types of waste, using lean techniques and reducing environmental impact through the utilization of green practices. Such practices extend beyond the organization to include inbound and outbound logistics along with reverse logistics (Kleindorfer et al. 2005).

Although lean and green practices are often seen compatible because of their intrinsic aim at reducing waste, some lean strategies such as just-in-time delivery may require increased transportation, packaging, and handling, which contradicts the green approach. Moreover, high initial cost of adopting green initiatives may hinder the companies to adopt it, yet it can be justified and compensated by the savings obtained by the elimination of non-value-adding activities through lean initiatives across the SC. Also, long-term benefits of implementing green practices will eventually start to materialize in terms of savings obtained from reduced energy consumption, recycling and recovering of materials, improved company image and market share and reduced waste treatment/discharge cost. Thus, companies will recognize synergetic benefits of implementing lean and green initiatives across the SC.

## 2.4 Literature Review

First, some review studies published in the domain of GSC, or Lean SC, or both were investigated to express the need for this study. Earlier works and reviews have a limited focus and narrow perspective. Table 2 illustrates these review studies, their area, scope, year and the number of papers they reviewed.

Table 2 contains review articles involving either "green supply chain" or "lean supply chain" or both in their title, abstract or keywords. In the domain of Green SC, Alzaman (2014) enumerates literature within the domain of GSC modelling, themes and modelling approaches and attributes. Luthra et al. (2014) have examined the existing studies from several perspectives, i.e. key themes of GSCM, methodology/techniques used in research, type of models developed, research methods used, type of industry considered in research, GSCM implementation enablers/barriers and performance measurement in GSCM. Srivastava (2007) classified the literature on GSCM on the basis of the problem context in major influential areas of supply chains, and also methodology and approaches have been used as other classification categories. They have mapped various mathematical tools/techniques used in literature vis-à-vis the contexts of GSCM.

However, the number of review papers in the literature on Lean-Green SCM is insufficient. Five review articles have been found in the literature on Lean-Green SC, nevertheless, neither of them shared the same focus with this article. Li and Found (2016) summarizes the state-of-art researches on Product Service System, Lean and Green SCM and the connections between them. Garza-Reyes (2015) aimed at systematically collecting and analyzing the existing studies on lean, with particular interest in its application and linkage with environmental organizational aspects (in other words, green paradigms). Garza-Reyes (2015) identifies and structures, through a concept map, six main research streams that comprise both conceptual and empirical research conducted within the context of various organisational

**Table 2** Characteristics of earlier review/partial review studies

| Paper | Area | Scope and highlights | Year | Number of papers |
|---|---|---|---|---|
| Srivastava (2007) | GSCM | Problem context, mathematical modelling | 1990–2007 | 227 |
| Sarkis et al. (2011) | GSCM | Organizational theories | Until 2010 | – |
| Duarte and Machado (2011) | LARG | LARG manufacturing paradigms in SCM | 2000–2009 | – |
| Ugochukwu et al. (2012) | LSC | Lean applications in SC | 1990–2012 | 40 |
| Ahi and Searcy (2013) | GSC | Definitions | Until 2013 | 22 |
| Xu et al. (2013) | GSCM | Concept and connotation, operations, performance evaluation and the applications | Until 2012 | – |
| Duarte and Cruz-Machado (2013) | Lean-Green | Modelling lean and green: a review from business models | Until 2013 | – |
| Alzaman (2014) | GSC | Modelling, and its' approaches and attributes | Until 2014 | 44 |
| Luthra et al. (2014) | GSCM | Whole area in GSC, techniques/methodologies applied, models developed, research methods and industry | Until 2014 | – |
| Sulistio and Rini (2015) | GSC | Multi-stage GSC, Management system, practices, organizational performance and barriers | Until 2015 | 50 |
| Fahimnia et al. (2015) | GSC | Performance measurement, supplier selection/evaluation, analytical modeling efforts, and some others with broader areas of focus. | Until 2014 | – |
| Malviya and Kant (2015) | GSCM | Whole area in GSC | 1998–2003 | 177 |
| Dawei et al. (2015) | GSCM | Definition, green concepts, Green supplier and selection | Until 2015 | – |
| Garza-Reyes (2015) | Lean-Green | Existing studies on Lean, with particular interest on its application and linkage with green paradigm | 1997–2014 | 59 |
| Hallam and Contreras (2016) | Lean-Green | Integrating lean and green management | Until 2016 | – |
| Li and Found (2016) | LGSC | Product-Service System SC | Until 2016 | – |
| Our study | LGSC | All areas in Lean-Green SC | 2000–2017 | 41 |

functions and industrial sectors, in which the six main research streams were compatibility between lean-green, integration of lean and green as a consolidated approach, integration of lean-green with other approaches, proposal of a lean-green performance assessment method/indicator and lastly, lean-green impact on organisational performance and lean-green research or empirical application. The journal articles and conference papers they reviewed have included the deployment of lean and green initiatives at either supply chains or operation/processes. Duarte and Machado (2011) examined how different business models can contribute to modelling a lean and green approach for an organization and its supply chain. Their study reveals a number of categories that are common in most business models that help to achieve a successful lean-green transformation. They developed guidelines to model a lean-green organization by applying specific principles and tools of a lean and green culture. As the latest review article on lean and green paradigms, Hallam and Contreras (2016) presented a literature review of peer-reviewed journal articles investigating the relationship between lean and green management in light of developing an integrated management model. The keywords they use are "lean manufacturing/management", "green manufacturing/management", "lean and green", "lean and environment" and "pollution prevention".

As discussed earlier, no comprehensive review study can be found in the literature on lean and green SC, which investigates state-of-the-art recently published papers focusing on the implementation of lean and green practices in the context of SCM. As concurrent deployment of lean and green practices in SCM has been devoted increasing attention nowadays, it becomes a necessity to present a literature review study based on recent publications in the Lean-Green SCM. Consequently, the literature review by Garza-Reyes (2015) analyzed journal papers published in 1997–2014 on lean-green, also including "articles in press" that were published later in 2015. In the last two years, there have been published as many articles on Lean-Green SCM as there have been published before 2015. Also, literature review by Duarte and Machado (2011) focused more on lean SC, green SC, agile SC and resilient SC separately, since the number of papers combining more than two paradigms were very rare at that time. Based on this information, in the current paper, we aim to cover the latest/recent publications in the area to differentiate this literature review from Garza-Reyes (2015) and the others. The last line of Table 2 shows this paper covering the presented gap of the literature.

## 3 Research Methodology

An efficient literature review is a method to help identify, synthesize and evaluate the existing studies published in the literature. The aim of conducting a literature review is to enable the researchers and practitioners to assess the existing intellectual works and to specify a research question to develop the existing body of knowledge more deeply (Tranfield et al. 2003). In order to provide a high-quality analysis, only peer-reviewed journal articles are included in the scope, which represents scientifically validated knowledge (Moher et al. 2009). For material collection, a method called Preferred Reporting Items for Systematic Reviews and Meta-Analyses (PRISMA) described by Moher et al. (2009) is adopted to obtain the most relevant peer-reviewed journal articles systematically, using a checklist. This section involves content analysis and description of research methodology.

## *3.1 Material Collection*

The following electronic databases were searched for articles on lean and green paradigms in the supply chain: Emerald (emeraldinsight.com), Elsevier (sciencedirect.com), Web of Science (webofknowledge.com), Scopus (scopus.com), Springer Link (springerlink.com) and Google Scholar (scholar.google.com). The study was conducted in July 2017, and only journal articles published from 2000 to 2017 that discuss lean/green paradigms in relation to the supply chain were selected. We further delimited the material (research paper) to be collected as per our scope, so the reports, conference papers, books or other kind of documents were excluded from the scope of this research, as Sauders et al. (2003) stated that journal papers provide the most useful and reliable information.

The initial search for the articles was done using Google-scholar search engine using the keywords "lean", "green" and "supply" or "chain" in the titles, sorted by relevance. It is worth mentioning that searching with using exact phrases, such as "supply chain" or "supply chain management", might accelerate the phase of acquiring the articles, yet it might cause overlooking some publications on the aimed topic, for instance, "Improving Environmental and Economic Performance in the Food Chain; the Lean and Green Paradigm" by Zokaei et al. (2015) (which was excluded from the scope of this study for being a book chapter). Also, the keyword of LARG (acronym for "lean, agile, resilient and green") is also used in the search to not overlook any publications, such as Figueira et al. (2012). An individual search on each keyword resulted in hundreds of articles, most of which were not related to the scope of this study. Combinations of these keywords are used to narrow the search (Table 3).

The narrowed search yielded in 183 publications. In some cases, some of the search runs resulted in the same articles being found on different databases. However, this systematic search and selection approach was necessary to ensure the reliability of the process of finding and selecting articles. Out of 183, 35 were rejected as duplicates. Limiting the scope of this study to peer-reviewed journal articles, 44 articles were excluded for being either conference proceedings, reviews, book chapters, or master theses. In the evaluation process of selecting related state-of-the-art papers in this area, all the articles were checked manually and carefully by their abstracts to maintain the inner-consistency of the literature review by ensuring that those were selected that address lean-green paradigms in relation to SCM. By doing so, 63 more articles are considered not applicable to our study. Thus we focused our detailed review on the remaining 41 relevant peer-reviewed journal articles published between 2000 and 2017, which are given in Table 5.

**Table 3** Summary of methodology used

| | |
|---|---|
| Unit of analysis | Relevant articles published on Lean and Green SCM are selected and reviewed. The reports/conference papers/books or other kind of documents were excluded. |
| Types of studies | Qualitative and quantitative, conceptual framework, survey/questionnaire/interview, case study, mathematical model, simulation |
| Search period | 2000 to July 2017 |
| Search engines | Emerald, Science Direct, Web of Science, Scopus, SpringerLink and Google Scholar |
| Keywords used in searches | Lean<br>Green<br>Lean and Green<br>Supply Chain<br>Chain<br>Supply chain management<br>Leanness<br>Greenness<br>LARG (Lean, Agile, Resilient and Green) |
| Total number of articles | 41 |

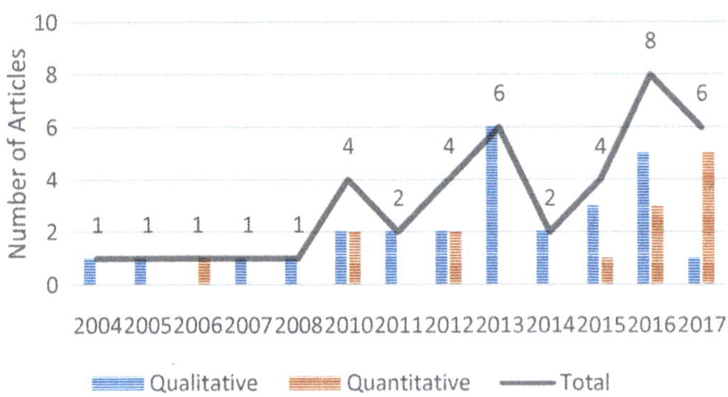


**Fig. 1** Distribution of publications per year across the period of the study (41 papers: 2000–2017)

## 3.2 *Descriptive Analysis*

This study attempts to analyze 41 articles published between 2000 and 2017 as illustrated in Fig. 1. Although, the scope of the literature review covered the articles since 2000, no publications are found until 2004. Figure 1 shows a trend over the last four years.

Figure 2 illustrates that most of the studies (66%) dealing with lean and green practices in SC context took qualitative approach, only 34% are quantitative.

Figure 3 illustrates that the vast majority of studies proposing a conceptual framework are qualitative studies, whereas the numbers of qualitative and quantitative case studies are almost the same.

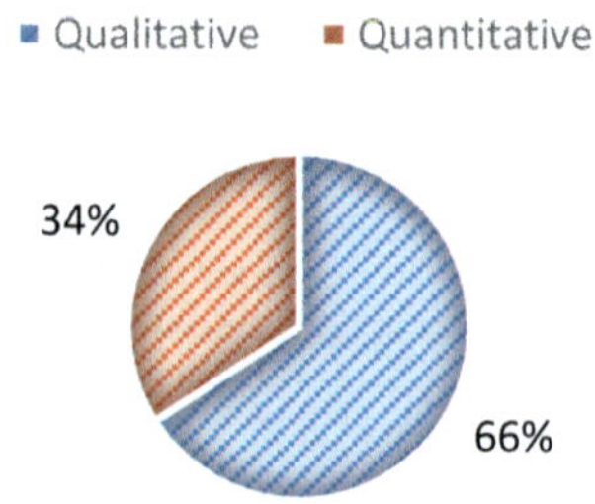


**Fig. 2** Distribution of papers by attribution

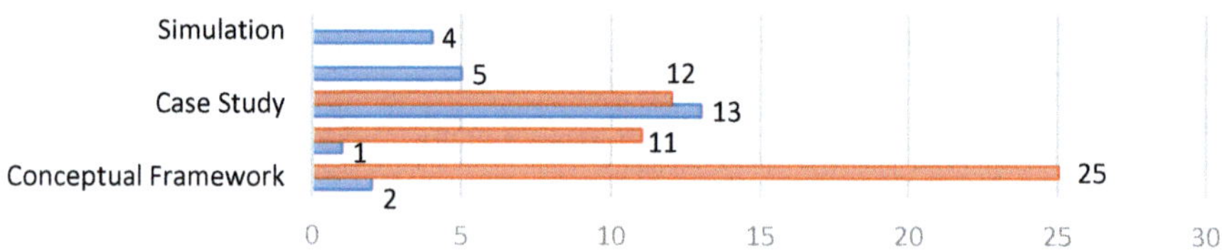


| | Conceptual Framework | Survey/Questionnaire/Interview | Case Study | Mathematical/ Decision Making Model | Simulation |
|---|---|---|---|---|---|
| Qualitative | 25 | 11 | 12 | 0 | 0 |
| Quantitative | 2 | 1 | 13 | 5 | 4 |

**Fig. 3** Distribution of papers by research types (Some papers have been classified under more than one type)

**Table 4** The most popular keywords and words in paper titles

| Keywords | Frequency | Words in paper titles | Frequency |
|---|---|---|---|
| Supply chain management | 11 | Lean | 35 |
| Green | 11 | Green | 35 |
| Green supply chain management | 5 | Supply | 32 |
| Lean | 5 | Chain | 26 |
| Lean production | 4 | Management | 18 |
| Resilient | 3 | Practices | 11 |
| Case study | 3 | Performance | 6 |
| Agile | 3 | Resilient | 5 |
| Green supply chain | 3 | Agile | 5 |
| Supply chain | 2 | Sustainability | 4 |

Table 4 gives an insight into the most frequently used keywords and words in paper titles. This information comes handy for prospective researchers interested in the domain of lean and green supply chain.

Table 5 includes the research papers on lean and green SC and classification by types of studies, i.e. qualitative, quantitative, conceptual framework, survey/questionnaire/interview, case study, mathematical/decision making model and simulation.

**Table 5** Articles included in the literature review and classification

| Authors | Qualitative | Quantitative | Conceptual framework | Survey/Questionnaire /Interview | Case study | Mathematical model | Simulation | Keywords |
|---|---|---|---|---|---|---|---|---|
| Zhu and Sarkis (2004) | • | | | • | | | | International management, GSCM, Moderated hierarchical regression, Empirical study |
| Simpson and Power (2005) | • | | • | | • | | | Supplier relations, Lean production, Environmental management |
| Kainuma and Tawara (2006) | | • | | | • | | • | SCM, Life cycle assessment, LGSC |
| Sawhney et al. (2007) | • | | • | • | • | | | Lean manufacturing, Green manufacturing, GSC, Chip forming, Metal cutting |
| Mason et al. (2008) | • | | • | | • | | | Industrial ecology, Lean thinking, SCM, Sustainable development, Value stream mapping, $CO_2$ emissions |
| Esmer et al. (2010) | | • | | | | | • | Green port, Lean port, SC, Simulation, Turkish ports |
| H. Carvalho et al. (2010) | • | | • | | | | | Lean paradigm, Green paradigm, Performance measurement, SCM |
| Mollenkopf et al. (2010) | • | | • | | | | | Globalization, Sustainable development, SCM, Environmental management, Lean production |

(continued)

**Table 5** (continued)

| Authors | Qualitative | Quantitative | Conceptual framework | Survey/Questionnaire /Interview | Case study | Mathematical model | Simulation | Keywords |
|---|---|---|---|---|---|---|---|---|
| Prentice et al. (2010) | | • | | | • | | | Environment, Cabotage, Containers, Freight |
| H. Carvalho et al. (2011) | • | | • | | | | | Lean production, Agile production, SCM |
| Azevedo et al. (2011) | • | | • | | | | | Lean, Agile, Resilient, Green, Performance measurement, SC |
| Azevedo et al. (2012) | • | | • | | • | | | Automotive industry, Green and lean supply chain practices, Performance measurement |
| Cabral et al. (2012) | | • | • | | • | • | | Lean, Agile, Resilient, Green, SCM, ANP |
| Jasti et al. (2012) | | • | | | • | | | GSCM, LSCM, Case study, Eco-efficiency |
| Figueira et al. (2012) | • | | • | | | | | SC, LARG paradigms, Conceptual framework, Human factors |
| Duarte and Cruz-Machado(2013) | • | | • | | | | | Organizational culture, SCM, Lean, Green, Modelling, Awards, Standards, Frameworks |
| Dües et al. (2013) | • | | • | | | | | Green, Sustainability, Environment, Lean, SCM |
| Govindan et al. (2013) | • | | • | | | | | Lean , Green, Resilient, SC Performance, Interpretive structural modeling, Automotive SC |
| Hajmohammad et al. (2013b) | • | | • | • | • | | | Environmental performance, Environmental management, Supply management, Lean management |

| | | | | | | | | |
|---|---|---|---|---|---|---|---|---|
| Liang and Wang (2013) | • | | • | | | | | Automobile enterprise, Lean logistics, GSC, Environment |
| Maleki and Cruz Machado (2013) | • | | • | • | | | | Automotive industry, Bayesian network, SC, Integration, SC practices |
| Duarte and Cruz-Machado (2014) | • | | • | | | | | Lean, Green, SC, Balanced scorecard |
| Govindan et al. (2014) | • | | • | • | • | | | Lean, Resilient, GSC, Sustainability, Case study |
| Fahimnia et al. (2015) | | • | | | • | • | | GSC, Environmental sustainability, Lean, Flexible, Agile, Nonlinear mathematical programming, Cross-Entropy method, Case study |
| Piercy and Rich (2015) | • | | • | • | • | | | Sustainability, Operations management, Lean, Green, Case study |
| Wiese et al. (2015) | • | | • | • | • | | | Lean, Green, Best practice, Integrated business model, Sustainable competitive advantage, Improved business performance |
| Wu et al. (2015) | • | | • | • | • | | | Lean, Green, Social, Sustainability, Triple bottom line |
| Bortolini et al. (2016) | • | | • | | | | | Environmental sustainability, GSC, Integrated framework, Lean thinking, SCM |

(continued)

**Table 5** (continued)

| Authors | Qualitative | Quantitative | Conceptual framework | Survey/Questionnaire /Interview | Case study | Mathematical model | Simulation | Keywords |
|---|---|---|---|---|---|---|---|---|
| Br et al. (2016) | • | | • | • | | | | Lean management, GSCM, Performance measurement system, Carbon footprint analysis |
| Campos and Vazquez-Brust (2016) | • | | • | • | • | | | Brazil, Lean production, GSC, SCM practices, Domestic appliance industry |
| A. L. Carvalho et al. (2016) | • | | | | | | | SCM, Multiple supply chains, GSCM |
| Garza-Reyes et al. (2016) | | • | | | • | | | Efficiency, Environmental performance, Environmental sustainability, Green, Lean, Road transportation, Waste elimination |
| Ugarte et al. (2016) | | • | | | • | | • | Lean logistics, Consumer goods, Sustainability, Retailing systems, Energy, Carbon |
| Zhan et al. (2016) | • | | | • | • | | | Green and lean practice, Environmental performance, Guanxi, Business performance, Chinese organizations, Sustainable development |
| Azevedo et al. (2016) | | • | | | • | | | Decision support systems, SCM, Agility, Lean production, Industrial performance, Lean, Green, Agile, Index, Automotive industry, Resilient |
| H. Carvalho et al. (2017) | | • | | • | • | • | | Eco-efficiency, LGSC, Linear programming model |

| | | | | | | | | |
|---|---|---|---|---|---|---|---|---|
| Colicchia et al. (2017) | | • | | | • | | | SCM, GSC, Lean systems, Intermodal transport, Fast moving consumer goods |
| Duarte and Cruz Machado (2017) | • | | • | • | • | | | Lean, Green, Case study, Assessment |
| Rachid et al. (2017) | | • | | | • | | • | Risk approach, LSC, Agile SC, GSC, Resilient SC |
| Pandey et al. (2017) | | • | | | • | • | | Environmental sustainability, Multi-objective supplier selection, Fuzzy set theory, Goal programming |
| Golpira et al. (2017) | | • | | | • | • | | GSC, Conditional value at risk, Bi-level programming, Robust optimization, KKT-Conditions |

# 4 Detailed Analyses of the Literature

In this section, the papers included in our literature review are discussed and analyzed in order to shed light on the recent state-of-the-art studies in LGSC.

## 4.1 *Classification of Studies*

There are various types of studies, such as surveys, studies using causal/cause-effect diagrams, and studies providing conceptual framework.

**Survey**
Survey papers try to find practical solutions to scientific questions in an interactive study with participants through questionnaires/interviews using statistical analyses.

**Causal/Cause-Effect Diagram**
Causal diagrams are used to represent the causal influence between variables. Some studies investigated the cause-effect relationship between lean-green practices and supply chain attributes.

**Conceptual Framework**
These studies analyze some theoretical or practical factors to develop a framework that will ease the process of adopting practices across the SC.

**Quantitative Analyses**
Quantitative studies provide an analytical way to deal with the problems related to lean and green in the SC context.

## 4.2 *Surveys in Lean and Green Supply Chain*

Zhu and Sarkis (2004), using survey instruments, examined the relationships between GSCM practices (internal environmental management, external GSCM, investment recovery, and eco-design or design for environment practices) and environmental/economic performance and how two types of management operations philosophies, namely quality management and just-in-time (JIT) (or lean) manufacturing principles, impact the relationship between GSCM practices and performance. They used modified hierarchical regression methodology to test the various hypotheses developed. Hajmohammad et al. (2013a) sought to answer the question whether lean and supply chain management are directly related to environmental performance by suggesting research propositions. Lean management practices comprise of JIT, lead time reduction, set-up time reduction, and TQM. Supply management practices include suppliers' evaluation, certification, recognition and development programs and auditing of suppliers' facilities. For environmental practices; ISO 14001 certification, pollution prevention, recycling of materials and waste reduction were considered. Lastly, for environmental performance criteria, air emissions, waste water generation, solid wastes disposal, consumption of hazardous/harmful/toxic materials and energy consumption were considered. Govindan et al. (2014) developed a conceptual model to examine the impact of lean, resilient and green practices on the degree of the sustainability of supply chains using a deductive research approach by suggesting eighteen research propositions. Case studies were implemented to test the proposed propositions. The findings indicated that the waste elimination, SC risk management and cleaner production were the most significant drivers for better sustainability of SCs, while flexible transportation, flexible sourcing, ISO 14001 certification and reverse logistics were found non-significant in favor of sustainability. Piercy and Rich (2015) sought to evaluate the relationship between lean operations at strategic, supply chain and workplace activities level and sustainability outcomes. The authors highlighted that lean tools, including supply monitoring, transparency, workforce treatment, and community engagement, along with lean tools into workplace, such as 5S, staff training etc., meet several sustainability outcomes and environmental benefits. Wu et al. (2015) offered a conceptual framework of achieving sustainability through lean, green and social practices selected from several management systems. Obtaining data from questionnaires, they empirically attempted to analyze the impact of selected lean, green and social practices on auto-part suppliers' economic, environmental and social performance (mostly referred to as Triple Bottom Line (3BL)). Campos and Vazquez-Brust (2016) empirically analyzed the synergic results that lean and green supply chain activities can bring when implemented in a synergic way, using a case study of the Brazilian subsidiary of a multinational company. Their findings indicate that 26 out of 31 integrated lean and green practices, obtained from the extensive literature review, show synergic results in lean and green performance. Besides, they pointed out that the waste reduction in lean and green practices differ in some aspects, and it is found as the key point of integration of lean and green practices that yields the most synergic result amongst the other lean and green practices. Four items that have been found non-synergic and that present trade-offs were JIT delivery, JIT philosophy, inventory reduction, Value Steam Mapping (VSM)/Sustainable VSM and green technology. Zhan et al. (2016), by using questionnaires administered to all the representatives and organizations, empirically analyzed how lean and green practices affect the organizations' environmental and business performance and how this relationship is moderated by guanxi, i.e. an organisational practice deeply rooted in 5000 years of Chinese culture. Nineteen identified lean and green practices in their study were rather different than usual lean and green tools that are listed in Table 6. Mindset and attitude of people responsible for lean and green practices, leadership and management, employee involvement, integrated approach, and tools and techniques were used to classify those nineteen lean and green practices.

We need to highlight that most of the survey studies collected the participants' expectations and beliefs on the relationships between practices and their impact on triple bottom line performances, rather than using actual financial numbers for

**Table 6** Lean and green practices used in SC context

| Practices | | 1 | 2 | 3 | 4 | 5 | 6 | 7 | 8 | 9 | 10 | 11 | 12 | 13 | 14 | 15 | 16 | 17 | 18 | 19 | 20 | 21 | 22 | 22 |
|---|---|---|---|---|---|---|---|---|---|---|---|---|---|---|---|---|---|---|---|---|---|---|---|---|
| Lean and green tools | JIT philosophy/Pull flow/Cellular man. | • | | • | | • | | • | • | • | | • | | | | • | • | | | | | | • | • |
| | Lead/Set-up/Total time reduction | | | | | • | | • | | • | | | | | | • | | | | | | | | |
| | Kaizen/5S/Kanban/Poka yoke/6 sigma/Mistake proofing/Quick changeover | | | • | | | | | | | | | | | | | | • | | | | | | |
| | Total quality management/Quality management system | • | | | | • | | • | • | | | • | | | | • | • | • | | | | | | |
| | Total productive management | | | • | | • | | | | | | | | | | • | | | | | | | | |
| | Waste mngmt., Elimination/Reduction | | | | | • | | • | | | • | • | | • | | • | | | | | | | | • |
| | JIT delivery, Lot sizing, Order consolidation | | | • | | • | | | | | | | • | | • | • | | | | | | | | |
| | Courier and transport modes opt. | | | | | | | | | | | | | | • | | | | | | • | | | |
| | Inventory control/Reduction | | | | | | | | | | | | | | | | | | • | | | | | |
| | Energy efficiency improvement/Emission reduction | | | | | | | | | | | | | | • | | • | | | | | | | |
| | Environmental management system (ISO 14001) | • | | | | • | | • | • | | | • | | | | | • | • | | • | | | • | |
| | External GSCM | • | | | | | | | | | | | | | | | | | | | | | | |
| | Pollution prevention | | | | | | | • | | | | | | | | • | | | | | | | | |
| | 3R (Reduce, reuse and recycle)/Reverse logistics | | | | | • | | • | | | | • | | | • | • | • | | | | | | | • |
| | Cleaner production | | | | | | | | | | | • | | | | | | • | | | | | | • |
| | Green tech./Eco-design/Green packaging | • | | | | • | | | • | | | | | | • | | • | • | | • | | | • | • |
| | Green purchasing | | | | | • | | | | | | | | | | | | | | • | | | | |

| | | | | | | | | | | | | | | | | | | | | | | | | |
|---|---|---|---|---|---|---|---|---|---|---|---|---|---|---|---|---|---|---|---|---|---|---|---|---|
| | Reduction of hazardous/Harmful/Toxic materials | | | | | • | | | | | | | | | | | | | | | | | | |
| | Life cycle assessment | • | • | | | • | | | | | | | | | | | | • | | | | | | |
| | Supplier management/Long term relationship | | | • | | • | | | | | • | | | • | • | | • | | | | | | | |
| | Geographic concentration | | | | | | | | | | | | | | | | | | | | | | • | |
| | Environmental risk sharing/Cooperation with suppliers/customers | | | | | • | | | | | | | | | | | | | | • | • | | | |
| | Information management | | | | | | | | | | • | | | • | | | • | | | | • | | | |
| | Human resources mngmt./Employee ed. and satisfaction | | | • | | | | | | | • | | | • | | • | | | | | | • | | |
| | Customer relationship | | | | | • | | | | | | | | | | | | | | | | | | |
| | Distribution network configuration | | | | | | | | | | | | | | | | | | • | | | | | |
| | (Sustainable) Value stream mapping | | | | • | | | | | | | | | | | | | | | | • | | | |

References: Zhu and Sarkis (2004), Kainuma and Tawara (2006), Sawhney et al. (2007), Mason et al. (2008), H. Carvalho et al. (2010, 2017), Azevedo et al. (2012, 2016), Hajmohammad et al. (2013a), Govindan et al. (2013, 2014), Maleki and Cruz Machado (2013), Duarte and Cruz-Machado (2014), Fahimnia et al. (2015), Piercy and Rich (2015), Wiese et al. (2015), Wu et al. (2015), Bortolini et al. (2016), Br et al. (2016), A. L. Carvalho et al. (2016), Garza-Reyes et al. (2016), Zhan et al. (2016), Pandey et al. (2017)

evaluation. However, whether lean and green practices have positive or negative performance on 3BL performance, can only be measured using actual financial numbers, if available. However, Zhu and Sarkis (2004) claimed that the expectations of management in organizations that implement any organizational practices will play a role in lessening the barrier to GSCM practice adoption.

### 4.2.1 Survey Papers Using Causal Diagrams

Duarte and Cruz-Machado (2014) highlighted that the cause-and-effect balanced score card can be an effective tool to measure the performance of LGSCs. H. Carvalho et al. (2010) introduced one of the first conceptual frameworks investigating the impact of lean and green paradigms on SCs performances. They used a causal diagram to visualize how lean and green practices interact with each other and overall SC performance metrics. In the conceptual framework, they employed supplier relationships, JIT philosophy, lot-size reduction, total quality management, total productive maintenance, cycle time reduction, and customer relationships as lean practices, while as to green practices, environmental collaboration with suppliers, environmental friendly practices in purchasing, working with designer and suppliers, minimization of waste, decrease the consumption of hazards and toxic materials, ISO 14001, reverse logistics, environmentally friendly packaging, and working with customers to change product specifications were employed. The set of measures implemented in the model was classified into three main categories; i.e. operational (inventory levels, quality, customer satisfaction and time), economic (cost, efficiency, return on assets and C2C cycle) and environmental (environmental cost, life-cycle assessment and business wastage). It should be noted that the proposed model was developed using anecdotal and empirical evidences, and has not yet been tested by case studies or real world data. H. Carvalho et al. (2011) further improved the previous work carried out by H. Carvalho et al. (2010) so that the models (cause-effect diagrams) developed represent the synergies, divergences and tradeoffs among not just lean and green, as in H. Carvalho et al. (2010), but lean, agile, resilient and green practices implemented in supply chains. Maleki and Cruz Machado (2013) sorted by importance the customer values in automobile industry, namely quality, cost, respect for the environment, know-how, customization and time, using a Bayesian network and data acquired through questionnaires. After analyzing the data, the order by importance through customers' eyes was identified as quality, cost, respect for the environment, know-how, customization and time. Afterwards, they identified the causal relationship between LARG practices and customer values aforementioned. For LARG practices, cycle time reduction, setup time reduction, batch sizing, lead time reduction, manufacturing transparency to customers, modularization, flexibility to demand change, reduction of raw materials variety, demand-based management, cooperation with product/production designers to decrease environmental impacts, visibility of upstream and downstream inventories, information sharing with customers, modularization, JIT, flexibility, supplier relationships and lead time reduction were used. Finally, Duarte and

Cruz-Machado (2014) explored the linkage between the integration of lean/green paradigms and several SC performance measures, namely profitability and revenues, cost, market share, customer satisfaction, productivity, waste reduction, supplier side management, information management, employee morale and satisfaction, and employee education and training, using a balanced score card. A causal diagram was used to represent the linkages between the aforementioned items.

## 4.3 Conceptual Frameworks Developed for Lean and Green Supply Chains

A framework refers to a set of prescriptive initiatives that guides organizations in achieving an organizational transformation. Simpson and Power (2005) developed a conceptual framework for investigating the relationship between a supplier firm's level of environmental management activity and the structure of the customer-supplier manufacturing relationship in automotive supply chain. Using the conceptual framework they conducted further research by proposing a number of hypothesis as follows; "the lean performance of a supplier firm is positively related to the environmental management practices of the supplier firm", "The customer's level of involvement in improvements to the lean performance of a supplier firm is positively related to the environmental management practices of the supplier firm" and so forth. Kainuma and Tawara (2006) proposed the multiple attribute utility theory method, which consists of the three single-attribute utility functions supply chain return of assets, customer satisfaction, and life cycle assessment (LCA—which is a measure of the effect on the environment and the contribution to the society) for assessing of a lean-green supply chain not only from a managerial viewpoint but also from an environmental performance aspect. Sawhney et al. (2007) proposed a framework (namely, En-Lean) to assist in developing the relationship between a comprehensive set of lean principles and their environmental impact for specific processes. The first phase of the methodology to develop a matrix to quantify the relationship between lean principles and specific environmental factors affected by them. Second phase defines the methodology for data collection, using questionnaires, for the specific type of process family, application or industry. In the third phase of the methodology, the data collected is analyzed to calculate the overall environmental impact score for each lean principle using a sensitivity analysis. Mason et al. (2008) proposed a Value Steam Mapping (VSM) tool, which is a lean-management tool used to illustrate the flow of materials and information through the supply chain from supplier to customer, and modified the tool to make it capable of delivering environmental benefits in terms of $CO_2$ in an attempt to improve the supply chain's economic and ecological performance. Their "Sustainable VSM" intends to maximize added value and at the same time, minimize $CO_2$ emission level through a supply chain. Mollenkopf et al. (2010) examined the relationship, synergies and divergences among lean, green and global supply chain strategies with an emphasis on concurrent deployment of these three paradigms, and addressed drivers, barriers, complementary and contradictory issues across green/lean, green/global and lean/global supply chain strategies. Azevedo et al. (2011) proposed the first conceptual model to help manufacturing SCs become more simultaneously LARG, in an attempt to improve their operational, economic and environmental performance. They considered the following LARG practices: "just in time", "supplier relationships" and "cycle/setup time reduction" as lean practices; "speed in improving responsiveness to changing market needs", "to produce in large or small batches", "ability to change delivery times of supplier's orders" as agile practices; "developing visibility to a clear view of upstream inventories and supply conditions", "lead time reduction" and "demand-based management" as resilient practices; and lastly as green practices, "to work with product designers and suppliers to reduce and eliminate product environmental impacts" and "reduction in the variety of materials employed in products manufacturing". Their model examines how these LARG practices impact the firms' inventory levels, quality, customer satisfaction and time (operational measures), cost, environmental cost and cast-to-cast cycle (economic measures) and business wastage (environmental measures). Figueira et al. (2012) presented a conceptual framework that allows integrating ergonomic and safety design principles during the implementation of LARG practices in SCs, which tends to increase the physiological and psychological strain on workers. This study is one of a kind dealing with LARG practices' impact on human health. For example, lead time reduction, one of the possible consequences of the lean and resilient practice, can lead to a reduction in the duration of work breaks that would affect the recovery period of workers. However, the authors also highlighted that not all LARG practices have a negative effect on workers, such as the decrease in the quantity and variety of hazardous and toxic materials decreases the physiological strain on workers. To the authors, organizations should take into account the impact that LARG practices in SCs might have on workers' health and that workplaces should be organized accordingly, considering the ergonomic and safety principles. Azevedo et al. (2012) are one of the first to carry out a study proposing a theoretical framework to find out the impact of green and lean supply chain management (SCM) on the sustainable development of businesses. They used a set of performance measures of economic (operational, environmental and inventory cost), environmental (wastage, green image and $CO_2$ emission) and social (corruption risk, supplier screening and local supplier). Duarte and Cruz-Machado (2013) analyzed several business models, awards, standards and frameworks that would help companies integrate lean and green paradigms into their supply chains and improve SC performance by providing a number of guidelines. Leadership, people, strategic planning, stakeholders, processes and results were defined as the key opponents of lean-green transformation frameworks. Dües et al. (2013) tried to explore which lean practices are synergic with green practices implemented in SCM and vice versa, and to help companies achieve green SCM using their own lean framework by analyzing several explanatory framework. In the light of the extensive literature review on published works on lean and green, they discussed in detail the distinguishing attributes, similarities and differences of the lean and green paradigms and how to get the best out of the synergic combination of lean and green SCM practices. Govindan et al. (2013) identified top critical lean, green and resilient practices and their simultaneous influence on automotive SC performances. The inter-relationships among lean, green and resilient practices and SC performances were drawn through an interpretive structural modeling approach. JIT and TQM were used as lean, ISO 14001 certification and environmentally friendly packaging as green, and strategic stock and flexible transportation as resilient tools, out of which JIT, flexible transportation and environmentally friendly packaging were observed to be the most significant practices affecting positively the performance of automobile supply chains, assessed by the operational cost, business wastage, environmental cost and customer satisfaction. Liang and Wang (2013) offered a strategy for automobile enterprises to achieve development of lean logistics under green SCM environment. Another conceptual framework to integrate lean, green and best practices into the business model carried out by Wiese et al. (2015), which offers an opportunity for organizations to become more competitive by integrating them concurrently. The proposed framework has been supported by a case study conducted on Toyota South African Motors to illustrate the synergic benefits stemming from concurrent integration of lean, green and best practices into a SC organization. Br et al. (2016) reviewed the previous works to search for supply chain elements, their leanness applicability and greening effects and to offer a strategic framework for managers to help them convert their ordinary SC to a GSC. A. L. Carvalho et al. (2016) sought to analyze the synergies between GSCM practices and LSC, along with other various types of SCs. It is found that three out of five predetermined GSCM practices, i.e. the internal environmental management, green purchasing and recovery of investment have strong alignment with LSC, meaning that the LSC would allow for the full incorporation of those practices, without restriction or obstacles; while the remaining two practices, i.e. cooperation with consumers and eco-design, have moderate alignment, meaning that partial incorporation exists, with some restrictions or obstacles. Another framework to integrate lean and green initiatives into a SC or its components, such as the distribution network or the assembly system, is offered by Bortolini et al. (2016). Their framework can be applied to any lean and green integrated supply chain components, such as the distribution network, the assembly system, the job-shop, the storage system etc. Colicchia et al. (2017) addressed critical issues, namely planning and management, assets, train services, collaboration, legal issues and incentive schemes, for lean and green SCs to adopt the intermodal transportation in their distribution network. The case study results confirm that the average cost and $CO_2$ emission in road-rail intermodal transport was found lower than all-road delivery mode, on the other hand, on time delivery completed within the agreed time window slightly reduced from 98% to 96%. The managers, as they are advised to adopt intermodal transportation, can achieve significant increase in sustainability and efficiency in their SCs with the opportunities offered by intermodal transportation that would allow them to reduce $CO_2$ emissions and optimize transportation costs. Duarte and Cruz Machado (2017) proposed a conceptual framework for evaluating lean and green implementation in an organization's SC. An explanatory case study that reviewed data through interviews was performed in the automotive sector to validate the credibility of the method proposed.

### 4.4 Quantitative Analyses

Quantitative studies provide an analytical way to deal with the problems related to lean and green in the SC context. Esmer et al. (2010) proposed a model for ports and simulated a container terminal system in which lean and green practices were employed, yet due to the lack of proper data, only cargo handling operations were considered among lean practices in the ports (JIT, postponement and cross-docking) and green practices (vessel operations, services of marine facilities and cargo handling). Their main objective in the simulation was to minimize the number of electric vehicles used for inland cargo transport, which would greatly help lean and green causes in the ports through minimizing the environmental footprint and increasing effectiveness, which increase lean capability. Prentice et al. (2010) investigated the environmental impact of a cabotage regulation in Canada domestic services on making use of empty marine containers, if owned by Canadian carrier, after an import load is discharged and how these regulations influenced supply chain efficiencies. The need for such regulation stems from the fact that repositioning the transportation equipment without a payload (in industry terms, this is called empty backhaul, or "deadhead") is somehow unavoidable, and it creates waste and generates unnecessary GHG emission. They examined the environmental impact through three cases studies, and reported that the eliminating empty backhauls could lead to a dramatic decrease in fuel consumption in transport. Cabral et al. (2012) proposed an analytic network process (ANP) decision-making model to help choose the most appropriate lean, agile, resilient and green practices affecting key performance indicators that companies are willing to implement in their SC, and they support their case by providing a case study in an automaker SC. Strategic stock, systems for rapid response in case of emergency and special demands, and reusing materials and packages were selected as LARG practices, and inventory cost, order fulfilment rate and responsiveness to urgent deliveries as key performance indicators. Jasti et al. (2012) tried to reorganize the current state of a steel manufacturing company to achieve green goals with the help of lean practices by changing the layout of the plant, providing a separate exit gate, new weigh bridges to facilitate faster movement of vehicles and better communication between inter-departments and with suppliers to reduce waiting time and faster unloading at destination, and more parking sites to reduce congestion of vehicles. By maximizing space and better material utilization, results from this case study indicate that the conjunction of the lean tools with green SCM practices helps to reduce the carbon emission by 34% and the pollutant level by more than 30%. Fahimnia et al. (2015) investigated the trade-off between lean operations and green outcomes by proposing a non-linear mathematical model for LGSC network. The proposed model and solution method was supported using a real-world data case study. Outdated and up-to-date set of machinery, the size of warehouses, three different truck sizes used in direct transportation differing in the cost of operating, energy consumption and carbon emission level, altogether creates a trade-off in their model, in which total cost in manufacturing, transport, inventory holding and backlogging, total carbon generated in manufacturing, transport and holding, total consumed energy in manufacturing and inventory holding and total waste generated in manufacturing and inventory holding in plants and warehouses were to be minimized simultaneously. The last three objectives, emission, energy and consumption values are expressed as dollar amounts in the overall objective function to convert the multi-objective into single. They developed a Nested Integrated Cross-Entropy method to deal with their mixed-integer non-linear mathematical model and the findings suggested that the distribution emission level in the leanest scenario is much higher than in other flexible situations and the centralized situation. Using smaller trucks helps making the situation leaner, but definitely not greener, as it causes more and smaller deliveries with less emission efficient vehicles on a per unit basis. Garza-Reyes et al. (2016) proposed a variant of VSM, namely Sustainable Transportation VSM, to integrate the lean and green paradigms in organizations' transport operations to improve both their operational efficiency and environmental performance. After mapping the road transportation process and analyzing possible causes of waste due to inefficiency and environmental waste, they offered a set of improvement strategies, which resulted in increased performance efficiency, reduced unnecessary travels, and decreased fuel consumption and therefore GHG emissions. Azevedo et al. (2016) presented an assessment tool for automotive companies and corresponding SCs that would enable them to evaluate their degree of leanness, agility, resilience and greenness. A LARG index provided by the tool for measuring the LARG behavior of companies can be seen as a benchmarking evaluation by managers that can be used to compare the corporations' performances. The Delphi method is used to compute the weight for each predetermined SCM practice. The method is tested using a case study involving several companies and was found an effective and useful tool to evaluate the companies' and the SCs' performance regarding the LARG practices. Ugarte et al. (2016) sought to evaluate the difference in GHG emission that each of the lean-oriented inventory management methodologies, i.e. JIT, product postponement and vendor managed inventory, have compared to traditional (economic order quantity) inventory policy, by using a two-echelon discrete-event simulation model. It is reported that JIT causes a dramatic raise in product shipment frequency, therefore it results in more GHG emission level than baseline policy, yet lean practices lead to reduced GHG in inventory and retail operations. On the other hand, product postponement and vendor managed inventory policy perform better than the baseline scenario regarding the GHG emission. H. Carvalho et al. (2017) proposed a mathematical model to assist managers willing to improve eco-efficiency of companies and supply chains by determining a set of practices that minimizes the supply chain's overall negative environmental and economic impacts. Their findings support the general consensus on JIT production creating a trade-off between economic benefits and environmental impacts, as it contributes to a reduction in the inventory of raw materials, works in process, and finished goods, however, the frequent delivery of materials leads to a higher level of emissions, worsening the environmental impact of SCs. Also, they highlighted that the geographic concentration with suppliers is another potential source of trade-off between economic and environmental goals. Rachid et al. (2017) mapped the agility and resilience risks created by lean, green and achievement goals using a risk management approach. It was found that the LGSC is perfectly agile and resilient, except when the demand variability is evident. Pandey et al. (2017) provided a framework for the selection of lean, green and sustainable suppliers. They proposed fuzzy goal programming to be used in the evaluation of suppliers on the basis of economic, lean, services and environmental criteria and the model was tested using a real world case. Golpira et al. (2017) developed an integrated model to design Green Opportunistic SC networks considering leanness, agility and robustness. The model incorporates the uncertainties of the market-related information, namely the demand, transportation and shortage costs. They adopted bi-level programming methods to deal with the uncertainties pertaining to demand, transportation and shortage costs. Karush-Kuhn Tucker (KKT) conditions are implemented to transform the model into a single-level mixed integer linear programming problem, and the resulting problem is finally solved by CPLEX.

# 5 Discussion, Future Directions and Conclusion

## *5.1 Discussion: Synergies and Trade-Offs Between Lean and Green*

There is a long-lasting debate as to whether the lean and green paradigms are somehow targeting the same goals. Campos and Vazquez-Brust (2016) pointed out that the waste reduction in lean and green practices differ in some aspects and Tice et al. (2005) reported that the environmental wastes are not explicitly included in the wastes targeted by lean paradigm. Thanki et al. (2016) addressed the conceptual difference between lean and green as follows: “lean has the goal of reducing operational waste through non-value adding activities, while green manufacturing targets environmental waste reduction”. However, lean practices contribute to reduce some of these environmental impacts because of their intrinsic focus on waste elimination. Lean paradigm can help green efforts by connecting green practices to stronger financial drivers and improving the effectiveness of green procedures (Tice et al. 2005). Also, Piercy and Rich (2015) highlighted that one of the goals of lean operations is to use fewer resources to generate the same outcome, which is inherently environmentally friendly, as it means less materials used in production, higher quality with reduced rework, scrap, resource consumption, and pollution costs. In addition to those, Wiese et al. (2015) mentioned that there is a strong correlation between lean and green thinking as both principles focus on reducing waste, decreasing costs and enhancing value for all stakeholders and partners within a supply chain. Hanson et al. (2004) also reported that while lean practices can lead to environmental benefits, environmental practices can lead to improved lean practices. Mollenkopf et al. (2010) highlighted the trade-off between JIT philosophy and green approach, yet they stated that "when lean initiatives enable only demanded volumes to flow through the SC (and not the safety stock and extra inventory associated with non-lean supply chains), a reduced amount of inventory needs to be sourced, produced, transported, packaged and handled, which also minimizes the negative environmental impact of the SC". These studies show that, by definition, there is consensus on that lean thinking can help reduce environmental impact and comply with green efforts.

There are numerous studies that take the view opposite to the aforementioned arguments, citing some lean practices such as JIT logistics or faster delivery methods that may hurt environmental efforts by creating a trade-off (Campos and Vazquez-Brust 2016; H. Carvalho et al. 2017; Ugarte et al. 2016; Zhu and Sarkis 2004). For example, Zhu and Sarkis (2004) reported that JIT has no effect for positive or negative economic performance when implementing GSCM practices, however, strong relationship exists between GSCM practice and positive economic and environmental performance. This argument is also supported by H. Carvalho et al. (2010), who found that lean practices barely contribute to the SC environmental performance. Some studies, for instance, Campos and Vazquez-Brust (2016), reported trade-offs associated with JIT delivery, JIT philosophy, inventory reduction and VSMs/SVSMs when aiming for green outcomes. They claimed that when a company is decisive in adopting lean philosophy which dominates its operational strategy and vision, trade-offs are more likely to occur instead of synergies between lean and green practices. Mason et al. (2008) concurred with the argument that JIT logistics and the demand for faster delivery methods contribute to higher emission rates, because smaller quantities of goods are delivered more frequently in smaller vehicles. However, Azevedo et al. (2011) claimed that JIT were one of the two LARG practices that improve the SC performance most. Also, Ugarte et al. (2016) added to that by claiming that lean practices lead to reduced GHG in inventory and retail operations, even so JIT increases GHG emission in transportation. In addition to mentioning trade-off between economic benefits of implementing JIT and its environmental impact, H. Carvalho et al. (2017) highlighted that not all companies belonging to the same supply chain can be absolutely lean or green, some must make concessions. JIT also leads to small batch size production, which may lead to higher amount of waste generated in the production (King and Lenox 2001). Another example to a trade-off between lean and green paradigms is that a supply chain may be lean when it is geographically wide-spread, yet increased transportation, packaging and handling lead to more carbon emission, which means that it is not necessarily green (Venkat and Wakeland 2006). Hereby, these conflicts show that we need to bear in mind that when discussing whether lean and green practices are targeting the same goal or conflicting with each other, the discussion should be taken in hand at each individual practice level.

When it comes to synergy, Dües et al. (2013) described it with "the equation 1+1=3", meaning that "combined practices have greater results than the sum of the separate performances". Dhingra et al. (2014) sought to answer the question of whether there are synergic benefits of simultaneously adopting both, or adoption of one leads to rapid adoption of the other. They reiterated and confirmed that lean leads to green, but not necessarily vice versa. Bergmiller and McCright (2009) identified the correlation between green and lean paradigms by studying Shingo prize winners and finalists, based on a complete assessment of an organization's culture and how well it drives world-class results. They found that lean companies which include green practices achieve better lean results than those companies which do not. Their findings can be interpreted as such, when both paradigms are implemented together, the benefits will be greater for the companies than when implemented separately. Also, Dües et al. (2013) explored the relationship and links between lean and green SCM practices, and their conclusion is similar to the studies mentioned above, that lean is beneficial for green practices and the implementation of green practices in turn also has a positive impact on existing lean business practices.

Concurrent deployments of lean and green practices were reported in the literature to have synergic benefits. Jasti et al. (2012) claimed that the conjunction of the lean tools with green SCM practices helps in reducing the carbon emission and pollutant level. Wiese et al. (2015) reported that they found synergic benefits stemming from concurrent integration of lean, green and best practices into a SC organization. Bortolini et al. (2016) reported that the combination of lean and green practices leads to greater results, compared to separate implementation of each practice. Campos and Vazquez-Brust (2016) reached to a conclusion that the majority of the practices (26 out of 31) brings synergic results to lean and green performance. A. L. Carvalho et al. (2016) found that lean supply chains present a strong alignment with the GSCM concepts. Br et al. (2016) reported that lean practices are positively and moderately interrelated with green practices. Zhan et al. (2016) pointed out that the lean and green practices have win-win relationship with environmental and business performance of organizations. Galeazzo et al. (2014) suggested managers to implement lean and green practices simultaneously to achieve synergic results. However, they also point out the importance of timing in the implementation of lean and green practices, as it affects the interdependencies (sequential and reciprocal), the search for supplier partners, and the results in terms of performance.

## *5.2 Future Directions and Conclusion*

This paper presents a comprehensive literature review of recent and state-of-the-art journal papers that have considered lean and green practices/initiatives in a SCM context. Totally, 41 published papers between January 2000 and July 2017 are selected, reviewed, categorized and analyzed to find the gaps in current literature and highlight future directions and opportunities in LGSC.

We recognize that most of the studies dealing with lean and green practices in SC context took qualitative approach and offered a conceptual framework. Besides, the possible impacts of lean and green practices on specified key performance indicators caught academicians' and practitioners' attention. In this context, Table 6 includes different combinations of lean and green practices implemented in SC context, which is quite crucial for the future development of the lean and green SC and guides prospective researchers along with the information of selected key performance indicators given in Table 7. By combining different lean and green practices and performance indicators, researchers can investigate not only the impact of selected practices on selected performance indicators, but also the trade-offs or synergies between the selected practices (Table 8).

Mathematical models developed for supply chain networks utilizing lean and green practices in SC context are rare (Fahimnia et al. 2015; Golpira et al. 2017). It is still a challenge for researchers to utilize lean and green practices concurrently in supply chain models. Truck types and outdated and up-to-date machinery and warehouses varying in sizes were assumed to create a trade-off between leanness and greenness of a distribution network, while using smaller warehouses or smaller trucks is assumed to be a leaner situation by Fahimnia et al. (2015). JIT production philosophy was tested using a linear programming model by H. Carvalho et al. (2017) and a simulation model by Ugarte et al. (2016). We recognize that there is a gap in the literature that lean/green practices are assessed by their trade-off using a simulation or mathematical model. We recommend the interested researchers to evaluate other lean/green practices that can be implemented in SC context by their trade-offs between leanness and greenness using a simulation or mathematical model, rather than qualitative type of research that is based on subjective arguments. As few quantitative contributions exist in the literature, multi objective programming emerges as a gap, since real world problems are rarely single objective. It is necessary for researchers to pay more attention to the real world problems with multi objective functions instead of single ones, as the goals of lean and green practices often contradict each other. Additionally, the approaches for handling multi objective models and achieving Pareto optimal solutions need to be considered. Manufacturing, transporting, inventory holding, handling, operational and backlog costs, waste generation/disposal costs, material purchasing costs and energy consumption costs which are associated with lean goals can be utilized in objective functions. Environmental goals such as carbon and GHG emission, amount of waste generation and energy, water, hazardous/harmful/toxic/PET materials consumption, can be utilized in objective functions. It is recognized that, up to this time, carbon and GHG emission, amount of waste (water-solid) generation and disposal, operational costs, inventory handling and waste treatment/discharge costs have been the most frequently utilized performance measures in previous works. Researchers should find a way how to integrate the least used lean/green practices so far in mathematical models, such as kaizen, quick changeover, 6 sigma, TPM, optimization of courier and transport modes, inventory control and reduction,

**Table 7** Measures and indicators to evaluate the influence of lean and green paradigms on SC performance

| | | References | 1 | 2 | 3 | 4 | 5 | 6 | 7 | 8 | 9 | 10 | 11 | 12 | 13 | 14 | 15 | 16 | 17 | 18 | 19 | 20 | 21 | 22 | 22 |
|---|---|---|---|---|---|---|---|---|---|---|---|---|---|---|---|---|---|---|---|---|---|---|---|---|---|
| SC performance measures (3PL) | Operational | Inventory levels | | | | | • | | | | | | | | | | | | | | | | | | |
| | | Quality | | | | | • | | | | | | | | | | | | | | | | | | |
| | | Customer satisfaction | | • | | | • | | | | | • | | | | | | | | | | | | | |
| | | Reliability | | | | | | | | | | | | | | | | | • | | | | | | |
| | | Order fulfilment rate/time | | | | | • | | | | | | | | | | | | | | | | | | |
| | Environmental | Carbon/GHG emission/C2C | • | | • | • | • | | • | | | | | • | | • | | | • | | | • | • | • | |
| | | Waste (water–solid) generation/disposal | • | | • | | • | | • | • | | | | • | | | | | | | | • | • | • | |
| | | Hazardous/harmful/toxic /PET materials consumption | • | | | | | | • | | | | | | | | | | • | | | • | • | | |
| | | Energy/water consumption | • | | • | | | | | | | | | • | | • | | | | | | | | • | |
| | | Environmental situation/accidents | | | • | | | | | | | | | | | | | | | | | | • | | |
| | Economic | Sustainability (social, environmental and economic) | | | | | | | | | | | | | • | | | | | | | | | | |
| | | Operational cost | • | | | | • | | | | | • | | | | | | | • | | | | | • | • |
| | | Inventory/handling cost | | | | | | | | | | | | | | • | | | • | • | | | | • | |
| | | Material purchasing cost | • | | | | | | | | | | | | | | | | | • | | | | | |
| | | Transportation cost | | | | | | | | | | | | | | | | | | • | | | | | • |
| | | Waste treatment/discharge cost | • | | | | | | | | | • | | | | | | | | • | | | | | |
| | | Energy consumption cost | • | | | | | | | | | | | | | • | | | | | | | | | |
| | | Profitability and revenues, return of asset, investment recovery | • | • | | | • | | | | | • | | | | | | | | | • | | • | | |
| | | Productivity/Efficiency | | | | | • | | | | | • | | | | | | | | | | | | | |
| | | Market share/Sales improvement | | | | | | | | | | • | | | | | | | | | | | • | | |

References: Zhu and Sarkis (2004), Kainuma and Tawara (2006), Sawhney et al. (2007), Mason et al. (2008), H. Carvalho et al. (2010, 2017), Azevedo et al. (2012, 2016), Hajmohammad et al. (2013a), Govindan et al. (2013, 2014), Maleki and Cruz Machado (2013), Duarte and Cruz-Machado (2014), Fahimnia et al. (2015), Piercy and Rich (2015), Wiese et al. (2015), Wu et al. (2015), Bortolini et al. (2016), Br et al. (2016), A. L. Carvalho et al. (2016), Garza-Reyes et al. (2016), Zhan et al. (2016), Pandey et al. (2017)

**Table 8** Synergies, trade-offs and contribution of studies

| Authors | Lean and green | | |
|---|---|---|---|
| | Synergies | Trade-offs | Findings |
| Zhu and Sarkis (2004) | | JIT may hurt the environmental performance associated with GSCM internal management practice. | JIT has no effect for positive or negative economic performance when implementing GSCM practices. Strong relationship exists between GSCM practice and positive economic and environmental performance. |
| Kainuma and Tawara (2006) | | | Information sharing, one of the lean SC strategies, were observed to reduce unnecessary production, transportation and stock holding, reducing environmental impacts. |
| H. Carvalho et al. (2010) | | | Lean practices considered in the model barely contribute to the SC environmental performance. |
| Azevedo et al. (2011) | | | Amongst suggested LARG practices, the ones that improve the SC performance most were found as the practices of "just in time" and "supplier relationships". |
| Azevedo et al. (2012) | | | The implementation of green upstream SCM practices will result in the improvement of business' economic and environmental performance. Lean practices in SCM will improve businesses' economic and environmental performance. Finally, lean and green practices in SCM improve businesses' social performance. |
| Jasti et al. (2012) | The conjunction of the lean tools with green SCM practices helps in reducing the carbon emission and pollutant level. | | |
| Dües et al. (2013) | | | Lean is beneficial for Green practices and the implementation of Green practices in turn also has a positive influence on existing Lean business practices. |

(continued)

**Table 8** (continued)

| Authors | Lean and green | | |
| --- | --- | --- | --- |
| | Synergies | Trade-offs | Findings |
| Hajmohammad et al. (2013a) | | Lean management must be accompanied by environmental practices to be a driver in reaching better environmental performances. | Environmental practices (such as management systems or pollution prevention) were found to be significantly improving the environmental performances. |
| Fahimnia et al. (2015) | | Between cost and environmental degradation including carbon emissions, energy consumption and waste generation. | Some lean practices were reported not to be resulting in green outcome. |
| Mollenkopf et al. (2010) | "When lean initiatives enable only demanded volumes to flow through the SC (and not the safety stock and extra inventory associated with non-lean supply chains), a reduced amount of inventory needs to be sourced, produced, transported, packaged and handled, which also minimizes the negative environmental impact of the SC" | JIT delivery of small lot sizes can require increased transportation, packaging, and handling that may contradict a green approach | |
| Piercy and Rich (2015) | | | Clear mutual benefit for lean and sustainability in every area of operation. |
| Wiese et al. (2015) | Synergic benefits stems from concurrent integration of lean, green and best practices into a SC organization. | | Strong correlation between lean and green thinking as both principles focus on reducing waste, decreasing costs and enhancing value for all stakeholders and partners within a supply chain. |
| Wu et al. (2015) | | | The implementation of lean, green and social practices have positive impacts on triple bottom line (Economic, environmental and social), but the integration of these practices have stronger impact. |

| | | | |
|---|---|---|---|
| Bortolini et al. (2016) | Combination of lean and green practices leads to greater results, compared to separate implementation of each practice. | | |
| Campos and Vazquez-Brust (2016) | The majority of the practices (26 out of 31) bring synergic results to lean and green performance. | Between JIT and green outcomes. Lean & green practices and production cost | Practices related to suppliers and customers are found to be the strongest synergic ones. |
| A. L. Carvalho et al. (2016) | Lean supply chain presented a strong alignment with the GSCM concepts. | | |
| Prasad et al. (2016) | Lean practices are positively and moderately interrelated with green practices. | | |
| Ugarte et al. (2016) | | JIT increases product shipment frequency, therefore it results in more GHG emission level, yet lean practices lead to reduced GHG in inventory and retail operations. | |
| Zhan et al. (2016) | The lean and green practices have win–win relationship with environmental and business performance of organizations. | | Lean and green practices were found to improve organization's environmental and business performance. |
| H. Carvalho et al. (2017) | | Between JIT and green outcomes, and geographic concentration with suppliers creates a trade-off between economic and environmental goals. | Not all companies belonging to the same supply chain can be absolutely lean or green. |

external GSCM, pollution prevention, cleaner production, green purchasing, reduction of hazardous, harmful, toxic materials or geographic concentration.

Qualitative studies using surveys, questionnaires or interviews to find out the practical implications of lean and green practices and their impacts on hypothesized key performance indicators are widely spread in the literature. As stated earlier, whether lean and green practices have positive or negative performance on 3BL performance indicators should be measured using actual financial numbers.

Finally, the deployment of lean and green practices in supply chains, either sequential or consolidated, has been attracting attention among academicians and practitioners recently. This attention is evident by the growing number of publications on lean and green supply chain management which have been published in scientific journals in recent years. This paper presents a comprehensive literature review of recent and state-of-the-art journal papers that have considered lean and green practices/initiatives in a SCM context to highlight future directions and opportunities in LGSC.

## References

Ahi, P., & Searcy, C. (2013). A comparative literature analysis of definitions for green and sustainable supply chain management. *Journal of Cleaner Production, 52*, 329–341. https://doi.org/10.1016/j.jclepro.2013.02.018.

Alzaman, C. (2014). Green supply chain modelling: Literature review. *International Journal of Business Performance and Supply Chain Modelling, 6*(1), 16–39.

Azevedo, S., Carvalho, H., & Cruz-Machado, V. (2011). A proposal of LARG supply chain management practices and a performance measurement system. *International Journal of e-Education, e-Business, e-Management and e-Learning*, 7–14. https://doi.org/10.7763/ijeeee.2011.v1.2.

Azevedo, S., Carvalho, H., Duarte, S., & Cruz-Machado, V. (2012). Influence of green and lean upstream supply chain management practices on business sustainability. *IEEE Transactions on Engineering Management, 59*(4), 753–765. https://doi.org/10.1109/tem.2012.2189108.

Azevedo, S., Pati, N., Carvalho, H., & Cruz-Machado, V. (2016). LARG index: A benchmarking tool for improving the leanness, agility, resilience and greenness of the automotive supply chain. *Benchmarking: An International Journal, 23*(6), 1472–1499. https://doi.org/10.1108/bij-07-2014-0072.

Bergmiller, G. G., & McCright, P. R. (2009). *Lean manufacturers' transcendence to green manufacturing*. Paper presented at the Proceedings of the 2009 industrial engineering research conference.

Bortolini, M., Ferrari, E., Galizia, F., & Mora, C. (2016). A reference framework integrating lean and green principles within supply chain management. *International Journal of Social, Behavioral, Educational, Economic, Business and Industrial Engineering, 10*(3), 859–864.

Br, R. K., Agarwal, A., & Sharma, M. K. (2016). Lean management – A step towards sustainable green supply chain. *Competitiveness Review, 26*(3), 311–331. https://doi.org/10.1108/cr-05-2015-0040.

Cabral, I., Grilo, A., & Cruz-Machado, V. (2012). A decision-making model for lean, agile, resilient and green supply chain management. *International Journal of Production Research, 50*(17), 4830–4845. https://doi.org/10.1080/00207543.2012.657970.

Campos, L. M. S., & Vazquez-Brust, D. A. (2016). Lean and green synergies in supply chain management. *Supply Chain Management: An International Journal, 21*(5), 627–641. https://doi.org/10.1108/scm-03-2016-0101.

Carvalho, H., Azevedo, S. G., & Cruz-Machado, V. (2010). Supply chain performance management: Lean and green paradigms. *International Journal of Business Performance and Supply Chain Modelling, 2*(3-4), 304–333.

Carvalho, H., Duarte, S., & Cruz Machado, V. (2011). Lean, agile, resilient and green: Divergencies and synergies. *International Journal of Lean Six Sigma, 2*(2), 151–179. https://doi.org/10.1108/20401461111135037.

Carvalho, H., Govindan, K., Azevedo, S. G., & Cruz-Machado, V. (2017). Modelling green and lean supply chains: An eco-efficiency perspective. *Resources, Conservation and Recycling, 120*, 75–87. https://doi.org/10.1016/j.resconrec.2016.09.025.

Carvalho, A. L., Ignacio, L. R., Esposto, K. F., & Ometto, A. R. (2016). Synergy between the multiple supply chain and green supply chain management (GSCM) approaches: An initial analysis aimed at fostering supply chain sustainability. *European Journal of Sustainable Development, 5*(3), 119–132. https://doi.org/10.14207/ejsd.2016.v5n3p119.

Colicchia, C., Creazza, A., & Dallari, F. (2017). Lean and green supply chain management through intermodal transport: Insights from the fast moving consumer goods industry. *Production Planning & Control, 28*(4), 321–334. https://doi.org/10.1080/09537287.2017.1282642.

Corbett, C. J., & Klassen, R. D. (2006). Extending the horizons: Environmental excellence as key to improving operations. *Manufacturing & Service Operations Management, 8*(1), 5–22.

Dawei, Z., Hamid, A. B. A., Chin, T. A., & Leng, K. C. (2015). Green supply chain management: A literature review. *Sains Humanika, 5*(2).

Dhingra, R., Kress, R., & Upreti, G. (2014). Does lean mean green? *Journal of Cleaner Production, 85*, 1–7. https://doi.org/10.1016/j.jclepro.2014.10.032.

Duarte, S., & Cruz Machado, V. (2017). Green and lean implementation: An assessment in the automotive industry. *International Journal of Lean Six Sigma, 8*(1), 65–88. https://doi.org/10.1108/ijlss-11-2015-0041.

Duarte, S., & Cruz-Machado, V. (2013). Modelling lean and green: A review from business models. *International Journal of Lean Six Sigma, 4*(3), 228–250. https://doi.org/10.1108/ijlss-05-2013-0030.

Duarte, S., & Cruz-Machado, V. (2014). Investigating lean and green supply chain linkages through a balanced scorecard framework. *International Journal of Management Science and Engineering Management, 10*(1), 20–29. https://doi.org/10.1080/17509653.2014.962111.

Duarte, S., & Machado, V. C. (2011). Manufacturing paradigms in supply chain management. *International Journal of Management Science and Engineering Management, 6*(5), 328–342. https://doi.org/10.1080/17509653.2011.10671181.

Dües, C. M., Tan, K. H., & Lim, M. (2013). Green as the new Lean: How to use Lean practices as a catalyst to greening your supply chain. *Journal of Cleaner Production, 40*, 93–100. https://doi.org/10.1016/j.jclepro.2011.12.023.

Esmer, S., Ceti, I. B., & Tuna, O. (2010). A simulation for optimum terminal truck number in a Turkish port based on lean and green concept. *The Asian Journal of Shipping and Logistics, 26*(2), 277–296. https://doi.org/10.1016/s2092-5212(10)80006-9.

Fahimnia, B., Sarkis, J., & Davarzani, H. (2015). Green supply chain management: A review and bibliometric analysis. *International Journal of Production Economics, 162*, 101–114. https://doi.org/10.1016/j.ijpe.2015.01.003.

Fahimnia, B., Sarkis, J., & Eshragh, A. (2015). A tradeoff model for green supply chain planning: A leanness-versus-greenness analysis. *Omega, 54*, 173–190. https://doi.org/10.1016/j.omega.2015.01.014.

Figueira, S., Machado, V. C., & Nunes, I. L. (2012). Integration of human factors principles in LARG organizations – A conceptual model. *Work, 41*(Suppl 1), 1712–1719. https://doi.org/10.3233/WOR-2012-0374-1712.

Friedman, P. (2008). Leaning toward green: Green your supply chain with lean practices. *Outsourced Logistics, 1*, 16–17.

Galeazzo, A., Furlan, A., & Vinelli, A. (2014). Lean and green in action: Interdependencies and performance of pollution prevention projects. *Journal of Cleaner Production, 85*, 191–200.

Garza-Reyes, J. A. (2015). Lean and green – A systematic review of the state of the art literature. *Journal of Cleaner Production, 102*, 18–29. https://doi.org/10.1016/j.jclepro.2015.04.064.

Garza-Reyes, J. A., Villarreal, B., Kumar, V., & Ruiz, P. M. (2016). Lean and green in the transport and logistics sector – A case study of simultaneous deployment. *Production Planning & Control, 27*(15), 1221–1232. https://doi.org/10.1080/09537287.2016.1197436.

Golpira, H., Najafi, E., Zandieh, M., & Sadi-Nezhad, S. (2017). Robust bi-level optimization for green opportunistic supply chain network design problem against uncertainty and environmental. *Computers & Industrial Engineering, 107*, 301–312. https://doi.org/10.1016/j.cie.2017.03.029.

Govindan, K., Azevedo, S. G., Carvalho, H., & Cruz-Machado, V. (2013). Lean, green and resilient practices influence on supply chain performance: Interpretive structural modeling approach. *International Journal of Environmental Science and Technology, 12*(1), 15–34. https://doi.org/10.1007/s13762-013-0409-7.

Govindan, K., Azevedo, S. G., Carvalho, H., & Cruz-Machado, V. (2014). Impact of supply chain management practices on sustainability. *Journal of Cleaner Production, 85*, 212–225. https://doi.org/10.1016/j.jclepro.2014.05.068.

Hajmohammad, S., Vachon, S., Klassen, R. D., & Gavronski, I. (2013a). Lean management and supply management: Their role in green practices and performance. *Journal of Cleaner Production, 39*, 312–320. https://doi.org/10.1016/j.jclepro.2012.07.028.

Hajmohammad, S., Vachon, S., Klassen, R. D., & Gavronski, I. (2013b). Reprint of Lean management and supply management: Their role in green practices and performance. *Journal of Cleaner Production, 56*, 86–93. https://doi.org/10.1016/j.jclepro.2013.06.038.

Hallam, C., & Contreras, C. (2016). Integrating lean and green management. *Management Decision, 54*(9), 2157–2187. https://doi.org/10.1108/md-04-2016-0259.

Hanson, J. D., Melnyk, S. A., & Calantone, R. J. (2004). Core values and environmental management: A strong inference approach. *Greener Management International, 46*.

Hervani, A. A., Helms, M. M., & Sarkis, J. (2005). Performance measurement for green supply chain management. *Benchmarking: An International Journal, 12*(4), 330–353.

Jasti, N. V. K., Sharma, A., & Kodali, R. (2012). Lean to green supply chain management: A case study. *Journal of Environmental Research and Development, 6*.

Kainuma, Y., & Tawara, N. (2006). A multiple attribute utility theory approach to lean and green supply chain management. *International Journal of Production Economics, 101*(1), 99–108. https://doi.org/10.1016/j.ijpe.2005.05.010.

King, A. A., & Lenox, M. J. (2001). Lean and green? An empirical examination of the relationship between lean production and environmental performance. *Production and Operations Management, 10*(3), 244–256.

Kleindorfer, P. R., Singhal, K., & Wassenhove, L. N. (2005). Sustainable operations management. *Production and Operations Management, 14*(4), 482–492.

Li, A. Q., & Found, P. (2016). Lean and green supply chain for the product-services system (PSS): The literature review and a conceptual framework. *Procedia CIRP, 47*, 162–167. https://doi.org/10.1016/j.procir.2016.03.057.

Liang, D., & Wang, H. (2013). Study on the development strategy of lean logistics for automobile enterprises under green supply chain environment. *Applied Mechanics and Materials, 397–400*, 2677–2680. https://doi.org/10.4028/www.scientific.net/AMM.397-400.2677.

Luthra, S., Garg, D., & Haleem, A. (2014). Green supply chain management: Implementation and performance – A literature review and some issues. *Journal of Advances in Management Research, 11*(1), 20–46. https://doi.org/10.1108/JAMR-07-2012-0027.

Machado, C., & Davim, J. P. (2017). *Green and lean management*. Springer.

Maleki, M., & Cruz Machado, V. (2013). Generic integration of lean, agile, resilient, and green practices in automotive supply chain. *Review of International Comparative Management/Revista de Management Comparat International, 14*(2), 237–248.

Malviya, R. K., & Kant, R. (2015). Green supply chain management (GSCM): A structured literature review and research implications. *Benchmarking: An International Journal, 22*(7), 1360–1394.

Mason, R., Nieuwenhuis, P., & Simons, D. (2008). Lean and green supply chain mapping: Adapting a lean management tool to the needs of industrial ecology. *Progress in Industrial Ecology, An International Journal, 5*(4), 302–324.

Moher, D., Liberati, A., Tetzlaff, J., Altman, D. G., & Group, P. (2009). Preferred reporting items for systematic reviews and meta-analyses: The PRISMA statement. *PLoS Med, 6*(7), e1000097. https://doi.org/10.1371/journal.pmed.1000097.

Mollenkopf, D., Halldórsson, Á., Stolze, H., Tate, W. L., & Ueltschy, M. (2010). Green, lean, and global supply chains. *International Journal of Physical Distribution & Logistics Management, 40*(1/2), 14–41. https://doi.org/10.1108/09600031011018028.

Ohno, T. (1988). *Toyota production system: Beyond large-scale production.* London: CRC Press.

Pandey, P., Shah, B. J., & Gajjar, H. (2017). A fuzzy goal programming approach for selecting sustainable suppliers. *Benchmarking: An International Journal, 24*(5), 1138–1165. https://doi.org/10.1108/bij-11-2015-0110.

Piercy, N., & Rich, N. (2015). The relationship between lean operations and sustainable operations. *International Journal of Operations & Production Management, 35*(2), 282–315. https://doi.org/10.1108/ijopm-03-2014-0143.

Prasad, S., Khanduja, D., & Sharma, S. K. (2016). An empirical study on applicability of lean and green practices in the foundry industry. *Journal of Manufacturing Technology Management, 27*(3), 408–426. https://doi.org/10.1108/jmtm-08-2015-0058.

Prentice, B. E., Vido, E., & Kosior, J. (2010). Creating lean, green and agile supply chains: The benefits of cabotage liberalization. *Canadian Journal of Transportation, 3*(1).

Rachid, B., Roland, D., Sebastien, D., & Ivana, R. (2017). Risk management approach for lean, agile, resilient and green supply chain. *World Academy of Science, Engineering and Technology, International Journal of Social, Behavioral, Educational, Economic, Business and Industrial Engineering, 11*(4), 742–750.

Reichhart, A., & Holweg, M. (2007). Lean distribution: Concepts, contributions, conflicts. *International Journal of Production Research, 45*(16), 3699–3722.

Sarkis, J., Zhu, Q., & Lai, K.-h. (2011). An organizational theoretic review of green supply chain management literature. *International Journal of Production Economics, 130*(1), 1–15. https://doi.org/10.1016/j.ijpe.2010.11.010.

Sauders, M., Lewis, P., & Thornhill, A. (2003). *Research methods for business students* (4th ed.pp. 100–109). London: Pearson.

Sawhney, R., Teparakul, P., Bagchi, A., & Li, X. (2007). En-Lean: A framework to align lean and green manufacturing in the metal cutting supply chain. *International Journal of Enterprise Network Management, 1*(3), 238–260.

Simpson, D. F., & Power, D. J. (2005). Use the supply relationship to develop lean and green suppliers. *Supply Chain Management: An International Journal, 10*(1), 60–68. https://doi.org/10.1108/13598540510578388.

Srivastava, S. K. (2007). Green supply-chain management: A state-of-the-art literature review. *International Journal of Management Reviews, 9*(1), 53–80. https://doi.org/10.1111/j.1468-2370.2007.00202.x.

Sulistio, J., & Rini, T. A. (2015). A structural literature review on models and methods analysis of green supply chain management. *Procedia Manufacturing, 4*, 291–299. https://doi.org/10.1016/j.promfg.2015.11.043.

Thanki, S., Govindan, K., & Thakkar, J. (2016). An investigation on lean-green implementation practices in Indian SMEs using analytical hierarchy process (AHP) approach. *Journal of Cleaner Production, 135*, 284–298. https://doi.org/10.1016/j.jclepro.2016.06.105.

Tice, J., Ahouse, L., & Larson, T. (2005). Lean production and EMSs: Aligning environmental management with business priorities. *Environmental Quality Management, 15*(2), 1–12.

Tranfield, D., Denyer, D., & Smart, P. (2003). Towards a methodology for developing evidence-informed management knowledge by means of systematic review. *British Journal of Management, 14*(3), 207–222.

Ugarte, G. M., Golden, J. S., & Dooley, K. J. (2016). Lean versus green: The impact of lean logistics on greenhouse gas emissions in consumer goods supply chains. *Journal of Purchasing and Supply Management, 22*(2), 98–109. https://doi.org/10.1016/j.pursup.2015.09.002.

Ugochukwu, P., Engström, J., & Langstrand, J. (2012). Lean in the supply chain: A literature review. *Management and Production Engineering Review, 3*(4), 87–96.

United States Environmental Protection Agency. (2000). *The lean and green supply chain: A practical guide for materials managers and supply chain managers to reduce costs and improve environmental performance* (MC-7409). Washington, D.C. 20460: EPA.

Venkat, K., & Wakeland, W. (2006). *Is lean necessarily green?* Paper presented at the Proceedings of the 50th Annual Meeting of the ISSS-2006, Sonoma, CA, USA.

Wiese, A., Luke, R., Heyns, G. J., & Pisa, N. M. (2015). The integration of lean, green and best practice business principles. *Journal of Transport and Supply Chain Management, 9*(1). https://doi.org/10.4102/jtscm.v9i1.192.

Wu, L., Subramanian, N., Abdulrahman, M., Liu, C., Lai, K.-h., & Pawar, K. (2015). The impact of integrated practices of lean, green, and social management systems on firm sustainability performance—Evidence from Chinese fashion auto-parts suppliers. *Sustainability, 7*(4), 3838–3858. https://doi.org/10.3390/su7043838.

Xu, A., Hu, X., & Gao, S. (2013). Review of green supply chain management. *International Journal of Networking and Virtual Organisations, 12*(1), 27–39.

Zhan, Y., Tan, K. H., Ji, G., Chung, L., & Chiu, A. S. F. (2016). Green and lean sustainable development path in China: Guanxi, practices and performance. *Resources, Conservation and Recycling*. https://doi.org/10.1016/j.resconrec.2016.02.006.

Zhu, Q., & Sarkis, J. (2004). Relationships between operational practices and performance among early adopters of green supply chain management practices in Chinese manufacturing enterprises. *Journal of Operations Management, 22*(3), 265–289.

Zokaei, K., Manikas, I., & Reza, S. (2015). Improving environmental and economic performance in the food chain; the lean and green paradigm. In I. P. Vlachos & G. Malindretos (Eds.), *Markets, business and sustainability* (p. 173). Sharjah: Bentham Science Publishers.

# A New Model for Lean and Green Closed-Loop Supply Chain Optimization

**Turan Paksoy, Ahmet Çalik, Alexander Kumpf, and Gerhard Wilhelm Weber**

## 1 Introduction

In recent years, there has been a growing attention in Supply Chain Management (SCM) among academicians and practitioners because of its beneficial factors such as cost minimization, profit maximization and customer satisfaction. Besides of these factors, many researchers have integrated their environmental concerns, such as pollution, global warming, overpopulation, climate change, etc., into the SCM. In this context, companies are obliged to keep in mind environmental concerns in their operations to satisfy customer expectations and obligations (Rostamzadeh et al. 2015).

At the same time, new paradigms have emerged and integrate green issues with the growing pressure from customers and governments. Lean manufacturing also can be referred to the Toyota production system that focus on elimination of waste and customer satisfaction simultaneously (Vlachos 2015). Also, integration of Lean&Green (L&G) paradigms has gained more popularity from researchers and

T. Paksoy
Department of Industrial Engineering, Konya Technical University, Konya, Turkey

A. Çalik (✉)
Department of International Trade and Logistics, KTO Karatay University, Konya, Turkey

A. Kumpf
Department of Logistics Management, Applied Sciences University of Landshut, Munich, Germany

G. W. Weber
Faculty of Engineering Management, Chair of Marketing and Economic Engineering, Poznan University of Technology, Poznan, Poland

IAM, METU, Ankara, Turkey



T. Paksoy et al. (eds.), *Lean and Green Supply Chain Management*, International Series in Operations Research & Management Science 273,
https://doi.org/10.1007/978-3-319-97511-5_2

practitioners. Companies have started to adopt these paradigms simultaneously to achieve the objectives of efficiency, effectiveness and sustainability (Fercoq et al. 2016; Colicchia et al. 2017). This means that Lean and Green Closed-Loop Supply Chain Management (L&G-CLSCM) is a mandatory research field that should be handled.

Therefore, L&G issues have appeared in the literature for the last decade. For example, Wang et al. (2011) proposed a multi-objective mixed-integer formulation by considering the environmental investment decision in supply network design. They established a trade-off between total cost and environmental influence and conducted a comprehensive set of numerical experiments. Kannan et al. (2012) developed a mixed-integer model based on a carbon footprint for a reverse logistics network design. Fahimnia et al. (2015) focused on lean-and-green paradigms to provide practical insights to managers using a mixed-integer nonlinear programming model. This model includes different cost functions, carbon emissions, energy consumption and waste generation. The carbon emissions represent the generated carbon pollution in manufacturing, transportation and inventory holding. Talaei et al. (2016) proposed a robust fuzzy optimization method for a carbon-efficient CLSC model, minimizing the total cost and carbon dioxide emission in the objectives.

Mollenkopf et al. (2010) conducted a literature survey on the simultaneous implementation of the three strategies as green, lean and global supply strategies. Cabral et al. (2012) created a decision-making model considering lean agile, resilient and green paradigms that are fundamental of companies in supply chain management. They applied the ANP method in an automotive supply chain to measure the impact of these concepts on the performance of companies and to select the best green supply chain application. Duarte and Cruz-Machado (2013) examined how to reach benefits with L&G approaches for any organization and its supply chain. For this purpose, 12 business models are examined and the differences and similarity relations between the models are investigated. Martínez-Jurado and Moyano-Fuentes (2014) established a linkage between Lean Management, Supply Chain Management and Sustainability through a literature search. Fifty eight articles published between the years 1990 and 2013 were divided into two categories and results were reported. Govindan et al. (2014) investigated the effects of lean, flexible and green supply chain management practices on sustainable supply chain issues. Eighteen research samples were tested with empirical data and the case study was conducted in the Portuguese automotive supply chain to determine which lean, flexible and green applications effect to sustainable supply chain. Govindan et al. (2015) identified critical, lean, green, and flexible applications that will help managers to improve the performance of their automotive supply chains. An application was conducted on the automotive supply chain in Portugal and concluded that just-in-time, flexible transportation and environmentally friendly packaging were found to be very important for the automotive supply chain. Ugarte et al. (2016) set up three different hypotheses to investigate the environmental impact of existing lean logistics applications and tested these hypotheses with the simulation model of a manufacturing-retailer supply chain.

Golpîra et al. (2017) constructed a bi-level optimization under risky environment. They handled both effects of the $CO_2$ emission and retailer's risk aversion by using the $\varepsilon$-constraint method. Carvalho et al. (2017) discussed green and lean practices of companies and eco-efficiency with a mathematical model. A case study in the automotive supply chain is selected to figure out the best green and lean practices by a real context. They stressed that the companies place in the whole network cannot be absolutely lean or green. Verrier et al. (2016) presented a state of the art of L&G topics through detailed best-practices. Garza-Reyes et al. (2016) proposed a new tool taking L&G paradigms into account in Mexico. Duarte and Cruz Machado (2017) investigated coordination of L&G initiatives in the automotive industry.

The paper is organized as follows. Following the introduction, the structure of the proposed model and problem definition is given in Sect. 2. After that, some insights on L&G-CLSCM are presented. Subsequently, the proposed model formulation is presented. In Sect. 5, computational studies are examined. Finally, some considerations are drawn in the conclusion in which some future directions are also discussed.

## 2 Problem Definition

We study a supply chain network of an anonymous automotive corporate group which consists of (*i*) raw material suppliers; (*j*) the corporate group that has subsidiaries such as suppliers, (*k*) assemblers, (*l*) plants, (*n*) collection/disassembly centre, and (*m*) customers. The schematic view of the developed model is shown in Fig. 1. In the network, suppliers buy raw material and transform into parts under capacity constraints. Assemblers provide the necessary parts from suppliers and assemble them into semi-finished products. Finished product consists of raw material, semi-finished product and components/parts, as seen in Fig. 2. The assembling of end-product is carried out in the potential plants by getting components, semi-finished product and raw material. The components of an end-product can be purchased from different suppliers, the semi-finished product of the end-product is provided by different assemblers and the necessary raw material (steel plate) is supplied from the raw material suppliers. In the proposed model, a single end-product and multiple components have been taken into consideration. As mentioned above, suppliers provide different components for plants and assemblers. In the manufacturing process, suppliers send the first type of the components (*C1*, *C2*, *C3*, and *C4*) to the plants and second type of the components (*P1*, *P2*, and *P3*) to the assemblers. Assembled parts in plants are transported to the customers to satisfy customer demands. After the used products are collected from customers, the reverse flow begins. After one period used products are collected from customers and send to the collection centres. The final step is shipping the used products to assemblers or plants according to a certain percentage.

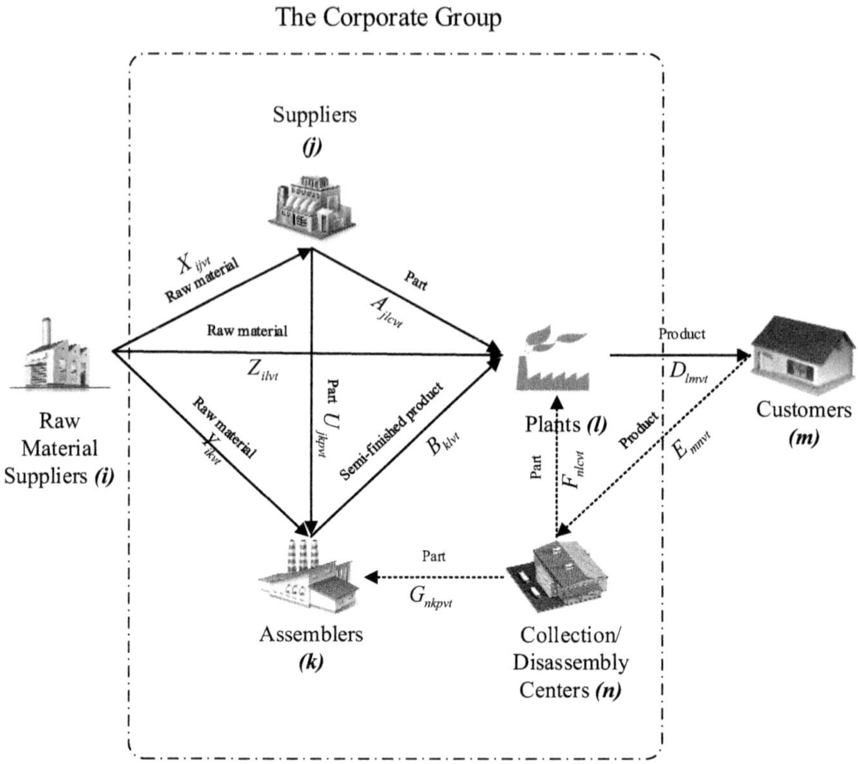


**Fig. 1** The structure of the developed model

## 3 Drivers of Lean&Green Supply Chain Management

Managers of the companies need to know which issues are important to improve their L&G performance and their competitiveness. L&G issues provide an opportunity to the managers for a further development of their supply chain management by enhancing collaboration among the business partners, managing investments, minimizing inventories, reducing $CO_2$ emissions, reusing and recycling of the packaging materials, etc. The aim of this section is to explain the drivers of the LG-SCM for a better understanding.

### *3.1 Lean&Green Construction*

According to the investment decisions, plants can be constructed with different sizes: Small, Medium and Large sizes are the construction options for the managers of the companies. As a pioneering paper in LG-SCM and optimization, Fahimnia

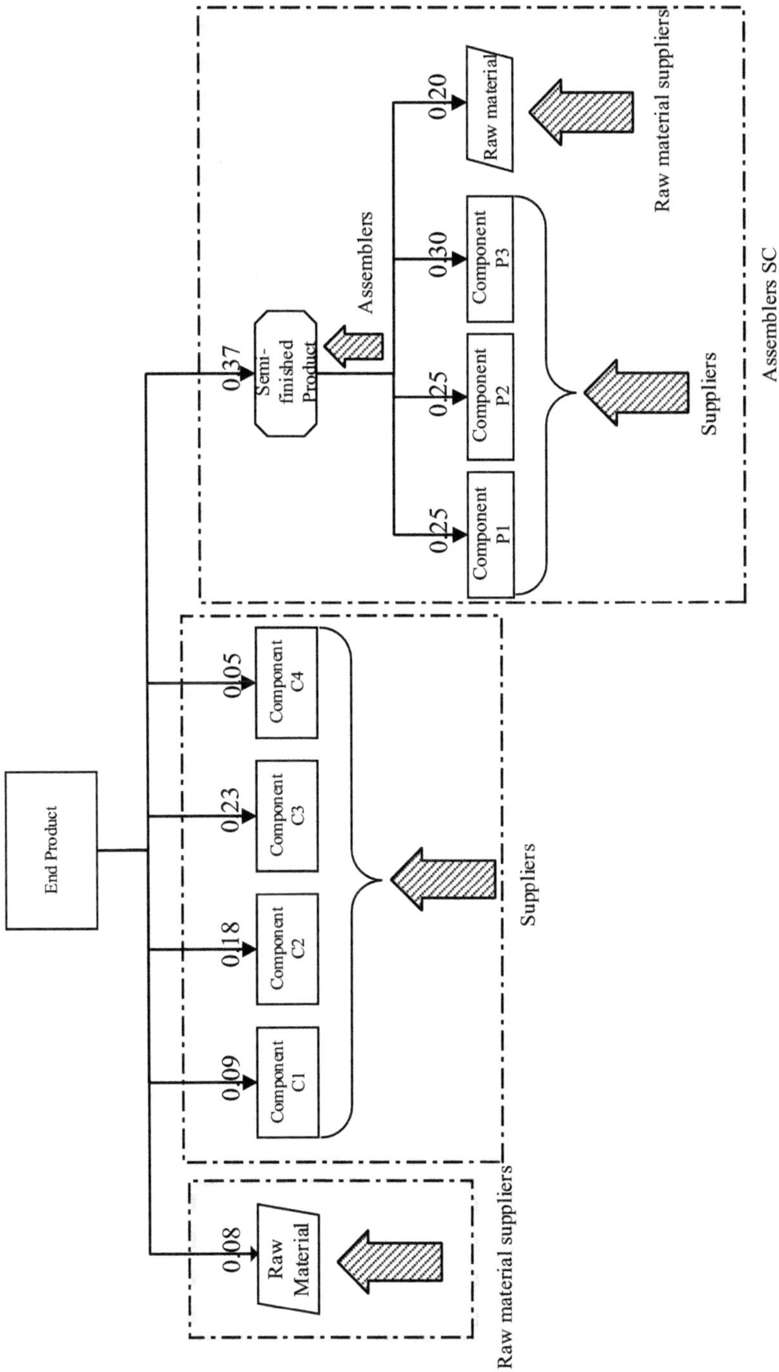


**Fig. 2** The structure of the end-product

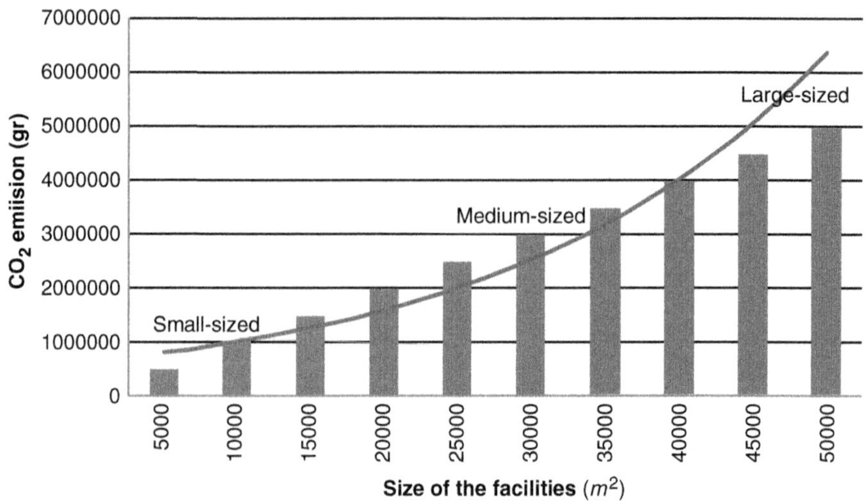


**Fig. 3** The amount of $CO_2$ emission emitted during the construction of the facilities (adapted from Klufallah et al. 2014)

et al. (2015) proposed a scale of structure for warehouses. They classified the warehouses according to three sizes. Following this approach, we offered an opportunity to the managers for making a trade-off between construction cost and $CO_2$ emission levels. The size of a construction has an influence on the emissions of a project. It is obvious that the $CO_2$ emissions during the construction of a building or facility are very important. The managers are increasingly interested in possibilities to reduce the $CO_2$ emissions in the construction process. The relationship between the construction of a facility and the $CO_2$ emissions during the construction can be seen in Fig. 3. The values are adapted from Klufallah et al. (2014). The amount of emitted $CO_2$ depends on the size of the facility. From this viewpoint, we handled this situation in the SC network with reducing the $CO_2$ emissions during the construction process.

## 3.2 *Lean&Green Production System*

Green Production should be an organizational strategy focusing on the bottom line by using environmental concerned production systems. The main purpose of green production is to reduce the environmental impact of the operating process at all the manufacturing systems by minimizing raw material usage, wastes, emissions and energy consumption. These aims also overlap with the aim of lean manufacturing directly (Emmett and Sood 2010). In this study, we focused on green production system with Environmental Protection Levels (EPL). The potential plants are separated according to the EPLs. The EPL concept was first proposed

**Fig. 4** The scale of Environmental Protection Level (EPL)

by Wang et al. (2011). At the beginning of the construction process, managers take this concept into account. Higher EPL means that a higher equipment or techniques should be employed in the production system with lower $CO_2$ emissions. According to the Wang et al. (2011): "The environmental investment can be used to purchase equipment or technology for environmental protection. The higher the environmental investment in facility is, the more sophisticated equipment or technology is installed."

Figure 4 depicts how the changes in the EPL can affect the investment with an illustrative scale. If this variable is not important (i.e., black and grey tones of protection level) for the manager, who does not pay any attention to the greenness, the investment will be cheaper, but this investment will lead to higher $CO_2$ emission levels.

When managers are willing to reduce $CO_2$ emissions but at the same time he/she wants to get a cheaper investment, he/she would choose white EPL. These managers are *hesitant managers* who have some awareness about the environmental protection however; they are not decisive. If the cost of investment is the same or a little bit higher, but not too much larger than the grey tones of EPL, they would choose a green EPL.

In contrast to the behaviour of the aforementioned managers, the *green managers* are interested in environmental protection and willing to expend extra amount of money.

The relationship between the fixed facility cost and environmental protection level can be seen in Fig. 5. If a facility is set up with higher equipment or technology, then the fixed facility cost will be higher. Facility-location decisions are fixed and unchangeable in the short-term period. However, in the long-term, this is a very critical decision for the managers. According to the environmental protection level, the change at the $CO_2$ emission can be seen in Fig. 5. The highest environmental protection level is the most green production system with the lowest $CO_2$ emission.

Manufacturers and their customers are aware of the fact that using high-tech equipment in production systems decreases the consumption of materials and resources (e.g., water and electricity), reduces wastes, and minimizes scraps. All these results obtained by usage of green/high-tech production systems are strongly overlapping with the aims of the lean manufacturing approach. Consequently, it is safe to say that the greener a supply chain is, the leaner it is as well.

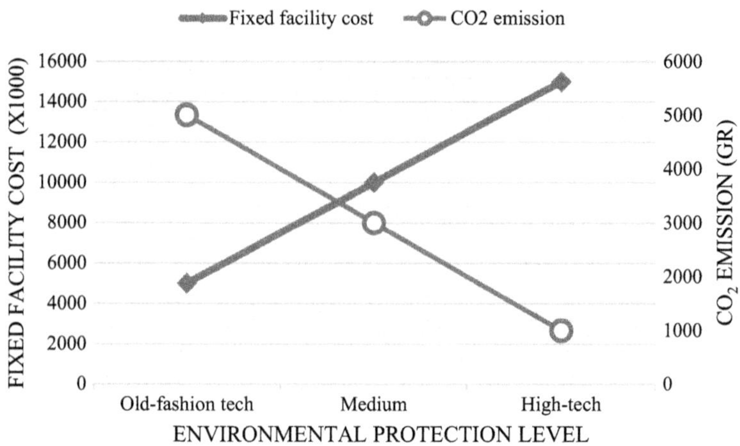


**Fig. 5** The fixed facility cost and the $CO_2$ emission change according to the environmental protection level (according to Wang et al. 2011)

### 3.3 Lean&Green Handling

Another way to reduce the $CO_2$ emissions in the plants can be achieved by using *greener technologies* for handling that forklifts are most used vehicles in handling. When forklifts are operating, they produce greenhouse gases. The usage of proper forklift is an important decision to increase the productivity and reduce the $CO_2$ emissions. With the type of proper forklift utilization, companies can achieve visible outcomes on the handling cost and on environmental protection. The most used forklifts are electric powered, battery operated, diesel engine and hybrid ones.

If the managers have an awareness about the green issues, then they can use lithium-ion or lead battery forklifts emitting significantly lower $CO_2$ emissions and spending lower fuel consumption. The usage of proper forklifts means another important choice for the managers (Fig. 6).

### 3.4 Lean&Green Transportation

According to the latest report issued by the U.S. Environmental Protection Agency (2014), the total transportation $CO_2$ emissions rose by 16% from 1990 to 2014. "The largest sources of transportation $CO_2$ emissions in 2014 were passenger cars (42.5%), medium- and heavy-duty trucks (23.1%), light-duty trucks, which include sport utility vehicles, pickup trucks, and minivans (17.8%) (U.S. Environmental Protection Agency 2014)".

From this viewpoint, we considered that all the transportation in the network is outsourced from a logistics firm where three different types of vehicles are used for

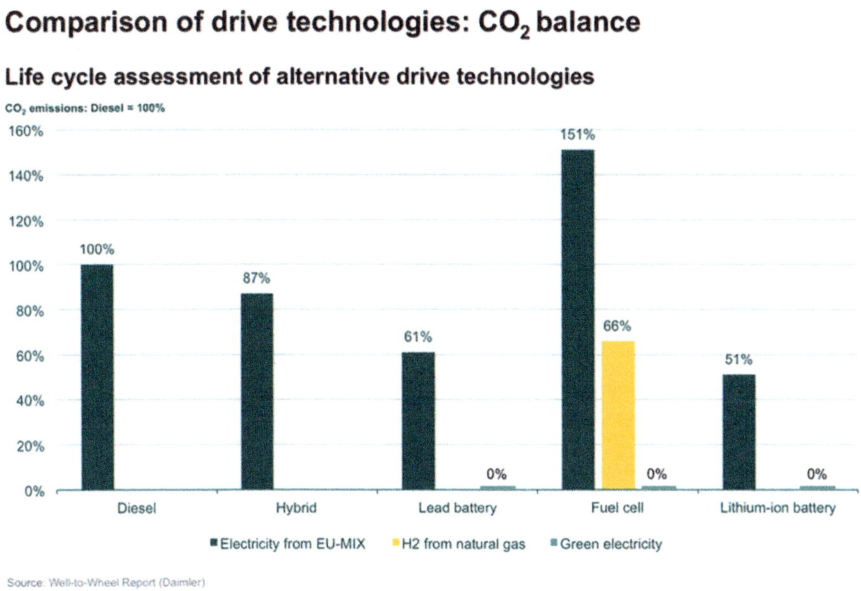


**Fig. 6** Comparison of forklift types with respect to $CO_2$ emission (http://www.jungheinrich.com/en/fork-lift-trucks-at-a-glance/drive-systems/co2-footprint/)

transportation. Small, medium, and heavy trucks are the transportation options in the network. "Smaller trucks are leaner (smaller lot size deliveries) requiring less storage space in warehouses", at the same time, it is environmentally efficient to use as well. On the other hand, it may not be economically efficient (Fahimnia et al. 2015).

In this study, we develop three different transportation strategies: L&G, mixed, and traditional strategies. In the L&G strategy, small size trucks are used and the lot sizes are as smaller as possible to support the lean manufacturing strategy. In the traditional strategy, large size trucks are used to carry material/parts/products between the partners of SC. This strategy offers an opportunity for saving less transportation cost. On the other hand, it does not consider environmental concerns. A comparison of these strategies can be seen in Table 1.

## 4 Model Formulation

The network can be formulated as a mixed-integer linear programming model. Sets, parameters and variables, objective functions and constraints are as follows:

**Table 1** Transportation strategies

| | Strategy | | |
|---|---|---|---|
| |  |  |  |
| Cost | Lean&Green | Mixed | Traditional |
| Environmental | Low | Medium | High |
| Transportation | High | Medium | Low |
| Late delivery | Low | Medium | High |

## 4.1 Indices

$I$ set of raw material suppliers $i \in I$,
$J$ set of suppliers $j \in J$,
$K$ set of assemblers $k \in K$,
$L$ set of plants $l \in L$,
$M$ set of customers $m \in M$,
$N$ set of collection centres $n \in N$,
$Q$ set of environmental protection level of production $q \in Q$,
$R$ set of forklifts $r \in R$,
$S$ set of size of plants $s \in S$,
$V$ set of vehicles $v \in V$,
$C$ set of parts of finished product $c \in C$,
$P$ set of parts of semi-finished product $p \in P$,
$T$ set of periods $t \in T$.

## 4.2 Parameters

### Distances and Unit Shipping Costs

$d_{ij}$ distance between raw material supplier '$i$' and supplier '$j$' (km),
$d_{ik}$ distance between raw material supplier '$i$' and assembler '$k$' (km),
$d_{il}$ distance between raw material supplier '$i$' and plant '$l$' (km),
$d_{jk}$ distance between supplier '$j$' and assembler '$k$' (km),
$d_{jl}$ distance between supplier '$j$' and plant '$l$' (km),
$d_{kl}$ distance between assembler '$k$' and plant '$l$' (km),
$d_{lm}$ distance between plant '$l$' and customer '$m$' (km),
$d_{mn}$ distance between customer '$m$' and collection centre '$n$' (km),
$d_{nl}$ distance between collection centre '$n$' and plant '$l$' (km),

$d_{nk}$ distance between collection centre '$n$' and assembler '$k$' (km),
$utc_v$ unit transportation cost of vehicle '$v$' (($/ton.km);

### Capacities

$caprm_{it}$ steel plate capacity of raw material supplier '$i$' at time period '$t$' (ton),
$capsc_{jct}$ part '$c$' capacity of supplier '$j$' at time period '$t$' (ton),
$capsp_{jpt}$ part '$p$' capacity of supplier '$j$' at time period '$t$' (ton),
$cappl_{sqt}$ finished product capacity of plant '$l$' with size '$s$' and environmental protection level '$q$' at time period '$t$' (ton),
$capa_{kt}$ semi-finished product capacity of assembler '$k$' at time period '$t$' (ton),
$capcc_{nt}$ used product capacity of collection centre '$n$' at time period '$t$' (ton),
$capce_{lsqt}$ capacity of $CO_2$ emission of plant '$l$' with size '$s$' and environmental protection level '$q$' at time period '$t$' (gr),
$vcap_v$ maximum capacity of vehicle '$v$' (ton);

### Demand

$de_{mt}$ demand of customer '$m$' at time period '$t$' (ton);

### CO$_2$ Emissions and Related Parameters

$cec_{lsqt}$ estimated $CO_2$ emissions for construction of plant '$l$' with size '$s$' and environmental protection level '$q$' at time period '$t$' (gr),
$cep_{lsqt}$ estimated $CO_2$ emissions to produce a ton of finished product at plant '$l$' with size '$s$' and environmental protection level '$q$' at time period '$t$' (gr/ton),
$cesc_{jct}$ estimated $CO_2$ emissions to produce a ton of part '$c$' at supplier '$j$' at time period '$t$' (gr/ton),
$cesp_{jpt}$ estimated $CO_2$ emissions to produce a ton of part '$p$' at supplier '$j$' at time period '$t$' (gr/ton),
$cea_{kt}$ estimated $CO_2$ emissions to produce a ton of semi-finished product at assembler '$k$' at time period '$t$' (gr/ton),
$cedc_{nct}$ estimated $CO_2$ emissions to disassembly a ton of part '$c$' at collection centre '$n$' at time period '$t$' (gr/ton),
$cedp_{npt}$ estimated $CO_2$ emissions to disassembly a ton of part '$p$' at collection centre '$n$' at time period '$t$' (gr/ton),
$ceh_{lrt}$ estimated $CO_2$ emissions for handling a ton of unit finished product with '$r$' type forklift at plant '$l$' at time period '$t$' (gr/ton),
$C_{CO_2}$ unit cost of $CO_2$ emissions ($/gr),
$CO_2^v$ amount of $CO_2$ emissions of vehicle '$v$' (gr/km),
$CO_2^r$ amount of $CO_2$ emissions of forklift '$r$' (gr/km);

**Fixed Costs**

$fcp_{lsqt}$ fixed opening cost of plant '*l*' with size '*s*' and environmental protection level '*q*' at time period '*t*' ($),
$fcc_{nt}$ fixed opening cost of collection centre '*n*' at time period '*t*' ($);

**Handling Costs**

$hc_{lrt}$ unit handling cost of forklift type '*r*' in plant '*l*' at time period '*t*' ($);

**Purchasing Costs**

$pcp_{jc}$ manufacturing cost of part '*c*' at supplier '*j*' ($/ton),
$pcs_{jp}$ manufacturing cost of part '*p*' at supplier '*j*' ($/ton),
$pcrm_i$ purchasing cost of raw material (steel plate) from raw material supplier '*i*' ($/ton),
$pccc_{nc}$ disassembly cost of part '*c*' at collection centre '*n*' (used part) ($/ton),
$pccp_{np}$ disassembly cost of part '*p*' at collection centre '*n*' (used part) ($/ton),
$pca_k$ assembly cost of semi-finished product at assembler '*k*' ($/ton),
$pcm_{lsqt}$ manufacturing cost of finished product at plant '*l*' with size '*s*' and environmental protection level '*q*' at time period '*t*' ($/ton),
$ccm_m$ collection cost of used product from customer '*m*' ($/ton);

**Ratios and Percentages**

$r$ weight ratio of semi-finished product in final product,
$r_c$ weight ratio of part '*c*' in finished product,
$r'$ weight ratio of raw material in finished product,
$rp_p$ weight ratio of part '*p*' in semi-finished product,
$rp'$ weight ratio of raw material in semi-finished product,
$\pi$ percentage of collected amount which is re-sent to plants (%),
$\mu$ maximum allowable percentages of late deliveries for plants (%),
$ldel_i$ late delivery percentage of raw material supplier '*i*' (%).

## 4.3 Variables

$X_{ijvt}$ amount of steel plate shipped from raw material supplier '*i*' to supplier '*j*' with vehicle '*v*' at time period '*t*' (ton),
$Y_{ikvt}$ amount of steel plate shipped from raw material supplier '*i*' to assembler '*k*' with vehicle '*v*' at time period '*t*' (ton),

$Z_{ilvt}$ amount of steel plate shipped from raw material supplier '*i*' to plant '*l*' with vehicle '*v*' at time period '*t*' (ton),

$U_{jkpvt}$ amount of part '*p*' shipped from supplier '*j*' to assembler '*k*' with vehicle '*v*' at time period '*t*' (ton),

$A_{jlcvt}$ amount of part '*c*' shipped from supplier '*j*' to plant '*l*' with vehicle '*v*' at time period '*t*' (ton),

$B_{klvt}$ amount of semi-finished product shipped from assembler '*k*' to plant '*l*' with vehicle '*v*' at time period '*t*' (ton),

$D_{lmvt}$ amount of finished product shipped from plant '*l*' to customer '*m*' with vehicle '*v*' at time period '*t*' (ton),

$E_{mnvt}$ amount of used product shipped from customer '*m*' to collection centre '*n*' with vehicle '*v*' at time period '*t*' (ton),

$F_{nlcvt}$ amount of part '*c*' shipped from collection centre '*n*' to plant '*l*' with vehicle '*v*' at time period '*t*' (ton),

$G_{nkpvt}$ amount of part '*p*' shipped from collection centre '*n*' to assembler '*k*' with vehicle '*v*' at time period '*t*' (ton),

$XX_{ijvt}$ if raw material shipped with vehicle '*v*' from raw material supplier '*i*' to supplier '*j*' at time period '*t*' 1; otherwise, 0,

$YY_{ikvt}$ if raw material shipped with vehicle '*v*' from raw material supplier '*i*' to assembler '*k*' at time period '*t*' 1; otherwise, 0,

$ZZ_{ilvt}$ if raw material shipped with vehicle '*v*' from raw material supplier '*i*' to plant '*l*' at time period '*t*' 1; otherwise, 0,

$UU_{jkpvt}$ if part '*p*' shipped with vehicle '*v*' from supplier '*j*' to assembler '*k*' at time period '*t*' 1; otherwise, 0,

$AA_{jlcvt}$ if part '*c*' shipped with vehicle '*v*' from supplier '*j*' to plant '*l*' at time period '*t*' 1; otherwise, 0,

$BB_{klvt}$ if semi-finished product shipped with vehicle '*v*' from assembler '*k*' to plant '*l*' at time period '*t*' 1; otherwise, 0,

$DD_{lmvt}$ if finished product shipped with vehicle '*v*' from plant '*l*' to customer '*m*' at time period '*t*' 1; otherwise, 0,

$EE_{mnvt}$ if used product shipped with vehicle '*v*' from customer '*m*' to collection centre '*n*' at time period '*t*' 1; otherwise, 0,

$FF_{nlcvt}$ if part '*c*' shipped with vehicle '*v*' from collection centre '*n*' to assembler '*k*' at time period '*t*' 1; otherwise, 0,

$GG_{nkpvt}$ if part '*p*' shipped with vehicle '*v*' from collection centre '*n*' to plant '*l*' at time period '*t*' 1; otherwise, 0,

$OP_{lsqt}$ if plant '*l*' is open with size '*s*' and environmental protection level '*q*' at time period '*t*', 1; otherwise, 0,

$OC_{nt}$ if collection centre '*n*' is open at time period '*t*', 1; otherwise, 0,

$UF_{lrt}$ if forklift type '*r*' is used at plant '*l*' at time period '*t*', 1; otherwise, 0,

$XINT_{ijvt}$ number of tours from raw material supplier '*i*' to supplier '*j*' with vehicle '*v*' at time period '*t*' (unit),

$YINT_{ikvt}$ number of tours from raw material supplier '*i*' to assembler '*k*' with vehicle '*v*' at time period '*t*' (unit),

$ZINT_{ilvt}$ number of tours from raw material supplier '$i$' to plant '$l$' with vehicle '$v$' at time period '$t$' (unit),

$UINT_{jkpvt}$ number of tours from supplier '$j$' to assembler '$k$' with vehicle '$v$' at time period '$t$' (unit),

$AINT_{jlcvt}$ number of tours from supplier '$j$' to plant '$l$' with vehicle '$v$' at time period '$t$' (unit),

$BINT_{klvt}$ number of tours from assembler '$k$' to plant '$l$' with vehicle '$v$' at time period '$t$' (unit),

$GINT_{nkpvt}$ number of tours from collection centre '$n$' to assembler '$k$' with vehicle '$v$' at time period '$t$' (unit),

$DINT_{lmvt}$ number of tours from plant '$l$' to customer '$m$' with vehicle '$v$' at time period '$t$' (unit),

$EINT_{mnvt}$ number of tours from customer '$m$' to collection centre '$n$' with vehicle '$v$' at time period '$t$' (unit),

$FINT_{nlcvt}$ number of tours from collection centre '$n$' to plant '$l$' with vehicle '$v$' at time period '$t$' (unit).

## 4.4 Objective Functions

The objective function ($Z$) is formulated as a sum of six expressions for the corporate group as follows:

$$\begin{aligned}\text{minimize } Z = {} & W_{TC} \cdot TC + W_{POPC} \cdot POPC + W_{FFC} \cdot FFC + W_{CEC} \cdot CEC \\ & + W_{HC} \cdot HC + W_{LD} \cdot LD,\end{aligned} \tag{1}$$

where the coefficients denoted by $W$ represent the weights of each objective function.

The transportation cost between the facilities can be calculated by multiplying the unit transportation cost with sum of distances:

$$\begin{aligned}TC = utc_{v} \cdot \Big[ & \sum_{i}\sum_{j}\sum_{v}\sum_{t} d_{ij} X_{ijvt} + \sum_{i}\sum_{k}\sum_{v}\sum_{t} d_{ik} Y_{ikvt} \\ & + \sum_{i}\sum_{l}\sum_{v}\sum_{t} d_{il} Z_{ilvt} + \sum_{j}\sum_{k}\sum_{p}\sum_{v}\sum_{t} d_{jk} U_{jkpvt} \\ & + \sum_{j}\sum_{l}\sum_{c}\sum_{v}\sum_{t} d_{jl} A_{jlcvt} + \sum_{k}\sum_{l}\sum_{v}\sum_{t} d_{kl} B_{klvt} \\ & + \sum_{l}\sum_{m}\sum_{v}\sum_{t} d_{lm} D_{lmvt} + \sum_{m}\sum_{n}\sum_{v}\sum_{t} d_{mn} E_{mnvt} \\ & + \sum_{n}\sum_{l}\sum_{c}\sum_{v}\sum_{t} d_{nl} F_{nlcvt} + \sum_{n}\sum_{k}\sum_{p}\sum_{v}\sum_{t} d_{nk} G_{nkpvt} \Big].\end{aligned} \tag{2}$$

Equation (3) expresses the purchasing and operational costs, such as purchasing cost of raw material, manufacturing cost of parts, assembly cost of semi-finished product, manufacturing cost of finished product, collection cost of used products and disassembly cost of parts. These terms are aggregated as follows:

$$POPC = \Big[\sum_i\sum_j\sum_v\sum_t pcrm_i X_{ijvt} + \sum_i\sum_k\sum_v\sum_t pcrm_i Y_{ikvt}$$
$$+\sum_i\sum_l\sum_v\sum_t pcrm_i Z_{ilvt} + \sum_j\sum_k\sum_p\sum_v\sum_t pcs_{jp} U_{jkpvt}$$
$$+\sum_j\sum_l\sum_c\sum_v\sum_t pcp_{jc} A_{jlcvt} + \sum_k\sum_l\sum_v\sum_t pca_k B_{klvt}$$
$$+\sum_l\sum_s\sum_q\sum_m\sum_v\sum_t pcm_{lsqt} D_{lmvt} + \sum_m\sum_n\sum_v\sum_t ccm_m E_{mnvt}$$
$$+\sum_n\sum_l\sum_c\sum_v\sum_t pccc_{nc} F_{nlcvt} + \sum_n\sum_k\sum_p\sum_v\sum_t pccp_{np} G_{nkpvt}\Big]. \quad (3)$$

The fixed-opening cost of plant '$l$' with size '$s$' and environmental protection level '$q$' and fixed-opening cost of collection centre '$n$' can be formulated as follows:

$$FFC = \sum_l\sum_s\sum_q\sum_t fcp_{lsqt} OP_{lsqt} + \sum_n\sum_t fcc_{nt} OC_{nt}. \quad (4)$$

The generated $CO_2$ emission costs which are caused by construction, transportation, manufacturing and handling, can be calculated as follows:

$$CEC = C_{CO_2}\cdot \Big\{\Big[\sum_l\sum_s\sum_q\sum_t cec_{lsqt}\, OP_{lsqt}\Big]$$
$$+\Big[\sum_i\sum_j\sum_v\sum_t CO_2^v d_{ij} XINT_{ijvt} + \sum_i\sum_k\sum_v\sum_t CO_2^v d_{ik} YINT_{ikvt}$$
$$+\sum_i\sum_l\sum_v\sum_t CO_2^v d_{il} ZINT_{ilvt} + \sum_j\sum_k\sum_p\sum_v\sum_t CO_2^v d_{jk} UINT_{jkpvt}$$
$$+\sum_j\sum_l\sum_c\sum_v\sum_t CO_2^v d_{jl} AINT_{jlcvt} + \sum_k\sum_l\sum_v\sum_t CO_2^v d_{kl} BINT_{klvt}$$
$$+\sum_n\sum_k\sum_p\sum_v\sum_t CO_2^v d_{nk} GINT_{nkpvt} + \sum_l\sum_m\sum_v\sum_t CO_2^v d_{lm} DINT_{lmvt}$$
$$+\sum_m\sum_n\sum_v\sum_t CO_2^v d_{mn} EINT_{mnvt} + \sum_n\sum_l\sum_c\sum_v\sum_t CO_2^v d_{nl} FINT_{nlcvt}\Big]$$
$$+\Big[\sum_l\sum_s\sum_q\sum_t cep_{lsqt} \sum_l\sum_s\sum_q\sum_t lin1_{lsqt}\Big]$$
$$+\Big[\sum_l\sum_r\sum_t ceh_{lrt} \sum_l\sum_r\sum_t lin2_{lrt}\Big]\Big\}$$
$$+C_{CO_2}\cdot\Big(\sum_j\sum_k\sum_p\sum_v\sum_t cesp_{jpt} U_{jkpvt} + \sum_j\sum_l\sum_c\sum_v\sum_t cesc_{jct} A_{jlcvt}$$
$$+\sum_k\sum_l\sum_v\sum_t cea_{kt} B_{klvt} + \sum_n\sum_l\sum_c\sum_v\sum_t cedc_{nct} F_{nlcvt}$$
$$+\sum_n\sum_k\sum_p\sum_v\sum_t cedp_{npt} G_{nkpvt}\Big), \quad (5)$$

where $lin1_{lsqt} = OP_{lsqt}\cdot \sum_l \sum_m \sum_v \sum_t D_{lmvt}$. To guarantee that the new variable operates properly, the following constraints are added to the model:

$$lin1_{lsqt} \le M\cdot OP_{lsqt} \qquad \forall_{l,s,q,t}, \quad (6)$$

$$lin1_{lsqt} \le \sum_m\sum_v D_{lmvt} \qquad \forall_{l,s,q,t}, \quad (7)$$

$$lin1_{lsqt} \geq \sum\nolimits_{m}\sum\nolimits_{v} D_{lmvt} - \left(1 - OP_{lsqt}\right) \cdot M \qquad \forall_{l,s,q,t}, \tag{8}$$

$$lin1_{lsqt} \geq 0 \qquad \forall_{l,s,q,t}, \tag{9}$$

where $M$ is a sufficiently large number.

The handling cost of plant '$l$' with forklift type '$r$' can be formulated as follows:

$$HC = \sum\nolimits_{l}\sum\nolimits_{r}\sum\nolimits_{t} hc_{lrt} UF_{lrt} \sum\nolimits_{l}\sum\nolimits_{m}\sum\nolimits_{v}\sum\nolimits_{t} D_{lmvt}. \tag{10}$$

Equation (10) is a nonlinear constraint which is given as a multiplication of a binary variables and continuous variables. To linearize the nonlinear effect of multiplication, we define a new continuous variable and reformulate the nonlinear term as follows:

$$HC = \sum\nolimits_{l}\sum\nolimits_{r}\sum\nolimits_{t} hc_{lrt} \cdot lin2_{lrt} \tag{11}$$

We shortened the multiplication of variables with a continuous variable, $lin2_{lrt} = UF_{lrt} \sum_{l} \sum_{m} \sum_{v} \sum_{t} D_{lmvt}$. To guarantee that the new variables operate properly, the following constraints are added to the model:

$$lin2_{lrt} \leq M \cdot UF_{lrt} \ \forall_{l,r,t}, \tag{12}$$

$$lin2_{lrt} \leq \sum_{m} \sum_{v} D_{lmvt} \ \forall_{l,r,t}, \tag{13}$$

$$lin2_{lrt} \geq \sum_{m} \sum_{v} D_{lmvt} - (1 - UF_{lrt}) \cdot M \ \forall_{l,r,t}, \tag{14}$$

$$lin2_{lrt} \geq 0 \ \forall_{l,r,t}, \tag{15}$$

where $M$ is a sufficiently large number.

The late delivery percentages of raw material suppliers can be calculated as follows:

$$LD = \sum\nolimits_{i}\sum\nolimits_{j}\sum\nolimits_{k}\sum\nolimits_{l}\sum\nolimits_{v}\sum\nolimits_{t} ldel_{i} \left(X_{ijvt} + Y_{ikvt} + Z_{ilvt}\right). \tag{16}$$

## 4.5 Constraints

**Capacity Constraints**
The total quantity of raw material sent from raw material suppliers to the suppliers, plants and assemblers should be less than or equal to the capacity of those raw material suppliers during any period:

$$\sum_j \sum_v X_{ijvt} + \sum_k \sum_v Y_{ikvt} + \sum_l \sum_v Z_{ilvt} \leq caprm_{it} \ \forall_{i,t}. \tag{17}$$

The total quantity of part *'c'* sent from suppliers to the plants should be less than or equal to the capacity of part *'c'* of those suppliers during any period:

$$\sum_l \sum_v A_{jlcvt} \leq capsc_{jct} \ \forall_{j,c,t}. \tag{18}$$

The total quantity of part *'p'* sent from suppliers to the assemblers should be less than or equal to capacity of part *'p'* of those suppliers during any period:

$$\sum_k \sum_v U_{jkpvt} \leq capsp_{jpt} \ \forall_{j,p,t}. \tag{19}$$

The total quantity of semi-finished products sent from assemblers to the plants should be less than or equal to the capacity of those assemblers during any period:

$$\sum_l \sum_v B_{klvt} \leq capa_{kt} \ \forall_{k,t}. \tag{20}$$

The total quantity of finished products sent from plants to customers should be less than or equal to capacity of end products those plants during any period:

$$\sum_m \sum_v D_{lmvt} \leq \sum_s \sum_q capp_{lsqt} OP_{lsqt} \ \forall_{l,t}. \tag{21}$$

The generated $CO_2$ emissions to produce a ton of finished product and handling should be less than or equal to the capacity of $CO_2$ emissions of plants during any period:

$$\begin{aligned}\left[\sum_s \sum_q cep_{lsqt} OP_{lsqt} \sum_m \sum_v D_{lmvt} + \sum_r ceh_{lrt} UF_{lrt} \sum_m \sum_v D_{lmvt}\right] \\ \leq \sum_s \sum_q capce_{lsqt} OP_{lsqt} \forall_{l,t}.\end{aligned} \tag{22}$$

The same linearization procedure is implemented for Eq. (22) and is given below:

$$\left[\sum_s \sum_q cep_{lsqt} lin1_{lsqt} + \sum_r ceh_{lrt} lin2_{lrt}\right] \leq \sum_s \sum_q capce_{lsqt} OP_{lsqt} \ \forall_{l,t}. \tag{23}$$

Constraint (24) ensures that at most one type size and one type environmental protection level can be selected at location '*l*' in each period.

$$\sum_s \sum_q OP_{lsqt} \leq 1 \; \forall_{l,t}. \tag{24}$$

Constraint (25) ensures that forklift types are chosen if corresponding plants are opened:

$$\sum_r UF_{lrt} \geq \sum_s \sum_q OP_{lsqt} \; \forall_{l,t}. \tag{25}$$

The total quantity of used products collected from customers should be less than or equal to capacity of product of common collection centres during any period:

$$\sum_m \sum_v E_{mnvt} \leq capcc_{nt} OC_{nt} \; \forall_{n,t}. \tag{26}$$

**Demand Constraints**
The demands of all customers are fully satisfied and the total quantity of products should be greater than or equal to the customers' demand during any period:

$$\sum_l \sum_v D_{lmvt} \geq dem_{mt} \; \forall_{m,t}. \tag{27}$$

**Balance Constraints (Kirchoff Law)**
According to the principle of conservation, Kirchoff equalities ensure that the sum of flows coming into that node is equal to the sum of flows going out that node in the CLSC network.

Constraint (28) ensures the flow balance at supplier '*j*' for both part '*c*' and part '*p*':

$$\sum_i \sum_v X_{ijvt} - \sum_l \sum_c \sum_v A_{jlcvt} - \sum_k \sum_p \sum_v U_{jkpvt} = 0 \; \forall_{j,t}. \tag{28}$$

Constraint (29) ensures the flow balance of raw material of semi-finished product at assembler '*k*':

$$\sum_i \sum_v Y_{ikvt} - rp' \cdot \sum_l \sum_v B_{klvt} = 0 \; \forall_{k,t}. \tag{29}$$

Constraint (30) ensures the flow balance at assembler '*k*' for part '*p*':

$$\sum_j \sum_v U_{jkpvt} + \sum_n \sum_v G_{nkpvt} - rp_p \cdot \left(\sum_l \sum_v B_{klvt}\right) = 0 \; \forall_{k,p,t}. \tag{30}$$

Constraint (31) ensures the flow balance of used products that are collected from customers after one period usage in SC:

$$\sum_l \sum_v D_{lmvt} - \sum_n \sum_v E_{mnv(t+1)} = 0 \; \forall_{m,t}. \tag{31}$$

Constraint (32) and (33) ensure the flow balance of part '$c$' and part '$p$' at collection centre '$n$':

$$r_c \cdot \left(\pi \cdot \sum_m \sum_v E_{mnvt}\right) - \sum_l \sum_v F_{nlcvt} = 0 \ \forall_{n,c,t} . \tag{32}$$

$$r \cdot rp_p \cdot \left((1-\pi) \cdot \sum_m \sum_v E_{mnvt}\right) - \sum_k \sum_v G_{nkpvt} = 0 \ \forall_{n,p,t} . \tag{33}$$

Constraint (34) ensures the flow balance of semi-finished product at plant '$l$':

$$\sum_k \sum_v B_{klvt} - r \cdot \sum_m \sum_v D_{lmvt} = 0 \ \forall_{l,t} . \tag{34}$$

Constraint (35) ensures the flow balance of part '$c$' at plant '$l$':

$$\sum_j \sum_v A_{jlcvt} + \sum_n \sum_v F_{nlcvt} - r_c \cdot \sum_m \sum_v D_{lmvt} = 0 \ \forall_{l,c,t} . \tag{35}$$

Constraint (36) ensures the flow balance of raw material at plants '$l$':

$$\sum_i \sum_v Z_{ilvt} - r' \cdot \sum_m \sum_v D_{lmvt} = 0 \ \forall_{l,t} . \tag{36}$$

**Transportation Constraints**
The following constraints ensure the minimum level of load for each type of vehicle ($M$ being a large number):

$$X_{ij1t} - M \cdot XX_{ij1t} \leq 0 \ \forall_{i,j,t} , \tag{37}$$

$$X_{ij1t} - M \cdot XX_{ij1t} \geq 0.001 - M \ \forall_{i,j,t} , \tag{38}$$

$$X_{ij2t} - M \cdot XX_{ij2t} \leq 0 \ \forall_{i,j,t} , \tag{39}$$

$$X_{ij2t} - M \cdot XX_{ij2t} \geq 10 - M \ \forall_{i,j,t} , \tag{40}$$

$$X_{ij3t} - M \cdot XX_{ij3t} \leq 0 \ \forall_{i,j,t} , \tag{41}$$

$$X_{ij3t} - M \cdot XX_{ij3t} \geq 20 - M \ \forall_{i,j,t} , \tag{42}$$

$$Y_{ik1t} - M \cdot YY_{ik1t} \leq 0 \ \forall_{i,k,t} , \tag{43}$$

$$Y_{ik1t} - M \cdot YY_{ik1t} \geq 0.001 - M \ \forall_{i,k,t} , \tag{44}$$

$$Y_{ik2t} - M \cdot YY_{ik2t} \leq 0 \ \forall_{i,k,t} , \tag{45}$$

$$Y_{ik2t} - M \cdot YY_{ik2t} \geq 10 - M \ \forall_{i,k,t} , \tag{46}$$

$$Y_{ik3t} - M \cdot YY_{ik3t} \leq 0 \ \forall_{i,k,t} , \tag{47}$$

$$Y_{ik3t} - M \cdot YY_{ik3t} \geq 20 - M \ \forall_{i,k,t} , \tag{48}$$

$$Z_{il1t} - M \cdot ZZ_{il1t} \leq 0 \ \forall_{i,l,t} , \tag{49}$$

$$Z_{il1t} - M \cdot ZZ_{il1t} \geq 0.001 - M \ \forall_{i,l,t} , \tag{50}$$

$$Z_{il2t} - M \cdot ZZ_{il2t} \leq 0 \ \forall_{i,l,t} , \tag{51}$$

$$Z_{il2t} - M \cdot ZZ_{il2t} \geq 10 - M \ \forall_{i,l,t} , \tag{52}$$

$$Z_{il3t} - M \cdot ZZ_{il3t} \leq 0 \ \forall_{i,l,t} , \tag{53}$$

$$Z_{il3t} - M \cdot ZZ_{il3t} \geq 20 - M \ \forall_{i,l,t} , \tag{54}$$

$$U_{jkp1t} - M \cdot UU_{jkp1t} \leq 0 \ \forall_{j,k,p,t} , \tag{55}$$

$$U_{jkp1t} - M \cdot UU_{jkp1t} \geq 0.001 - M \ \forall_{j,k,p,t} , \tag{56}$$

$$U_{jkp2t} - M \cdot UU_{jkp2t} \leq 0 \ \forall_{j,k,p,t}\,, \tag{57}$$

$$U_{jkp2t} - M \cdot UU_{jkp2t} \geq 10 - M \ \forall_{j,k,p,t}\,, \tag{58}$$

$$U_{jkp3t} - M \cdot UU_{jkp3t} \leq 0 \ \forall_{j,k,p,t}\,, \tag{59}$$

$$U_{jkp3t} - M \cdot UU_{jkp3t} \geq 20 - M \ \forall_{j,k,p,t}\,, \tag{60}$$

$$A_{jlc1t} - M \cdot AA_{jlc1t} \leq 0 \ \forall_{j,l,c,t}\,, \tag{61}$$

$$A_{jlc1t} - M \cdot AA_{jlc1t} \geq 0.001 - M \ \forall_{j,l,c,t}, \tag{62}$$

$$A_{jlc2t} - M \cdot AA_{jlc2t} \leq 0 \ \forall_{j,l,c,t}, \tag{63}$$

$$A_{jlc2t} - M \cdot AA_{jlc2t} \geq 10 - M \ \forall_{j,l,c,t}\,, \tag{64}$$

$$A_{jlc3t} - M \cdot AA_{jlc3t} \leq 0 \ \forall_{j,l,c,t}\,, \tag{65}$$

$$A_{jlc3t} - M \cdot AA_{jlc3t} \geq 20 - M \ \forall_{j,l,c,t}\,, \tag{66}$$

$$B_{kl1t} - M \cdot BB_{kl1t} \leq 0 \ \forall_{k,l,t}\,, \tag{67}$$

$$B_{kl1t} - M \cdot BB_{kl1t} \geq 0.001 - M \ \forall_{k,l,t}\,, \tag{68}$$

$$B_{kl2t} - M \cdot BB_{kl2t} \leq 0 \ \forall_{k,l,t}\,, \tag{69}$$

$$B_{kl2t} - M \cdot BB_{kl2t} \geq 10 - M \ \forall_{k,l,t}\,, \tag{70}$$

$$B_{kl3t} - M \cdot BB_{kl3t} \leq 0 \ \forall_{k,l,t}\,, \tag{71}$$

$$B_{kl3t} - M \cdot BB_{kl3t} \geq 20 - M \ \forall_{k,l,t}, \tag{72}$$

$$D_{lm1t} - M \cdot DD_{lm1t} \leq 0 \ \forall_{l,m,t}, \tag{73}$$

$$D_{lm1t} - M \cdot DD_{lm1t} \geq 0.001 - M \ \forall_{l,m,t}, \tag{74}$$

$$D_{lm2t} - M \cdot DD_{lm2t} \leq 0 \ \forall_{l,m,t}, \tag{75}$$

$$D_{lm2t} - M \cdot DD_{lm2t} \geq 10 - M \ \forall_{l,m,t}, \tag{76}$$

$$D_{lm3t} - M \cdot DD_{lm3t} \leq 0 \ \forall_{l,m,t}, \tag{77}$$

$$D_{lm3t} - M \cdot DD_{lm3t} \geq 20 - M \ \forall_{l,m,t}, \tag{78}$$

$$E_{mn1t} - M \cdot EE_{mn1t} \leq 0 \ \forall_{m,n,t}, \tag{79}$$

$$E_{mn1t} - M \cdot EE_{mn1t} \geq 0.001 - M \ \forall_{m,n,t}, \tag{80}$$

$$E_{mn2t} - M \cdot EE_{mn2t} \leq 0 \ \forall_{m,n,t}, \tag{81}$$

$$E_{mn2t} - M \cdot EE_{mn2t} \geq 10 - M \ \forall_{m,n,t}, \tag{82}$$

$$E_{mn3t} - M \cdot EE_{mn3t} \leq 0 \ \forall_{m,n,t} \tag{83}$$

$$E_{mn3t} - M \cdot EE_{mn3t} \geq 20 - M \ \forall_{m,n,t}, \tag{84}$$

$$F_{nlc1t} - M \cdot FF_{nlc1t} \leq 0 \ \forall_{n,l,c,t}, \tag{85}$$

$$F_{nlc1t} - M \cdot FF_{nlc1t} \geq 0.001 - M \ \forall_{n,l,c,t}, \tag{86}$$

$$F_{nlc2t} - M \cdot FF_{nlc2t} \leq 0 \; \forall_{n,l,c,t} \,, \tag{87}$$

$$F_{nlc2t} - M \cdot FF_{nlc2t} \geq 10 - M \; \forall_{n,l,c,t} \,, \tag{88}$$

$$F_{nlc3t} - M \cdot FF_{nlc3t} \leq 0 \; \forall_{n,l,c,t} \,, \tag{89}$$

$$F_{nlc3t} - M \cdot FF_{nlc3t} \geq 20 - M \; \forall_{n,l,c,t} \,, \tag{90}$$

$$G_{nkp1t} - M \cdot GG_{nkp1t} \leq 0 \; \forall_{n,k,p,t} \,, \tag{91}$$

$$G_{nkp1t} - M \cdot GG_{nkp1t} \geq 0.001 - M \; \forall_{n,k,p,t} \,, \tag{92}$$

$$G_{nkp2t} - M \cdot GG_{nkp2t} \leq 0 \; \forall_{n,k,p,t} \,, \tag{93}$$

$$G_{nkp2t} - M \cdot GG_{nkp2t} \geq 10 - M \; \forall_{n,k,p,t} \,, \tag{94}$$

$$G_{nkp3t} - M \cdot GG_{nkp3t} \leq 0 \; \forall_{n,k,p,t} \,, \tag{95}$$

$$G_{nkp3t} - M \cdot GG_{nkp3t} \geq 20 - M \; \forall_{n,k,p,t} \,. \tag{96}$$

The following constraints calculate the number of tours at each arc regarding to vehicle types:

$$XINT_{ijvt} \geq \left(X_{ijvt}/vcap_v\right) \forall_{i,j,v,t} \,, \tag{97}$$

$$YINT_{ijvt} \geq \left(Y_{ikvt}/vcap_v\right) \forall_{i,k,v,t} \,, \tag{98}$$

$$ZINT_{ilvt} \geq \left(Z_{ilvt}/vcap_v\right) \forall_{i,l,v,t} \,, \tag{99}$$

$$UINT_{jkpvt} \geq \left(U_{jkpvt}/vcap_v\right) \forall_{j,k,p,v,t} \,, \tag{100}$$

$$AINT_{jlcvt} \geq \left(A_{jlcvt}/vcap_v\right) \forall_{j,l,c,v,t}, \tag{101}$$

$$BINT_{klvt} \geq \left(B_{klvt}/vcap_v\right) \forall_{k,l,v,t}, \tag{102}$$

$$GINT_{nkpvt} \geq \left(G_{nkpvt}/vcap_v\right) \forall_{n,k,p,v,t}, \tag{103}$$

$$DINT_{lmvt} \geq \left(D_{lmvt}/vcap_v\right) \forall_{l,m,v,t}, \tag{104}$$

$$EINT_{mnvt} \geq \left(\frac{E_{mnvt}}{vcap_v}\right) \forall_{m,n,v,t}, \tag{105}$$

$$FINT_{nlcvt} \geq \left(\frac{F_{nlcvt}}{vcap_v}\right) \forall_{n,l,c,v,t}. \tag{106}$$

The following constraint assures that total late delivery percentage of raw material supplied from raw material suppliers cannot exceed the plants' maximum allowable late delivery percentage:

$$\sum_i \sum_j \sum_v \sum_t ldel_i X_{ijvt} + \sum_i \sum_k \sum_v \sum_t ldel_i Y_{ikvt} + \sum_i \sum_l \sum_v \sum_t ldel_i Z_{ilvt} \leq \sum_l \sum_m \sum_v \sum_t \mu D_{lmvt}. \tag{107}$$

**Non-negativity Constraints**

The following constraints show non-negativity restrictions on the decision variables.

$$X_{ijvt} \geq 0 \ \forall_{i,j,v,t}, \tag{108}$$

$$Y_{ikvt} \geq 0 \ \forall_{i,k,v,t}, \tag{109}$$

$$Z_{ilvt} \geq 0 \ \forall_{i,l,v,t}, \tag{110}$$

$$U_{jkpvt} \geq 0 \ \forall_{j,k,p,v,t}, \tag{111}$$

$$A_{jlcvt} \geq 0 \; \forall_{j,l,c,v,t}, \tag{112}$$

$$B_{klvt} \geq 0 \; \forall_{k,l,v,t}, \tag{113}$$

$$D_{lmvt} \geq 0 \; \forall_{l,m,v,t}, \tag{114}$$

$$E_{mnvt} \geq 0 \; \forall_{m,n,v,t}, \tag{115}$$

$$F_{nlcvt} \geq 0 \; \forall_{n,l,c,v,t}, \tag{116}$$

$$G_{nkpvt} \geq 0 \; \forall_{n,k,p,v,t}. \tag{117}$$

**Binary Variables Constraints**
The following constraints show binary restrictions on the decision variables.

$$OP_{lsqt} \in \{0, 1\} \; \forall_{l,s,q,t}, \tag{118}$$

$$OC_{nt} \in \{0, 1\} \; \forall_{n,t}, \tag{119}$$

$$XX_{ijvt} \in \{0, 1\} \; \forall_{i,j,v,t}, \tag{120}$$

$$YY_{ikvt} \in \{0, 1\} \; \forall_{i,k,v,t}, \tag{121}$$

$$ZZ_{ilvt} \in \{0, 1\} \; \forall_{i,l,v,t}, \tag{122}$$

$$UU_{jkpvt} \in \{0, 1\} \; \forall_{j,k,p,v,t}, \tag{123}$$

$$AA_{jlcvt} \in \{0, 1\} \; \forall_{j,l,c,v,t}, \tag{124}$$

$$BB_{klvt} \in \{0, 1\} \; \forall_{k,l,v,t}\,, \tag{125}$$

$$DD_{lmvt} \in \{0, 1\} \; \forall_{l,m,v,t}\,, \tag{126}$$

$$EE_{mnvt} \in \{0, 1\} \; \forall_{m,n,v,t}\,, \tag{128}$$

$$GG_{nkpvt} \in \{0, 1\} \; \forall_{n,k,p,v,t}\,. \tag{129}$$

**Integer-Variables Constraints**

The following constraints show integer restrictions on the decision variables.

$$XINT_{ijvt} \; integer \; \forall_{i,j,v,t}\,, \tag{130}$$

$$YINT_{ikvt} \; integer \; \forall_{i,k,v,t}\,, \tag{131}$$

$$ZINT_{ilvt} \; integer \; \forall_{i,l,v,t}\,, \tag{132}$$

$$UINT_{jkpvt} \; integer \; \forall_{j,k,p,v,t}\,, \tag{133}$$

$$AINT_{jlcvt} \; integer \; \forall_{j,l,c,v,t}\,, \tag{134}$$

$$BINT_{klvt} \; integer \; \forall_{k,l,v,t}\,, \tag{135}$$

$$DINT_{lmvt} \; integer \; \forall_{l,m,v,t}\,, \tag{136}$$

$$EINT_{mnvt} \; integer \; \forall_{m,n,v,t}\,, \tag{137}$$

$$FINT_{nlcvt} \; integer \; \forall_{n,l,c,v,t}\,, \tag{138}$$

$$GINT_{nkpvt} \; integer \; \forall_{n,k,p,v,t}\,. \tag{139}$$

**Table 2** The trade-off table for membership functions

| Objective functions | $\mu = 1$ | $\mu = 0$ |
|---|---|---|
| $Z_{TC}$ | 15701616.05 | 395944597.50 |
| $Z_{POPC}$ | 50325976.93 | 204757836.92 |
| $Z_{FFC}$ | 33800000 | 126600000 |
| $Z_{CEC}$ | 9025162.06 | 937653793.51 |
| $Z_{HC}$ | 15288 | 222101.40 |
| $Z_{LD}$ | 346.05 | 4287.69 |

# 5 Computational Studies

In this section, we implement a numerical example to validate the proposed model based on randomly generated data. The parameters of the numerical example have been obtained using a uniform distribution, and all related data are given at the corresponding author's website (www.turanpaksoy.com/data), because of the page restrictions. The mixed-integer linear model for $Z_{CEC}$ is solved with GAMS 24.0.1 on a PC (I5 CPU 2.67 GHz with 3 GB RAM) and the computational time required to solve the model is 7200 CPU seconds with the optimality gap of 10%. The other models for the rest of the objective functions are found with the optimal solution.

## 5.1 Construction of Trade-Off Table

After the developed model is designed, six objective functions are optimized separately to obtain upper and lower bounds of each individual objective functions. The obtained results are given subsequently:

Using the values given in Table 2, membership functions for each individual objective functions are formulated as follows l:

$$\mu_{TC}\left(Z_{TC}(x)\right) = \begin{cases} 1, \ Z_{TC}(x) \leq 15701616.05, \\ \frac{395944597.50 - Z_{TC}(x)}{395944597.50 - 15701616.05}, \ 15701616.05 \leq Z_{TC}(x) \leq 395944597.50, \\ 0, \ Z_{TC}(x) \geq 4395944597.50, \end{cases} \tag{140}$$

$$\mu_{POPC}\left(Z_{POPC}(x)\right) = \begin{cases} 1, \ Z_{POPC}(x) \leq 50325976.93, \\ \frac{204757836.92 - Z_{POPC}(x)}{204757836.92 - 50325976.93}, \ 50325976.93 \leq Z_{POPC}(x) \leq 204757836.92, \\ 0, \ Z_{POPC}(x) \geq 204757836.92\ , \end{cases} \tag{141}$$

$$\mu_{FFC}\left(Z_{FFC}(x)\right)=\begin{cases}1, \; Z_{FFC}(x)\leq 33800000.00,\\ \frac{126600000.00-Z_{FFC}(x)}{126600000.00-33800000.00}, \; 33800000.00\leq Z_{FFC}(x)\leq 126600000.00,\\ 0, \; Z_{FFC}(x)\geq 126600000.00,\end{cases} \tag{142}$$

$$\mu_{CEC}\left(Z_{CEC}(x)\right)=\begin{cases}1, \; Z_{CEC}(x)\leq 9025162.06,\\ \frac{937653793.51-Z_{CEC}(x)}{937653793.51-9025162.06}, \; 9025162.06\leq Z_{CEC}(x)\leq 937653793.51,\\ 0, \; Z_{CEC}(x)\geq 937653793.51,\end{cases} \tag{143}$$

$$\mu_{HC}\left(Z_{HC}(x)\right)=\begin{cases}1, \; Z_{HC}(x)\leq 15288.00,\\ \frac{222248.84-Z_{HC}(x)}{222248.84-15288.00}, \; 15288.00\leq Z_{HC}(x)\leq 222248.84,\\ 222248.84,\end{cases} \tag{144}$$

$$\mu_{LD}\left(Z_{LD}(x)\right)=\begin{cases}1, \; Z_{LD}(x)\leq 346.05,\\ \frac{4287.69-Z_{LD}(x)}{4287.69-346.05}, \; 4346.05\leq Z_{LD}(x)\leq 4287.69,\\ 0, \; Z_{LD}(x)\geq 4287.69.\end{cases} \tag{145}$$

## 5.2 Determination of Weights Using Fuzzy AHP

In order to obtain the weights of objective functions, we use fuzzy AHP method developed by Buckley (1985). The linguistic scale (cf. Table 3) for a pairwise comparison matrix is adapted from Awasthi et al. (2010).

The decision makers of the anonymous corporate group expressed their opinion about the priority degrees of objective functions. Decision-makers consisted of three managers: the CEO, the operations and production manager (OPM), and the purchasing and supply chain manager (PSM). The ratings provided by the decision makers and aggregated weights of objective functions are given in Tables 4, 5, and 6.

**Table 3** Linguistic terms for objective functions ratings

| Linguistic term | Triangular fuzzy numbers |
|---|---|
| Equally importance (E) | (1, 1, 1) |
| Moderate importance (MI) | (1, 3, 5) |
| Essential or strong importance (SI) | (3, 5, 7) |
| Demonstrated importance (DI) | (5, 7, 9) |
| Extreme importance (EI) | (7, 9, 9) |

**Table 4** Rating of suppliers according to CEO

| | TC | POPC | FFC | CEC | HC | LD |
|---|---|---|---|---|---|---|
| TC | E | 1/DI | 1/MI | SI | SI | 1/DI |
| POPC | | E | SI | DI | EI | E |
| FFC | | | E | SI | DI | 1/MI |
| CEC | | | | E | MI | 1/DI |
| HC | | | | | E | 1/EI |
| LD | | | | | | E |
| Weights | 0.102 | 0.356 | 0.151 | 0.040 | 0.025 | 0.326 |

**Table 5** Rating of suppliers according to OPM

| | TC | POPC | FFC | CEC | HC | LD |
|---|---|---|---|---|---|---|
| TC | E | 1/MI | 1/MI | SI | SI | DI |
| POPC | | E | MI | DI | SI | DI |
| FFC | | | E | SI | DI | DI |
| CEC | | | | E | 1/MI | 1/MI |
| HC | | | | | E | MI |
| LD | | | | | | E |
| Weights | 0.198 | 0.375 | 0.279 | 0.037 | 0.067 | 0.043 |

**Table 6** Rating of suppliers according to PSM

| | TC | POPC | FFC | CEC | HC | LD |
|---|---|---|---|---|---|---|
| TC | E | 1/MI | 1/MI | MI | SI | DI |
| POPC | | E | MI | SI | SI | DI |
| FFC | | | E | SI | DI | EI |
| CEC | | | | E | MI | SI |
| HC | | | | | E | MI |
| LD | | | | | | E |
| Weights | 0.185 | 0.359 | 0.286 | 0.091 | 0.051 | 0.027 |

**Table 7** Weights of objective functions from three decision-makers

| | CEO | P-SC | PM | Weights |
|---|---|---|---|---|
| TC | 0.102 | 0.185 | 0.198 | 0.147 |
| POPC | 0.356 | 0.359 | 0.375 | 0.361 |
| FFC | 0.151 | 0.286 | 0.279 | 0.217 |
| CEC | 0.040 | 0.091 | 0.037 | 0.052 |
| HC | 0.025 | 0.051 | 0.067 | 0.042 |
| LD | 0.326 | 0.027 | 0.043 | 0.181 |

At first, we calculated all criterion weights by the FAHP method and then we aggregated the criteria according to criteria. In this aggregation process, the CEO's weights were multiplied by the coefficient of 2, while the other decision-makers' weights were multiplied by the coefficient of 1. Then, in order to find the objective functions weights, we calculated the average of all individual objective functions. As a result of the calculations, the weights of the criteria are given in Table 7.

Following Buckley's (1985) fuzzy AHP method, weights are obtained for each objective function as seen in Table 7. The purchasing and operational costs are selected as the most important driver with the rate of 0.361 in the developed model. It is observed that handling cost has the lowest weight for the developed model.

## *5.3 Fuzzy Weighted Additive Model*

The proposed model is solved with Tiwari et al.'s (1987) fuzzy weighted additive approach and transformed into the following model:

$$\text{minimize } Z = W_{TC}\cdot\lambda_1 + W_{POPC}\cdot\lambda_2 + W_{FFC}\cdot\lambda_3 + W_{CEC}\cdot\lambda_4 + W_{HC}\cdot\lambda_5 + W_{LD}\cdot\lambda_6$$

subject to

$$\lambda_1 \le \mu_{TC}\left(Z_{TC}(x)\right),$$

$$\lambda_2 \le \mu_{POPC}\left(Z_{POPC}(x)\right),$$

$$\lambda_3 \le \mu_{FFC}\left(Z_{FFC}(x)\right),$$

$$\lambda_4 \le \mu_{CEC}\left(Z_{CEC}(x)\right),$$

$$\lambda_5 \le \mu_{HC}\left(Z_{HC}(x)\right),$$

$$\lambda_6 \le \mu_{LD}\left(Z_{LD}(x)\right),$$

Constraints (17)–(139),

$$\lambda_1, \lambda_2, \lambda_3, \lambda_4, \lambda_5, \lambda_6 \in [0, 1]. \tag{146}$$

When this problem is solved using Tiwari's weighted additive approach, the costs and satisfaction degrees of the objective functions are calculated as seen in Table 8. The highest satisfaction degrees are obtained for fixed-opening cost and late delivery.

**Table 8** Optimal results according to the Tiwari's approach

| | |
|---|---|
| $\lambda_1$ | 0.995 |
| $\lambda_2$ | 0.996 |
| $\lambda_3$ | 1 |
| $\lambda_4$ | 0.995 |
| $\lambda_5$ | 0.994 |
| $\lambda_6$ | 1 |
| $Z_{TC}$ | 17479046.74 |
| $Z_{POPC}$ | 50925605.34 |
| $Z_{FFC}$ | 33800000 |
| $Z_{CEC}$ | 13433061.80 |
| $Z_{HC}$ | 16547 |
| $Z_{LD}$ | 347.66 |

**Table 9** The objective function's values and the percentages of the each objective functions

| | Performance criteria (PC) | Value ($) | Percentage of total cost (%) |
|---|---|---|---|
| PC1 | Total objective function | 115654260.9 | 100 |
| PC2 | Total transportation cost | 17479046.74 | 15.113 |
| PC3 | Total purchasing and operational cost | 50925605.34 | 44.033 |
| PC4 | Total fixed facility cost | 33800000 | 29.225 |
| PC5 | Total environmental cost | 13433061.80 | 11.615 |
| PC6 | Total handling cost | 16547 | 0.014 |
| PC7 | Late delivery | 347.66 | – |

When the developed model is solved with the given parameters, the total cost, which includes transportation, purchasing and operational, fixed facility, handling and generated $CO_2$ emission costs is found to be $115654260.9 for the weighted additive model. Table 9 provides the primary performance measures as a percentage of the overall value of the objective function. The results given in Table 9 show that the environmental costs account for a 11% share of the overall cost. In comparison, the maximum share is actualized by purchasing and operational costs with 44%, the minimum share is actualized by handling costs, approximately with 1%.

The optimal distribution networks of all three periods are illustrated in Fig. 7.

According to the distribution plan, in all three periods, 2589.48 ton of raw materials are sent from raw material suppliers to suppliers, assemblers and plants. At all periods, only first raw material supplier is chosen and only the first plant is opened. For this opened plant, EPL is 2 in first period, and 1 for the second and third period. The entire transportation process is actualized by using Vehicle 3 (heavy truck). Throughout the three periods, 1210, 1259, and 1218 tons of finished products are sent to the customers, respectively. The same amount of used products are delivered to the collection centres after one period of usage for disassembly process.

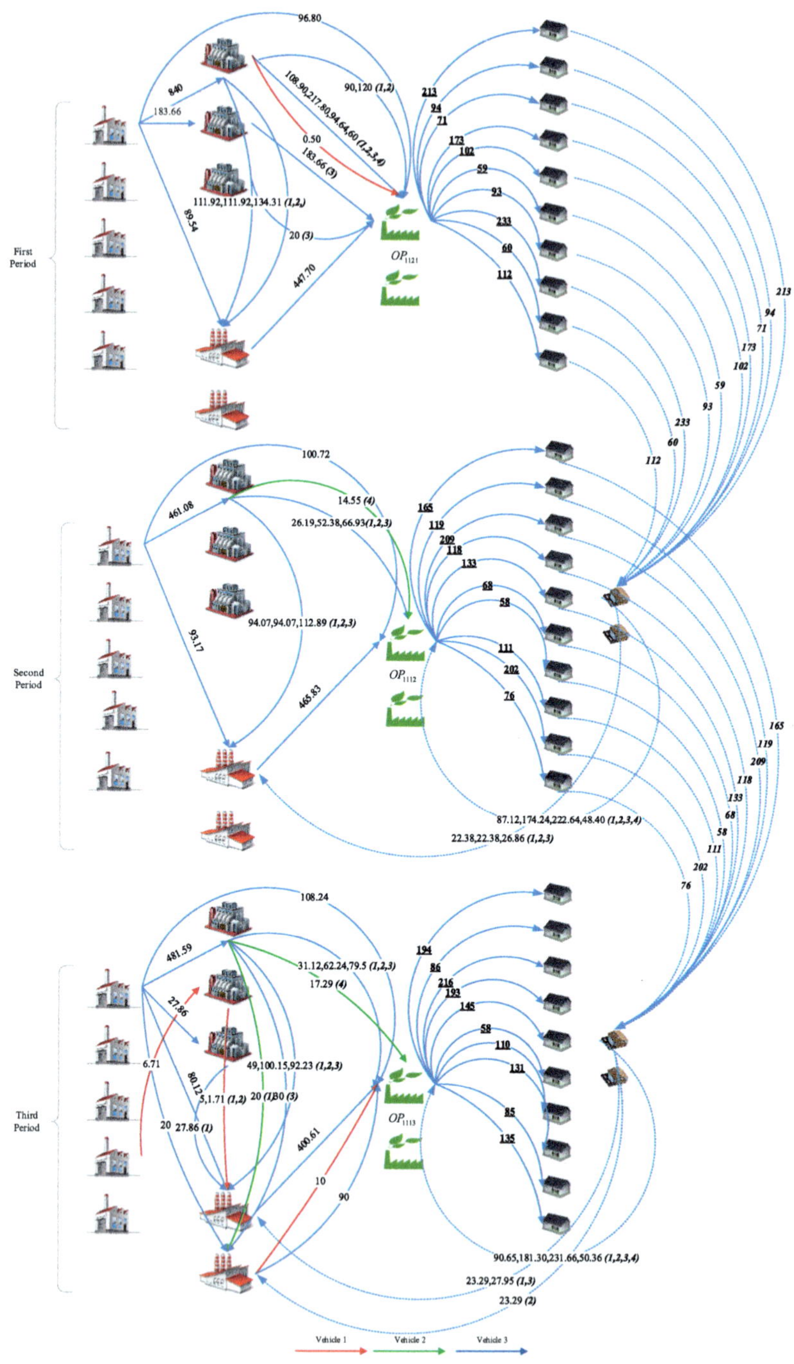


**Fig. 7** Optimal distribution plan

## 6 Concluding Remarks and Future Directions

Integration of L&G paradigms into the supply chain management provides a major benefits in several fields for companies. Some of the L&G activities have been explained in detail and investigated with the developed CLSC model. To determine the effects of L&G integration, a mixed-integer linear programming model is proposed for a multi-echelon closed-loop supply chain network consisting of raw material suppliers, suppliers, assemblers, plants, customers and collection centres. The objectives are minimization of the transportation cost, purchasing and operational costs, fixed-opening costs, handling cost, the generated $CO_2$ emission costs and the percentages of late deliveries, respectively.

As a result of the computational studies which we conducted, the manager of the anonymous corporate group tends to traditional strategy. They gave little priority to environmental issues in their judgements. Following the given priorities and weights, the model yields heavy trucks as a transportation option. Since large-size trucks are used to carry material/parts/products between the partners of SC, the manager considers a traditional strategy. On the other hand, same manager gives higher priority to the late deliveries. This contradictory point of view unveils a reality of the modern industry. Although the manager of the corporate group wants to implement lean manufacturing and maintain on time delivery, he does not want to compromise transportation costs with environmental concerns and chooses a traditional transportation strategy for the sake of lower costs.

To better understanding of L&G paradigms, a conceptual framework is constructed with several drivers. This study is a guide to identify how these drivers can be integrated into a model. Other drivers especially related with production system such as waste minimization technologies, pull production systems, delivery practices, and suppliers' performance can be included to this framework. The presented framework could be extended to a real case study to understand its application.

## References

Awasthi, A., Chauhan, S. S., & Goyal, S. K. (2010). A fuzzy multicriteria approach for evaluating environmental performance of suppliers. *International Journal of Production Economics, 126*(2), 370–378. https://doi.org/10.1016/j.ijpe.2010.04.029.

Buckley, J. J. (1985). Fuzzy hierarchical analysis. *Fuzzy Sets and Systems, 17*(3), 233–247. https://doi.org/10.1016/0165-0114(85)90090-9.

Cabral, I., Grilo, A., & Cruz-Machado, V. (2012). A decision-making model for Lean, Agile, Resilient and Green supply chain management. *International Journal of Production Research, 50*(17), 4830–4845. https://doi.org/10.1080/00207543.2012.657970.

Carvalho, H., Govindan, K., Azevedo, S. G., & Cruz-machado, V. (2017). Modelling green and lean supply chains: An eco-efficiency perspective. *Resources, Conservation and Recycling, 120*, 75–87. https://doi.org/10.1016/j.resconrec.2016.09.025.

Colicchia, C., Creazza, A., & Dallari, F. (2017). Lean and green supply chain management through intermodal transport: Insights from the fast moving consumer goods industry. *Production Planning & Control The Management of Operations, 28*(4), 321–334. https://doi.org/10.1080/09537287.2017.1282642.

Duarte, S., & Cruz Machado, V. (2017). Green and lean implementation: An assessment in the automotive industry. *International Journal of Lean Six Sigma, 8*(1), 65–88. https://doi.org/10.1108/IJLSS-11-2015-0041.

Duarte, S., & Cruz-Machado, V. (2013). Modelling lean and green: A review from business models. *International Journal of Lean Six Sigma, 4*(3), 228–250. https://doi.org/10.1108/IJLSS-05-2013-0030.

Emmett, S., & Sood, V. (2010). *Green supply chains: An action manifesto*. Chichester: Wiley.

Fahimnia, B., Sarkis, J., & Eshragh, A. (2015). A tradeoff model for green supply chain planning: A leanness-versus-greenness analysis. *Omega (United Kingdom), 54*, 173–190. https://doi.org/10.1016/j.omega.2015.01.014.

Fercoq, A., Lamouri, S., & Carbone, V. (2016). Lean/Green integration focused on waste reduction techniques. *Journal of Cleaner Production, 137*, 567–578. https://doi.org/10.1016/j.jclepro.2016.07.107.

Garza-Reyes, J. A., Villarreal, B., Kumar, V., & Molina, P. (2016). Lean and green in the transport and logistics sector – A case study of simultaneous deployment. *Production Planning & Control The Management of Operations, 27*(15), 1221–1232. https://doi.org/10.1080/09537287.2016.1197436.

Golpîra, H., Najafi, E., Zandieh, M., & Sadi-nezhad, S. (2017). Robust bi-level optimization for green opportunistic supply chain network design problem against uncertainty and environmental risk. *Computers & Industrial Engineering, 107*, 301–312. https://doi.org/10.1016/j.cie.2017.03.029.

Govindan, K., Azevedo, S. G., Carvalho, H., & Cruz-Machado, V. (2014). Impact of supply chain management practices on sustainability. *Journal of Cleaner Production, 85*, 212–225. https://doi.org/10.1016/j.jclepro.2014.05.068.

Govindan, K., Azevedo, S. G., Carvalho, H., & Cruz-Machado, V. (2015). Lean, green and resilient practices influence on supply chain performance: Interpretive structural modeling approach. *International Journal of Environmental Science and Technology, 12*(1), 15–34. https://doi.org/10.1007/s13762-013-0409-7.

Kannan, D., Diabat, A., Alrefaei, M., et al. (2012). A carbon footprint based reverse logistics network design model. *Resources, Conservation and Recycling, 67*, 75–79. https://doi.org/10.1016/j.resconrec.2012.03.005.

Klufallah, M. M. A., Nuruddin, M. F., Khamidi, M. F., & Jamaludin, N. (2014). Assessment of carbon emission reduction for buildings projects in Malaysia – A comparative analysis. *E3S Web Conferences*. https://doi.org/10.1051/e3sconf/20140301016.

Martínez-Jurado, P. J., & Moyano-Fuentes, J. (2014). Lean management, supply chain management and sustainability: A literature review. *Journal of Cleaner Production, 85*, 134–150. https://doi.org/10.1016/j.jclepro.2013.09.042.

Mollenkopf, D., Stolze, H., Tate, W. L., & Ueltschy, M. (2010). Green, lean, and global supply chains. *International Journal of Physical Distribution & Logistics Management, 40*(1/2), 14–41. https://doi.org/10.1108/09600031011018028.

Rostamzadeh, R., Govindan, K., Esmaeili, A., & Sabaghi, M. (2015). Application of fuzzy VIKOR for evaluation of green supply chain management practices. *Ecological Indicators, 49*, 188–203. https://doi.org/10.1016/j.ecolind.2014.09.045.

Talaei, M., Farhang Moghaddam, B., Pishvaee, M. S., et al. (2016). A robust fuzzy optimization model for carbon-efficient closed-loop supply chain network design problem: A numerical illustration in electronics industry. *Journal of Cleaner Production, 113*, 662–673. https://doi.org/10.1016/j.jclepro.2015.10.074.

Tiwari, R. N., Dharmar, S., & Rao, J. R. (1987). Fuzzy goal programming – An additive model. *Fuzzy Sets and Systems, 24*(1), 27–34. https://doi.org/10.1016/0165-0114(87)90111-4.

U.S. Environmental Protection Agency. (2014). *Draft U.S. Inventory of U.S. greenhouse gas emissions and sinks: 1990–2014*. Executive Summary. https://www.epa.gov/sites/production/files/2016-04/documents/us-ghg-inventory-2016-main-text.pdf

Ugarte, G. M., Golden, J. S., & Dooley, K. J. (2016). Lean versus green: The impact of lean logistics on greenhouse gas emissions in consumer goods supply chains. *Journal of Purchasing & Supply Management, 22*(2), 98–109. https://doi.org/10.1016/j.pursup.2015.09.002.

Verrier, B., Rose, B., & Caillaud, E. (2016). Lean and Green strategy: The Lean and Green House and maturity deployment model. *Journal of Cleaner Production, 116*, 150–156. https://doi.org/10.1016/j.jclepro.2015.12.022.

Vlachos, I. (2015). Applying lean thinking in the food supply chains: A case study. *Production Planning & Control The Management of Operations, 26*, 1351–1367. https://doi.org/10.1080/09537287.2015.1049238.

Wang, F., Lai, X., & Shi, N. (2011). A multi-objective optimization for green supply chain network design. *Decision Support Systems, 51*, 262–269. https://doi.org/10.1016/j.dss.2010.11.020.

# Risk Management in Lean & Green Supply Chain: A Novel Fuzzy Linguistic Risk Assessment Approach

**Turan Paksoy, Ahmet Çalik, Abdullah Yildizbaşi, and Sandra Huber**

## 1 Introduction

Increasing customer awareness on sustainability and environmental activities brings new and different responsibilities as well as risks to the firm (Blackburn 2007). Environmental, social, and beyond these, financial/economic dangers are some of these risks. Considering customer expectations, sustainability in carbon footprint, accidents in the workplace, and environmental changes, companies need to be more active in risk management, taking into account the costs that come with them. Classical supply chain management activities are not effective on issues such as carbon footprint and environmental awareness (Bazan et al. 2015), but directly affect the performance of firms in production, packaging and transportation. Environmental awareness brings with it an advantage for firms along with some risks (Dhingra et al. 2014). Activities such as pollution, inadequate waste management and recycling, $CO_2$ emissions (transport, packaging, and buildings), water and energy consumption during production are reported as serious concerns for complex supply chain operations (Dhingra et al. 2014). In addition, competitive environment and sustainable growth require firms to be more sensitive towards

T. Paksoy
Department of Industrial Engineering, Konya Technical University, Konya, Turkey

A. Çalik (✉)
Department of International Trade and Logistics, KTO Karatay University, Konya, Turkey

A. Yildizbaşi
Department of Industrial Engineering, Yıldırım Beyazıt University, Ankara, Turkey

S. Huber
Department of Logistics Management, Helmut Schmidt University, Hamburg, Germany
e-mail: sandra-huber@hsu-hh.de



T. Paksoy et al. (eds.), *Lean and Green Supply Chain Management*, International Series in Operations Research & Management Science 273,
https://doi.org/10.1007/978-3-319-97511-5_3

social and environmental issues. In other words, environmental wastes management should be provided to prevent the unnecessary or excessive use of resources and materials that can harm human health or the environment (EPA 2007).

However, recent developments have shown that natural disasters, and product recalls originating from customer security and environmental factors also affect the supply chain performance of firms, as well as the regular factors. For example, the product recalls made by Toyota, Audi, Ford, Volkswagen in the automobile industry, that of Samsung in the electronic products industry, Mattel in the toy industry, and Baxter in the pharmaceutical industry are some of these cases. The Tohoku earthquake, followed by the Tsunami in Japan in 2011, led to serious declines in Toyota's production and resulted in a $72 million daily cost (Pettit et al. 2013). The flood disaster in Thailand in 2011 negatively affected both the computer manufacturers and the supply chains of Japanese automotive manufacturers (Chopra and Sodhi 2014).

In addition to the risks stemming from unexpected disasters and distribution problems, different risks arise due to the long and complicated structures of the supply chain processes that grow with the global business world today (Nooraie and Mellat Parast 2015; Sodhi 2005; Chopra and Sodhi 2004). At this point, supply chain risk management has a strong role in preventing, defining, and reducing supply chain risks; so that chain failure operations will be recovered quickly after a breakdown or interruption (Zsidisin 2003).

In this context, although there is a significant increase in the number of studies investigating the relationship between sustainability and risk in supply chains, there are very few publications on risks arising from sustainable activities in supply chains (Cousins et al. 2013). While some of these studies focus on specific sectors, the rest focus solely on environmental factors (Teuscher et al. 2006). For example, Foerstl et al. (2010) investigated how the development of sustainable supplier management programs would provide competitive advantage. Hofmann et al. (2014) looked at the risks that may arise in the sustainable supply chain processes. Seuring and Müller (2008) investigated the relationship between sustainable supply chain management and supply risks and stressed the lack of research on this topic.

Due to increasing awareness of customers, the increase in $CO_2$ emissions, pollution, as well as the strict enforcement of rules and regulations on environmental performance, companies started to adopt the Lean & Green philosophy in their supply chain operations that considers the ecosystem, environmental aspect, corporate reputation, customer expectations, fiscal position, and consequences of complying with laws instead of focusing on distribution. Manufacturers also wish to maintain their lean & green strategy to create an environmentalist stance that leads to a reduction of costs and risks, revenue growth and improved brand image (Corbett and Klassen 2006).

However, it can be said that risks associated with the lean & green philosophy are different in many respects compared to classical supply chain risks. Therefore, the types of precautions and actions to be taken depending on the values to be obtained from risk management differ greatly. For example, the frequency of visits and the types of inspection that the key industry will make to a high risk level firm will vary from that of a lower one. In addition, the quality and competence of the teams to be assigned by the key industry for the solution of problems in sub-industries will also vary according to the risk level, and the measures to be taken against possible problems will be shaped according to these data.

Based on the literature search, it appears that there is a gap in the literature on risk management for lean & green supply chain operations. Therefore, an attempt has been made to fill this gap in this study. The main contribution of current research is developing a new fuzzy linguistic risk assessment approach based on the application fuzzy AHP method. This approach provides action plans to the decision makers according to the risk levels of suppliers. The study consists of five main sections. General information is provided in the first section. In section two, the studies conducted on risk management in supply chain operations, and risk types were examined. In the third section, the proposed fuzzy linguistic risk assessment methodology and the criteria to be used in the study are explained. In section four, the data related to case analysis is obtained and then the binary comparison matrix calculations made by taking expert opinions are presented and the managerial conclusions are defined. In the fifth and last section, the conclusions on the findings are presented.

## 2 Literature Review

Most of studies about lean and green supply chain management practices focus on the performance of supply chains (Azevedo et al. 2012; Carvalho et al. 2010; Carvalho et al. 2011) and analyze their impact on environmental performance (Carvalho et al. 2017; King and Lenox 2001; Mollenkopf et al. 2010; Simpson and Power 2005). Based on a literature survey ten different types of risk can be identified. A detailed classification is presented in Table 1.

The literature is analyzed and classified with respect to the risk types as well as the MCDM methods (Table 2).

When we look at the areas on which the abovementioned studies focus and the criteria they have identified, we see that (Bochao 2010) have established and classified a risk assessment index system based on supply chain risk sources. He then dealt with the risks in supply chain using AHP and Fuzzy comprehensive assessment method, and measured risk indicators with this approach. The aim of this study was to make supply chain risks measurable in the evaluation and decision making processes. The risks addressed in the study were addressed in two categories, as internal and external risk, and classified into different groups according to their risk levels and a correlation model was established. Both the systematic of the entire supply chain and the independence of the members were taken into account in this study where quantitative and qualitative approaches were employed.

Tang and Nurmaya Musa (2011) have identified a period of 14 years as a review period in their review on supply chain risk management and related studies and

**Table 1** Types of risks encountered in supply chain operations

| No | Risk | Sub-risk description and keywords |
|---|---|---|
| 1 | External risk (EXT) | • War and Terrorism, Natural disaster, Fire Accidents, Political Instability, Economic Downturns, Sovereign risk, Regional instability, Government regulations, Social and cultural grievances, Recovery risk |
| 2 | Flexibility and demand risk (FD) | • Inaccurate demand forecasts, Serious forecasting errors, Bullwhip effect or information distortion, Demand uncertainty, Sudden shoot-up demand, Demand variability, Customer fragmentation, High level of service required by customers, Customer dependency, Deficient or missing customer relation management function<br>• Short lead times, Short products' life cycle, On time delivery, Loss sales, Quick response, Flexibility to response demand changes, Competitor moves, Competition changes, Market changes, High competition in the market, Low in-house Production, Order fulfillment |
| 3 | Manufacturing/operational risk (M/O) | • Lean inventory, Labour disputes/strikes, Employee accidents, Operator absence, Dissatisfaction with work, Lack of experience or training, Insufficient breaks, Working conditions, Product obsolescence, Inventory holding cost, Inventory ownership, Production flexibility, Production capabilities/capacity, Products quality and safety, Technical/knowledge resources, Engineering and innovation, Shorter life time products, Linked phases in manufacturing, Warehouse and production disruption, Insufficient maintenance, Instable manufacturing process, Centralized storage of finished products |
| 4 | Supply risk (S) | • Inability to handle volume demand changes, Failures to make delivery requirements, not able to provide competitive pricing, technologically behind competitors, Inability to meet quality requirements, Supplier bankruptcy, Single supply sourcing, Small supply base, Suppliers' dependency, Supply responsiveness, High capacity utilization ay supply source, Global outsourcing, Narrow number of intermediate suppliers, Lack of integration with suppliers, Lack of suppliers' visibility |
| 5 | Information risk (I) | • Information infrastructure breakdown, System integration or extensive systems networking, E-commerce, Information delays, Lack of information transparency between logistics and marketing, Lack of compatibility in IT platforms among supply chain partners |
| 6 | Product recovery risk (PR) | Reverse logistics network design, Uncertainties related to viewing, inspection and screening of damaged and defective return products at the entry point of reprocessing stations |
| 7 | Transportation risk (T) | • Excessive handling due to border crossings or change in transportation modes, Lack of outbound effectiveness, Transport providers' fragmentation, No transport solution alternatives, On-time/on-budget delivery, Damages in transport, Accidents in transportation, Maritime pirate attack, Remote high-way theft, Stress on crew, Lack of training, Long working times, Negligently maintenance, Old technology, Transportation breakdowns, Port strikes, Global sourcing network, Supply chain complexity, Port capacity and congestion, Custom clearances at ports |

| | | |
|---|---|---|
| 8 | Financial risk (F) | • Exchange rate, Currency fluctuations, Interest rate level, Wage rate shifts, Financial strength of customers, Price fluctuations, Product cost, Financial and insurance issues, Loss of contract, Low profit margin, Market growth, Market size, Lead time for internal processing and the timing of its related cash outflows, Credit periods for accounts receivable to its customers and the pattern of early collection of accounts receivable, Credit periods for accounts payable from its suppliers and the pattern of early payment of accounts payable |
| 9 | Environmental risk (EN) | • $CO_2$ Emission, Elimination level of waste and hazardous materials, Environmental standards and certificates, Energy efficiency, Difficulty of disposal |
| 10 | Social and organizational related risk (SO) | • Management policy failures, Government policy risks, Lack in enterprise strategic goals, Legal risks, Partnership risk, External legal issues, Occupational health and safety risks |

**Table 2** Literature review of supply chain management studies which is related to risk management

| | Risk types | | | | | | | | | | Use of MCDM method | | | | | |
|---|---|---|---|---|---|---|---|---|---|---|---|---|---|---|---|---|
| Authors | EXT | FD | M/O | S | I | PR | T | F | EN | SO | AHP | Fuzzy AHP | PROMETHEE | TOPSIS | OM | Review |
| Bochao (2010) | ✓ | ✓ | ✓ | ✓ | ✓ | | | ✓ | | ✓ | ✓ | ✓ | | | | |
| Tang and Nurmaya Musa (2011) | ✓ | ✓ | ✓ | ✓ | ✓ | | | ✓ | | | | | | | | ✓ |
| Wang et al. (2012) | | ✓ | ✓ | ✓ | ✓ | | ✓ | ✓ | | | | ✓ | | | | |
| Tao (2012) | | | ✓ | ✓ | | | ✓ | ✓ | | | | ✓ | | | | |
| Venkatesan and Kumanan (2012) | ✓ | | ✓ | ✓ | | | | ✓ | | ✓ | ✓ | | ✓ | | | |
| Aggarwal and Sharma (2013) | ✓ | ✓ | ✓ | ✓ | | | | | | | | ✓ | | | | |
| Ganguly and Guin (2013) | | ✓ | ✓ | ✓ | | | ✓ | ✓ | | | | ✓ | | | | |
| Badea et al. (2014) | | ✓ | ✓ | ✓ | ✓ | | | | ✓ | | ✓ | | | | | |
| Kumar et al. (2015) | | ✓ | ✓ | ✓ | | ✓ | | ✓ | | ✓ | | ✓ | | | | |
| Nazam et al. (2015) | | ✓ | ✓ | ✓ | | | ✓ | | | | | ✓ | | ✓ | | |
| Ho et al. (2015) | ✓ | ✓ | ✓ | ✓ | ✓ | ✓ | ✓ | ✓ | | | | | | | | ✓ |
| Dong and Cooper (2016) | ✓ | ✓ | ✓ | ✓ | ✓ | | | ✓ | | | ✓ | | | | ✓ | |
| Chatterjee and Kar (2016) | ✓ | ✓ | ✓ | ✓ | | | | | ✓ | | | | | ✓ | | |
| Xiaoping (2016) | ✓ | | ✓ | ✓ | ✓ | | ✓ | | | ✓ | | ✓ | | | | |
| Radivojević and Gajović (2017) | ✓ | | ✓ | | ✓ | | | ✓ | | ✓ | ✓ | ✓ | | | | |
| Mital et al. (2017) | | | ✓ | ✓ | | | | ✓ | | | ✓ | | | | | |
| Proposed model | ✓ | ✓ | ✓ | ✓ | ✓ | ✓ | | | ✓ | ✓ | | ✓ | | | | ✓ |

developments. Based on the literature scan they conducted in terms of citation and co-citation, the most important risks were defined in three sub-titles; material, financial, and information flow risk. At the same time, they also categorized the studies in literature in terms of the use of qualitative and quantitative approaches. In this study, many risks encountered in supply chain management were classified and defined based on years.

Wang et al. (2012) proposed a two-stage Fuzzy AHP model to assess the risks that may arise during the application of green activities in their study dealing with demand fluctuations and uncertainties inherent in the clothing industry supply chain. In this study, a nine level evaluation scale and triangular fuzzy numbers were used, and risks were collected under nine main headings. These were; delivery, quality, procurement warranty, flexibility, cost, production, purchasing, logistics, and marketing, and the proposed criteria were evaluated in three different scenarios.

In his study dealing with increasing management risks in supply chain management approaches, (Tao 2012) proposed a Fuzzy AHP approach for a holistic evaluation of decision-making processes in risk management. In the risk index proposed for supply chain management; operation, technology, finance, and market risks were handled, and classified into 14 sub-criteria. As a result of the study, Tao (2012) found that operational risk has the highest priority and that relations with suppliers should be strengthened in order to reduce risk.

In their case analysis study on plastics industry (Venkatesan and Kumanan 2012), categorized the risks inherent in supply chains under six categories as organizational, supply chain, financial, political, social, and natural disasters, and used the Hybrid AHP PROMETHEE method to prioritize these risks.

In his study, Aggarwal and Sharma (2013) points out that defining and assessing the risks involved in supply chains is a multipurpose problem and suggested a multi-criteria assessment method. In this context, he identified procurement, process, demand, control and environmental risks and proposed the Fuzzy AHP approach to determine the importance levels and priority relations between these risks. As a result of the case analysis, demand fluctuations, preference changes, inconsistencies in suppliers and payment problems were identified as the most important criteria in evaluating supplier risks.

Ganguly and Guin (2013) used the Fuzzy AHP approach in their study to assess supply risks on a product basis. In order to determine the risks of procurement and the possible effects on the purchasers, five criteria were proposed. These criteria were defined as delivery, order completion, order correctness, defect-free delivery, and cost, and it the aim of the proposed approach was to prioritize these objectives. Five linguistic variables and triangular fuzzy numbers were used and evaluators working in different fields were included. Based on the results, it was stated that the proposed approach was an appropriate solution to allow for the handling of subjective judgments during the supply chain risk assessment process.

In their study, Badea et al. (2014) pointed out that risk factors impede the establishment of effective supply chains. The hierarchy of vertical and horizontal assistance was defined at the core of the study, and therefore, five alternatives were proposed in the fields of information sharing, decision synchronization, resource and skill sharing, adaptation promotion and information management for good communication and reducing risks. Afterwards, the most commonly used 16 risk factors were identified by a literature search and the AHP approach with a 9 level evaluation scale was used as the assessment method.

Kumar et al. (2015) identified and emphasized the risks that arise in the implementation of green supply chains. A two-stage approach was proposed in the study, and in the first stage they identified 25 sub-risks of green supply chain in 6 main categories. In the classification of the risks, the opinions of experts working in four different manufacturing firms in India were used. In the second stage, the classified risks were analyzed quantitatively and qualitatively by the Fuzzy AHP approach and the priority levels were determined. Talking about the importance of the use of Fuzzy AHP approach when incorporating subjective judgments into the risk analysis process, the study stated that the risks in the operational category were the most important risk group in terms of Green supply chain.

Nazam et al. (2015) stated that risks are a natural source of uncertainty, and they negatively affect green supply chain operations, reducing the success rate in the industry. The fuzzy multi-criteria decision making model (FMCGDM) was proposed in order to reduce these adverse effects and to assess the potential risks that are identified. In order for GSCM applications to be evaluated under fuzzy conditions and to determine the weights of each risk and sub-criteria, the fuzzy AHP method was proposed, and in order for the performance rankings to be made, an integrated model consisting of TOPSIS approaches was proposed. The proposed model was applied as a case study in a textile manufacturing industry and its results were evaluated. Five headings were identified as the main criteria; production, procurement, logistics, flexibility, and retail. Logistic risk factors were identified as the most important criteria in the study and flexibility criteria were determined as the least important criteria.

Ho et al. (2015) conducted a comprehensive literature review of the studies put forward in the field of supply chain risk management between 2003 and 2013. With this study, they initially made a presentation and categorization of publications related to supply chain risk management. Secondly, they analyzed risk management strategies, risk factors, risk definitions, strategies for risk types, and thirdly, they performed an analysis for identifying possible deficiencies when analyzing the literature on supply chain risk management. Within the scope of the study, they defined large and small factors and seven different criteria under these.

Dong and Cooper (2016) proposed the AHP (OM–AHP) model based on the importance of orders in order to solve the problems encountered especially in the analysis of abstract criteria and they discussed the difficulties in identifying the risks in supply chains. It was stated that evaluation of concrete and abstract criteria together was possible with this method. Risks are identified in the first stage of the three-stage model, risks are assessed in the second stage, and risks are ranked and analyzed in the final stage. In order to show the effectiveness of the results, a two-way risk matrix based on the likelihood and the significance of the result was used and sensitivity analysis was performed. In the case analysis, 31 different risks were identified and these risks were grouped and evaluated fewer than 7 different clusters.

The results showed that the grouping of risks resulted in a significant reduction in assessment and analysis time, and that the proposed OM–AHP approach had many advantages over classical AHP.

In their study emphasizing the importance of supplier choices with increasing competition conditions in recent years, (Chatterjee and Kar 2016) stated that more emphasis should be placed on suppliers with a strategic importance. In order to evaluate and rank the risks related to the suppliers, they used the TOPSIS approach from the MCDM methods. Uncertainties were modeled through linguistic variables and data obtained from decision makers was refined by this method. The effectiveness and feasibility of the work were evaluated by a case study dealing with electronic supply chain through six basic risks (environmental, demand, procurement, control, process, insurance) and four alternatives. As a result of the study, supply risk was determined as the highest risk criterion while the lowest risk criterion was found to be control risk.

Xiaoping (2016) addressed the food supply chain, and aimed to identify the supply chain safety risks associated with this industry and grade these risks. For this purpose, he identified eight main risk groups and divided them into various sub-criteria. As a solution method, an integrated Fuzzy AHP evaluation method that included a new scale was proposed.

According to Radivojević and Gajović (2017), the complex and dynamic structure of supply chains may not always be reliable, and for this reason supply chain risk management can be used as a crucial tool to reduce risks and eliminate ambiguities. For this reason, in this study where they defined the basic structure of supply chain, they used the AHP and Fuzzy AHP approaches to rank the risk categories and determine the level of each risk category within total risk. Within the scope of the study, the risks were dealt with in five basic categories as operational/technological, economic/competition, natural disasters, social risks, and laws/policies, and the knowledge and experiences of experts working in insurance companies were used during the evaluation process. Based on the results, it was determined that there was a great similarity between the results obtained with AHP and Fuzzy AHP, and the operational/technological risks were identified as the most important risk category.

Mital et al. (2017) mentioned that the basic philosophy of risk management is achieved by establishing a safety stock margin and a safe time limit. However, it was mentioned that changing circumstances reversed this situation and organizations worked in short time periods with less stock, and did not establish safe structures, and the importance of risk management for this purpose was emphasized. For this purpose, the different risks studied under four categories (milk, rice, mobile phone and cigarettes) were evaluated using cognitive mapping and AHP methods for managing and evaluating risks. Within the scope of the study, sub-criteria were identified under the five main criteria within the first level risk category: quality, cost, supply continuity, supplier service, and supplier customer partnership.

## 3 Proposed Fuzzy Linguistic Risk Assessment Approach

The proposed solution approach consists of four steps: In the first step, the risk criteria used for evaluating the suppliers are compared in a binary manner and the importance weights are obtained for each other. In the second stage, the fuzzy scale is used to evaluate existing suppliers according to various risk criteria and to obtain assessment scores. In the assessment process, decision makers use triangular fuzzy numbers to evaluate suppliers. In the third step, the level of risk for each supplier is determined using the scores and criteria weights obtained. In the final step, suppliers are assigned to risk groups according to their risk levels and a separate action plan is developed for each supplier. The steps of the proposed solution approach and the flow diagram are given in Fig. 1.

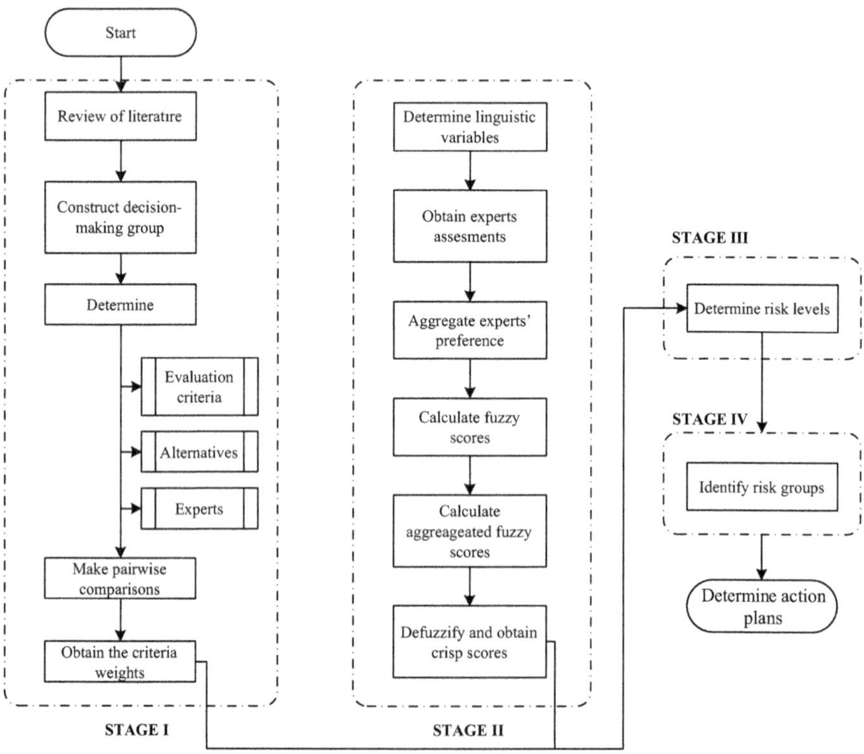


**Fig. 1** Framework of proposed approach

### 3.1 Identification of Problem and Binary Comparison of Criteria and Obtaining Relative Importance Weights

In this step, the Fuzzy AHP method to be used in determining the risk criteria and criteria weights, which is the aim of the research, is explained. As a result of the literature survey, nine criteria were selected for the risk assessment of suppliers, and the definitions of these criteria are presented as follows:

*Experience Level of Suppliers* The level of experience is determined by the year of establishment, sectoral contribution, specialization and the experience of the personnel employed. The risk level is low if the supplier is specialized in its job and delivers smooth shipments without defects. In other cases, if supplier causes significant production deficits and is having difficulty to supply the desired products, the risk level is high.

*Criticality Level of Parts Supplied* The risk level of the suppliers that produce the high importance and indispensable parts in the manufactured product is higher. The technical know-how level determines the risk levels according to whether a supplied part is a standard part, a simple part, or a more complex part related to module and system safety.

*Manufacturing Technical Requirements and Complexity of Parts Supplied* If the technology used by the supplier in the production of the supplied part is a new and non-established technology, or if the parts to be supplied are very complex, the risk level of the supplier will be high. In other cases, the risk level will be defined as low if the use of a newly proven and known production technology is needed.

*Effect of the Deviations on the Final Product Function* The effect of the deviations of the purchased part from the supplier on the function and visual quality of the final product determines the supplier risk level. For example, if the effect of the deviations on function and visual quality are small, the risk level will be low. If these deviations pose a risk to the final product, or affect/disrupt visual specifications such as gaps, surface roughness, color, etc. then the risk level of the supplier will be high.

*Flexibility and Demand Risk* Failure to comply with the current production plan and delays in production delivery times, together with uncertainties in the demand structure of the suppliers, instantaneous changes and misplaced demand forecasts, will result in lost sales. However, according to lean manufacturing, the level of continuing and responding to just-in-time deliveries determines the level of risk for suppliers. Despite all these uncertainties, the supplier must have a robust infrastructure to support on-time delivery.

*Cooperation Level of Suppliers* IT infrastructure of suppliers, their know-how sharing levels, information transmission channels being developed and the level of establishment of a fast and effective communication channel in solving problems determine this level.

*Green Activity Level of Suppliers and Environmental Risk* Criteria such as the $CO_2$ emission rate that the supplier attacks on the environment during production, the energy efficiency level of the equipment used and the production technology, whether supplier has the standards and certifications related to environmental management (ISO 14001, ecolabel, etc.), the level of utilization of renewable technology resources, the level of waste management, the capacity to dispose of hazardous materials used determine the risk level.

*Product Recovery Risk* The level of development of the recall system, the level of success in the checks and inspection of collected parts, and the competence of the recycling centers affect the risk level of the supplier. At the same time, risk level in this category is determined by whether the product produced by the supplier can be easily demounted or not, and the level of renewability of the product.

*Occupational Health Safety (OHS)* Interruptions or breakdowns in production occur depending on whether the suppliers take the necessary safety precautions in the context of occupational health and safety practices. These efforts and the occupational health and safety certifications obtained (ISO 18001) reveal the risk levels of suppliers.

The Fuzzy AHP method to be used in determining criteria weights once the criteria are defined is described along with its steps. The linguistic variables to be used in the binary comparison of the criteria are shown in Table 3.

**Table 3** The linguistic variables to be used in evaluating the criteria (Awasthi and Chauhan 2012)

| Linguistic terms | Crisp scale | Triangular fuzzy scale | Triangular fuzzy reciprocal scale |
|---|---|---|---|
| Equal importance (E) | 1 | (1, 1, 1) | (1/1, 1/1, 1/1) |
| Moderate importance (M) | 3 | (1, 3, 5) | (1/5, 1/3, 1/1) |
| Strong importance (S) | 5 | (3, 5, 7) | (1/7, 1/5, 1/3) |
| Very strong importance (VS) | 7 | (5, 7, 9) | (1/9, 1/7, 1/5) |
| Extreme importance (EX) | 9 | (7, 9, 9) | (1/9, 1/9, 1/7) |
| Intermediate values between two adjacent judgments (IV) | 2 | (1, 2, 3) | (1/3, 1/2, 1) |
| | 4 | (3, 4, 5) | (1/5, 1/4, 1/3) |
| | 6 | (5, 6, 7) | (1/7, 1/6, 1/5) |
| | 8 | (7, 8, 9) | (1/9, 1/8, 1/7) |

To obtain the importance weights of lower level DMs, we used the Buckley's FAHP method. The steps of the method can be summarized as follows:

**Step 1** A fuzzy pair-wise comparison matrix ($\tilde{A} = [\mathrm{a}_{ij}]$) is constructed as:

$$\tilde{A} = \begin{bmatrix} 1 & \cdots \tilde{a}_{1j} & \cdots \tilde{a}_{1n} \\ \vdots & \vdots & \vdots \\ \tilde{a}_{i1} & \cdots 1 & \cdots \tilde{a}_{in} \\ \vdots & \vdots & \vdots \\ \tilde{a}_{n1} & \tilde{a}_{nj} & \cdots 1 \end{bmatrix} = \begin{bmatrix} 1 & \cdots & \tilde{a}_{1j} & \cdots \tilde{a}_{1n} \\ \vdots & & \vdots & \vdots \\ 1/\tilde{a}_{i1} & \cdots & 1 & \cdots \tilde{a}_{in} \\ \vdots & & \vdots & \vdots \\ 1/\tilde{a}_{n1} & & 1/\tilde{a}_{nj} & \cdots 1 \end{bmatrix}$$

where; $\tilde{a}_{ij} = (l_{ij}, m_{ij}, u_{ij})$ is a triangular fuzzy numbers ($i = 1, 2, \ldots, n, j = 1, 2, \ldots, m$).

**Step 2** The fuzzy weight matrix is calculated as

$$\tilde{a}_i = (\tilde{a}_{i1} \otimes \tilde{a}_{i2} \otimes \cdots \otimes \tilde{a}_{in})^{1/n}$$

$m$ is the number of alternatives, $n$ is the number of criteria.

**Step 3** The fuzzy weights of each criterion/alternative is calculated by

$$w_i = \tilde{a}_i \otimes (\tilde{a}_1 \oplus \tilde{a}_2 \oplus \cdots \oplus \tilde{a}_n)^{-1}$$

where; $\oplus$ is a fuzzy addition operator, and $\otimes$ is a fuzzy multiplication operator (Buckley 1985).

## 3.2 *Obtaining Assessment Scores of Suppliers*

Since the evaluations of decision makers involve subjective judgments, they do not give exact numerical values and exact results. Therefore, fuzzy set theory has been incorporated into decision-making methods to overcome linguistic uncertainties. At this stage, a five-point scale was used to evaluate the alternatives under each criterion, that is, to obtain the assessment scores of the alternatives. Five linguistic expressions as Very poor (VP), Poor (P), Fair (F), Good (G) and Very good (VG) given in Table 4 were chosen. In a group of $K$ decision makers, each decision maker express their evaluation under each criterion using a triangular fuzzy number, and the aggregated evaluations of the decision makers were calculated according to the following equations:

$$\widetilde{x}_{kl} = (a_{kl}, b_{kl}, c_{kl}) = \frac{1}{N}\left[\widetilde{x}_{kl}^{1} + \widetilde{x}_{kl}^{2} + \cdots + \widetilde{x}_{kl}^{N}\right]$$

**Table 4** The linguistic variables to be used in evaluating the alternatives (Awasthi and Chauhan 2012)

| Linguistic variable | Corresponding triangular fuzzy number |
|---|---|
| Very poor (VP) | (1, 1, 3) |
| Poor (P) | (1, 3, 5) |
| Fair (F) | (3, 5, 7) |
| Good (G) | (5, 7, 9) |
| Very good (VG) | (7, 9, 9) |

Here, $\widetilde{x}_{kl}^{N}$ is the fuzzy number indicating the assessment of alternative $l$ under criterion $k$ by decision maker $N$. Afterwards defuzzification procedure was performed on the $\widetilde{x}_{kl}$ triangular fuzzy number according to Center of Area (gravity)—CoA,

$$d = \frac{(a_{kl} + b_{kl} + c_{kl})}{3}$$

using the equation.

## 3.3 Determining the Risk Levels of Suppliers

The risk levels of suppliers are calculated as follows using the weights obtained in Sect. 3.1 and the assessment scores obtained in Sect. 3.2:

$$RL_i = \sum\nolimits_{j=1}^{J} S_{ij} \cdot W_j$$

$S_{ij}$: Assessment score of supplier $i$ according to criteria $j$
$W_j$: Weight of criteria $j$
$RL_i$: Risk level of supplier $i$

## 3.4 Assigning Suppliers to Risk Groups and Determination of Action Plans

Pre-determination of the possible risks to be encountered in the supplier risk management process and taking necessary precautions for these risks have a critical importance. In this way, possible future problems will be defined in advance and the key industry will be able to take necessary measures against these problems. Action plans will be determined for the suppliers by using the risk levels obtained during the identification of risk groups. Suppliers with a risk level of 8 or above belong in group A, suppliers with a risk level of 5–7 belong in group B, suppliers with a risk level of 3–5 belong in group C, and lastly, suppliers with a risk level of less than three belong in group D. After the determination of risk groups, managerial impacts and different action plans such as effective sharing of information, developing Corrective–Preventive Action plan (CPA), more reliable cooperation, cost-effective implementations for each risk group will be developed and these processes will be continued until the risk levels are brought to an acceptable level.

# 4 Case Study

## *4.1 Data of the Company*

We have to take precautions against the risks that may arise in the network since supply chain management is responsible for the movement of materials, parts and products from procurement until delivery to customers. Risk modeling in supply chain management is a difficult process to deal with because of the complexity of the risk structure in the network and the large number of decision makers. The difficulty mentioned in this study was tried to be overcome by using Fuzzy AHP and multi-purpose linear programming models together. In order to investigate the performance of the proposed supplier risk assessment model, a case study that was influenced by an automobile company operating in Turkey was conducted. The aim of this case study was to determine how companies handle risk criteria as the supplier risk assessment model described in the conceptual framework is applied in the real world.

Case study company is located in Bursa, a region in Turkey. The ABC Company supplies the 70% of the automotive parts needed for production from different cities in Turkey, and the rest of the parts 15%, 9%, 4%, 2% along with countries such as Germany, Italy, Belgium, and India respectively.

In the risk assessment of ABC, it is aimed to determine the risk levels of the suppliers in Turkey that the company purchases different parts from. After a preliminary review with company managers, five potential auto parts suppliers ($A1$, $A2$, $A3$, $A4$, $A5$) were identified for risk assessment. Nine criteria considered in the assessment process are as follows: $C1$: (EXT) Experience level of suppliers; $C2$: (S) Criticality level of parts supplied; $C3$: (M/O) Manufacturing technical requirements and complexity of parts supplied; $C4$: (M/O) Effect of the deviations on the final product function; $C5$: (FD) Flexibility and demand risk; $C6$: (I) Cooperation level of suppliers; $C7$: (EN) Green activity level of suppliers and environmental risk; $C8$: (PR) Product recovery risk; $C9$: (SO) Occupational Health Safety (OHS). The hierarchical structure of the problem is given in Fig. 2. Three experts from the managers of the company with at least 5 years of experience were involved in the study to address the risk assessment. All specialists are people with different expertise who have worked in various departments within the company such as procurement, planning, production, finance, quality, and logistics. For the risk assessment of the suppliers, personal interviews were conducted in order to gather the data needed by these experts.

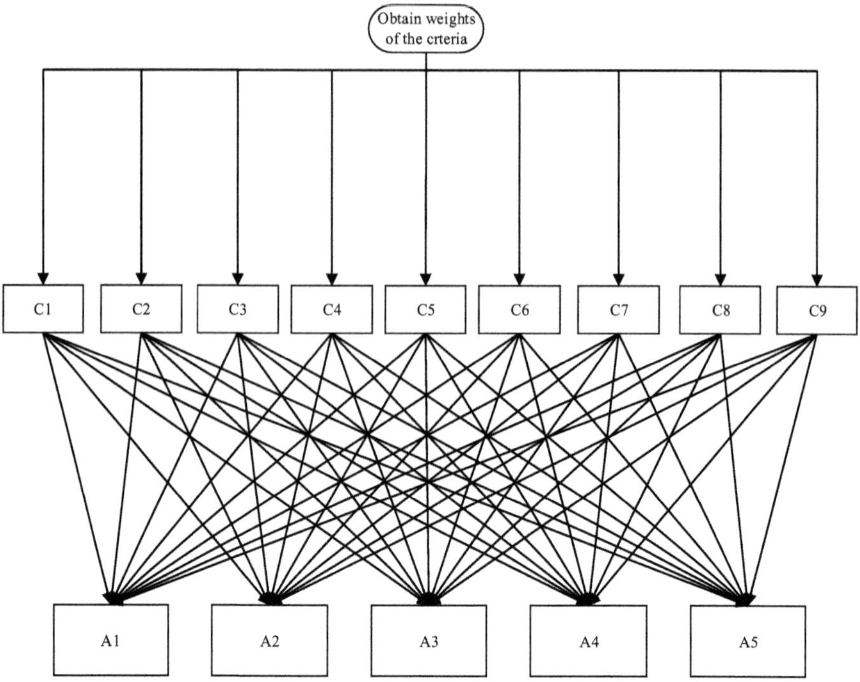


**Fig. 2** Hierarchical structure of problem

**Table 5** Pairwise comparison matrix for the criteria

| | C1 | C2 | C3 | C4 | C5 | C6 | C7 | C8 | C9 |
|---|---|---|---|---|---|---|---|---|---|
| C1 | – | IV | IV | IV | IV | M | IV | IV | S |
| C2 | | – | IV | IV | IV | S | IV | IV | VS |
| C3 | | | – | M | S | IV | VS | EX | IV |
| C4 | | | | – | IV | IV | S | EX | IV |
| C5 | | | | | – | IV | M | S | IV |
| C6 | | | | | | – | IV | IV | M |
| C7 | | | | | | | – | M | IV |
| C8 | | | | | | | | – | IV |
| C9 | | | | | | | | | – |

## 4.2 *Stages of Assessment*

**Step 1** In order to calculate the risk scores of the suppliers, the criteria weights must first be obtained. At this stage, weights were calculated for the determined risk criteria using the geometric mean method of Buckley (1985). The linguistic scale shown in Table 4 was used to determine the relative weights of the criteria. The binary comparison matrix and the fuzzy counterparts of the criteria are given in Tables 5 and 6.

**Table 6** Triangular fuzzy number for the pairwise comparison matrix

| | C1 | C2 | C3 | C4 | C5 | C6 | C7 | C8 | C9 |
|---|---|---|---|---|---|---|---|---|---|
| C1 | (1, 1, 1) | (1/5, 1/4, 1/3) | (1/5, 1/4, 1/3) | (1/3, 1/2, 1) | (1, 2, 3) | (1, 3, 5) | (3, 4, 5) | (5, 6, 7) | (3, 5, 7) |
| C2 | | (1, 1, 1) | (1/3, 1/2, 1) | (1, 2, 3) | (3, 4, 5) | (3, 5, 7) | (5, 6, 7) | (7, 8, 9) | (5, 7, 9) |
| C3 | | | (1, 1, 1) | (1, 3, 5) | (3, 5, 7) | (5, 6, 7) | (5, 7, 9) | (7, 9, 9) | (7, 8, 9) |
| C4 | | | | (1, 1, 1) | (1, 2, 3) | (3, 4, 5) | (3, 5, 7) | (5, 7, 9) | (5, 6, 7) |
| C5 | | | | | (1, 1, 1) | (1, 2, 3) | (1, 3, 5) | (3, 5, 7) | (3, 4, 5) |
| C6 | | | | | | (1, 1, 1) | (1, 2, 3) | (3, 4, 5) | (1, 3, 5) |
| C7 | | | | | | | (1, 1, 1) | (1, 3, 5) | (1, 2, 3) |
| C8 | | | | | | | | (1, 1, 1) | (1/3, 1/2, 1) |
| C9 | | | | | | | | | (1, 1, 1) |

**Table 7** The geometric mean of each criteria

| | |
|---|---|
| $\widetilde{a}_1$ | (0.9448, 1.4133, 1.9504) |
| $\widetilde{a}_2$ | (2.2660, 3.1057, 4.0569) |
| $\widetilde{a}_3$ | (2.8129, 4.1472, 5.1787) |
| $\widetilde{a}_4$ | (1.3511, 2.0200, 3.0027) |
| $\widetilde{a}_5$ | (0.6737, 1.0461, 1.5711) |
| $\widetilde{a}_6$ | (0.4538, 0.6853, 1.0584) |
| $\widetilde{a}_7$ | (0.3147, 0.4732, 0.7402) |
| $\widetilde{a}_8$ | (0.1931, 0.2411, 0.3555) |
| $\widetilde{a}_9$ | (0.2465, 0.3324, 0.4986) |
| **SUM** | (9.2565, 13.4644, 18.4125) |

**Table 8** Fuzzy weights of criteria

| | |
|---|---|
| $\widetilde{w}_1$ | (0.0513, 0.1050, 0.2107) |
| $\widetilde{w}_2$ | (0.1231, 0.2307, 0.4383) |
| $\widetilde{w}_3$ | (0.1528, 0.3080, 0.5595) |
| $\widetilde{w}_4$ | (0.0734, 0.1500, 0.3244) |
| $\widetilde{w}_5$ | (0.0366, 0.0777, 0.1697) |
| $\widetilde{w}_6$ | (0.0246, 0.0509, 0.1143) |
| $\widetilde{w}_7$ | (0.0171, 0.0351, 0.0800) |
| $\widetilde{w}_8$ | (0.0105, 0.0179, 0.0384) |
| $\widetilde{w}_9$ | (0.0134, 0.0247, 0.0539) |

According to decision makers' views, the consistency ratio of the pair-wise comparison matrix was found to be 0.0364. This value was found to be less than 10%. For this reason, the pair-wise comparison matrix can be regarded as consistent.

After the consistency of the binary comparison matrix was found, $\widetilde{a}_i$ values were calculated by taking the geometric mean of each criterion and the results are given in Table 7. Using the geometric means, the related fuzzy weights ($\widetilde{w}_i$) were calculated. Sample calculations were made for $\widetilde{a}_1$ and $\widetilde{w}_1$ and the general results were given below (Table 8):

$$\widetilde{a}_1 = \left(\widetilde{a}_{11} \times \widetilde{a}_{12} \times \cdots \times \widetilde{a}_{19}\right)^{\frac{1}{9}}$$

$$\widetilde{a}_1 = \left[\left(1 \times {}^1\!/_5 \times \cdots \times 3\right)^{\frac{1}{9}}, \left(1 \times {}^1\!/_4 \times \cdots \times 5\right)^{\frac{1}{9}}, \left(1 \times {}^1\!/_3 \times \cdots \times 7\right)^{\frac{1}{9}}\right]$$

$$\widetilde{a}_1 = (0.9448, 1.4133, 1.9503)$$

$$\widetilde{w}_1 = \left[\widetilde{a}_1 \otimes \left(\widetilde{a}_1 \oplus \widetilde{a}_2 \oplus \cdots \oplus \widetilde{a}_9\right)^{-1}\right]$$

$$\widetilde{w}_1 = (0.0513, 0.1049, 0.2107)$$

**Table 9** Relative weight of criteria

| | Defuzzified weights | Normalized weights |
|---|---|---|
| C1 | 0.1223 | 0.1051 |
| C2 | 0.2640 | 0.2268 |
| C3 | 0.3400 | 0.2922 |
| C4 | 0.1825 | 0.1569 |
| C5 | 0.0946 | 0.0813 |
| C6 | 0.0632 | 0.0544 |
| C7 | 0.0440 | 0.0379 |
| C8 | 0.0222 | 0.0191 |
| C9 | 0.0306 | 0.0263 |

Fuzzy criterion weights are refined by defuzzification to obtain exact values. The criteria weights are given in Table 9.

Based on the calculation results, "Manufacturing technical requirements and complexity of parts supplied (*C*3)", "Criticality level of parts supplied (*C*2)" and "effect of the deviations on the final product function (*C*4)" were identified as the most important criteria.

**Step 2** To obtain the risk levels, we use the five-point scale in the second step to determine the assessment scores of alternatives under each criterion. The three decision makers assessed the alternatives under the nine criteria using the linguistic variables shown in Table 4. The assessment scores obtained are shown in Table 10:

While the decision makers' assessment scores were obtained, the following steps were taken in Table 3, respectively. The sample calculations here are made for the evaluation of the 1st supplier (S1) under the 1st criterion (C1).

**Step 2.1:** Fuzzy numbers corresponding to the assessments of the decision makers in the study, in other words the linguistic variables, are obtained.

First decision makers assessed S1 as "F" while second and third decision makers assessed as "G". The corresponding fuzzy numbers for these linguistic variables are, respectively:

$$\widetilde{x}_{11}^{1} = (3, 5, 7)\,, \widetilde{x}_{11}^{2} = (5, 7, 9)\,, \widetilde{x}_{11}^{3} = (5, 7, 9)$$

**Step 2.2:** Fuzzy assessments of decision makers are added up and divided by the number of decision makers to obtain the added fuzzy assessment score.

$$\widetilde{x}_{kl} = \frac{1}{N}\left[\widetilde{x}_{kl}^{1} + \widetilde{x}_{kl}^{2} + \cdots + \widetilde{x}_{kl}^{N}\right]$$

$$\widetilde{x}_{11} = \frac{1}{3}\left[\widetilde{x}_{11}^{1} + \widetilde{x}_{11}^{2} + \widetilde{x}_{11}^{3}\right]$$

**Table 10** Suppliers' evaluation according to the criterion

| | Suppliers | Decision makers | | | Fuzzy scores | Crisp scores |
|---|---|---|---|---|---|---|
| | | DM1 | DM2 | DM3 | | |
| C1 | S1 | F | G | G | (4.33, 6.33, 8.33) | 6.33 |
| | S2 | F | F | P | (2.33, 4.33, 6.33) | 4.33 |
| | S3 | F | G | F | (3.66, 5.66, 7.66) | 5.66 |
| | S4 | VG | VG | G | (6.33, 8.33, 9.66) | 8.106 |
| | S5 | G | G | VG | (5.66, 7.66, 9.33) | 7.55 |
| C2 | S1 | F | F | P | (2.33, 4.33, 6.33) | 4.33 |
| | S2 | G | G | VG | (5.66, 7.66, 9.33) | 7.55 |
| | S3 | P | VP | P | (0.66, 2.33, 4.33) | 2.44 |
| | S4 | VG | G | VG | (6.33, 8.33, 9.66) | 8.106 |
| | S5 | G | G | VG | (5.66, 7.66, 9.33) | 7.55 |
| C3 | S1 | F | F | G | (3.66, 5.66, 7.66) | 5.66 |
| | S2 | G | G | F | (4.33, 6.33, 8.33) | 6.33 |
| | S3 | P | P | F | (1.66, 3.66, 5.66) | 3.66 |
| | S4 | VG | G | VG | (6.33, 8.33, 9.66) | 8.106 |
| | S5 | VG | G | G | (5.66, 7.66, 9.33) | 7.55 |
| C4 | S1 | P | P | P | (1, 3, 5) | 3 |
| | S2 | F | F | G | (3.66, 5.66, 7.66) | 5.66 |
| | S3 | VP | VP | P | (0.33, 1.66, 3.66) | 1.883 |
| | S4 | VG | VG | G | (6.33, 8.33, 9.66) | 8.106 |
| | S5 | G | G | VG | (5.66, 7.66, 9.33) | 7.55 |
| C5 | S1 | P | F | P | (1.66, 3.66, 5.66) | 3.66 |
| | S2 | F | F | G | (3.66, 5.66, 7.66) | 5.66 |
| | S3 | VP | VP | P | (0.33, 1.66, 3.66) | 1.883 |
| | S4 | VG | VG | G | (6.33, 8.33, 9.66) | 8.106 |
| | S5 | G | G | VG | (5.66, 7.66, 9.33) | 7.55 |
| C6 | S1 | F | F | F | (3, 5, 7) | 5 |
| | S2 | G | G | G | (5, 7, 9) | 7 |
| | S3 | P | VP | P | (0.66, 2.33, 4.33) | 2.44 |
| | S4 | G | VG | G | (5.66, 7.66, 9.33) | 7.55 |
| | S5 | G | G | G | (5, 7, 9) | 7 |
| C7 | S1 | VP | P | P | (0.66, 2.33, 4.33) | 2.44 |
| | S2 | P | F | F | (2.33, 4.33, 6.33) | 4.33 |
| | S3 | VP | P | P | (0.66, 2.33, 4.33) | 2.44 |
| | S4 | G | G | VG | (5.66, 7.66, 9.33) | 7.55 |
| | S5 | F | F | F | (3, 5, 7) | 5 |
| C8 | S1 | VP | P | P | (0.66, 2.33, 4.33) | 2.44 |
| | S2 | F | P | F | (2.33, 4.33, 6.33) | 4.33 |
| | S3 | VP | P | VP | (0.33, 1.66, 3.66) | 1.883 |
| | S4 | G | VG | G | (5.66, 7.66, 9.33) | 7.55 |
| | S5 | F | F | G | (3.66, 5.66, 7.66) | 5.66 |
| C9 | S1 | VP | F | P | (1.33, 3, 5) | 3.11 |
| | S2 | VG | G | VG | (6.33, 8.33, 9.66) | 8.106 |
| | S3 | F | G | F | (3.66, 5.66, 7.66) | 5.66 |
| | S4 | G | VG | G | (5.66, 7.66, 9.33) | 7.55 |
| | S5 | G | VG | VG | (6.33, 8.33, 9.66) | 8.106 |

$$\widetilde{x}_{11} = \frac{1}{3}(13, 19, 25)$$

$$\widetilde{x}_{11} = (4.33, 6.33, 8.33)$$

**Step 2.3:** Obtained fuzzy numbers are refined by the Center of gravity method.

$$a = \frac{(4.33 + 6.33 + 8.33)}{3}$$

$$a = 6.33$$

**Step 3** The risk level for the first supplier can be calculated as follows.

$$RS_1 = \sum\nolimits_{j=1}^{J} S_{ij} \cdot W_j = \sum\nolimits_{j=1}^{9} S_{1j} \cdot W_j = S_{11} \cdot W_1 + S_{12} \cdot W_2 + \cdots + S_{19} \cdot W_9$$

$$= 6.33 \bullet 0.1051 + 4.33 \bullet 0.2268 + \cdots + 3.11 \bullet 0.0263$$

$$= 4.56231$$

Risk levels of other suppliers can be calculated similarly, and the following values are obtained after the calculations (Table 11):

**Step 4** Determination of supplier risk levels

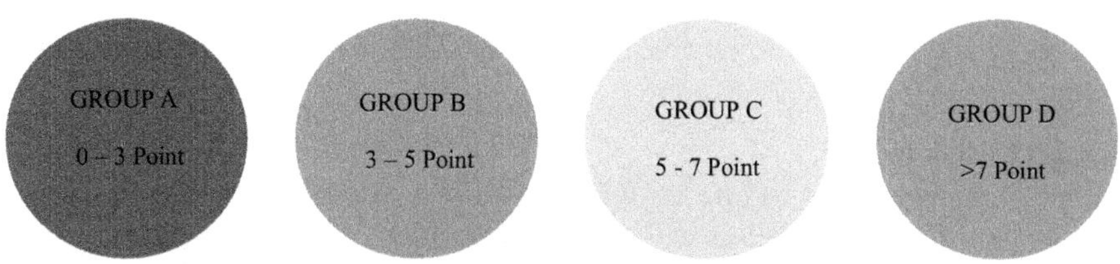


**Step 5** Determination of action plans according to Risk Levels:

**Group A Risk level**

- Corrective–Preventative Action (CPA) plan should be filled in for suppliers and kept under record after sharing with the key industry.

**Table 11** Risk scores of suppliers

| Suppliers | Risk levels | Risk group |
|---|---|---|
| S1 | 4.56231 | GROUP B |
| S2 | 6.20605 | GROUP C |
| S3 | 3.07627 | GROUP B |
| S4 | 8.02943 | GROUP D |
| S5 | 7.40195 | GROUP D |

- Supplier should prepare together with competent supervisors in the field of Emergency action plans under the supervision of the key industry for high risk activities and should be inspected every 3 months.
- Key industry and inspectors will determine whether the proposed plan meets the requirements to reduce Supplier Risks, and will provide necessary guidance if not.
- Inspectors competent in their fields should be assigned to minimize or eliminate the high risks identified during the process of guiding the suppliers.
- Effective sharing of information, personnel and expertise should be provided between the key industry and the suppliers until the suppliers' transition to green color class.
- Specialists should be appointed by the key industry to ensure that the limits and boundaries of the supplier companies are determined accurately and rationally.
- Cost-effective implementations should be developed with efficient use of data flow channels created with suppliers, which should enable suppliers to focus on reducing risks.
- The identified risks should be corrected according to the agreed Corrective–Preventive Action Plan (CPA) and monitored by the key industry on a monthly basis.
- Visits to be made by the key industry to the suppliers should be made on an informed – uninformed manner, and planned to include all departments.
- After the initial assessment, a monitoring-assessment date should be determined once every 3 months.
- The assessment process should continue until the suppliers enter the green color class or their agreement is terminated.
- Longer lasting and more reliable cooperation should be established with suppliers.

**Group B Risk level**

- A (CPA) plan should be filled in for suppliers and shared with the key industry.
- Key industry and inspectors will determine whether the proposed plan meets the requirements to reduce Supplier Risks, and will provide necessary guidance if not.
- Inspectors competent in their fields should be assigned to minimize or eliminate the high risks identified during the process of guiding the suppliers.
- Emergency action plans for high risk and risky activities should be prepared together with competent supervisors in their fields under the supervision of the key industry and should be inspected every 6 months.
- Key industry and inspectors will determine whether the proposed plan meets the requirements to reduce Supplier Risks, and will provide necessary guidance if not.
- Inspectors competent in their fields should be assigned to minimize or eliminate the high risks identified during the process of guiding the suppliers.
- The identified risks should be corrected as defined in the agreed CPA plan.

- After the initial assessment, a monitoring-assessment date should be determined within 6 months, and the modifications in the CPA plan should be investigated.
- The assessment process should continue until the suppliers enter the green color class.

**Group C Risk level**

- In order to join the green status that is the Group D Risk level, findings that are identified as high risk or risky should be corrected.
- An approval agreement should be made between the key industry and suppliers and a corrective preventive action report should be prepared.
- Focus groups should be formed by suppliers to eliminate risks, and the annual assessment results should be shared with the key industry.
- Feedback to the key industry and inspectors should be made with the necessary findings proving that corrective actions have been taken within the period specified in the corrective preventive action report.
- Suppliers should be assessed on an annual basis.

**Group D Risk level**

- No precautions are required.
- Suppliers should be re-assessed within 1–3 years.

## 5 Conclusion

In this study, a new solution method was proposed to determine the risk scores of the suppliers in supply chain management. With the proposed method, the risk levels of the suppliers are determined according to the risk scores obtained, and the risks of the suppliers can be deduced. Through identification, understanding, and prioritization of supplier risks, this approach will be beneficial for companies in terms of being more flexible and capable in issues that reduce the effectiveness of businesses.

The main steps of the proposed solution method are as follows: First, the relative weights of the criteria considered in the risk assessment method are determined by binary comparison matrices. In the second stage, risk-based assessments of suppliers were obtained using linguistic variables. At this stage, a five-point scale is used to obtain assessment scores. In the final stage, the risk assessment model is constructed using the values obtained in the first two stages.

The proposed method can be extended for future studies as follows: Uncertainties can be addressed by using different forms of fuzzy sets (type-2 fuzzy sets, Intuitionistic fuzzy sets, and hesitant fuzzy sets, etc.). Different criteria can be added to the risk assessment model and a new risk level calculation can be made when the risk groups are identified A decision support system can be developed so that decision makers can be supported in real life applications.

# References


Aggarwal, R., & Sharma, S. (2013). Prioritization of supply chain risk assessors using fuzzy analytic hierarchy process. In *International conference on machine intelligence and research advancement* (pp. 100–104).

Awasthi, A., & Chauhan, S. S. (2012). A hybrid approach integrating affinity diagram, AHP and fuzzy TOPSIS for sustainable city logistics planning. *Applied Mathematical Modelling, 36*(2), 573–584. https://doi.org/10.1016/j.apm.2011.07.033.

Azevedo, S. G., Carvalho, H., Duarte, S., & Cruz-Machado, V. (2012). Influence of green and lean upstream supply chain management practices on business sustainability. *IEEE Transactions on Engineering Management, 59*(4), 753–765.

Badea, A., Prostean, G., Goncalves, G., & Allaoui, H. (2014). Assessing risk factors in collaborative supply chain with the analytic hierarchy process (AHP). *Procedia – Social and Behavioral Sciences, 124*, 114–123. https://doi.org/10.1016/j.sbspro.2014.02.467.

Bazan, E., Jaber, M. Y., & Zanoni, S. (2015). Supply chain models with greenhouse gases emissions, energy usage and different coordination decisions. *Applied Mathematical Modelling, 39*(17), 5131–5151. https://doi.org/10.1016/j.apm.2015.03.044.

Blackburn, W. R. (2007). *The sustainability handbook: The complete management guide to achieving social, economic and environmental responsibility*. London: Earthscan.

Bochao, L. (2010). Supply chain risk assessment based on AHP and fuzzy comprehensive evaluation. In *International conference on management of e-commerce and e-government* (pp. 317–322).

Buckley, J. J. (1985). Fuzzy hierarchical analysis. *Fuzzy Sets and Systems, 17*(3), 233–247.

Carvalho, H., Azevedo, S. G., & Cruz Machado, V. (2010). Supply chain performance management: Lean and green paradigms. *International Journal of Business Performance and Supply Chain Modelling, 2*(3/4), 304–333.

Carvalho, H., Duarte, S., & Cruz Machado, V. (2011). Lean, agile, resilient and green: Divergencies and synergies. *International Journal of Lean Six Sigma, 2*(2), 151–179.

Carvalho, H., Govindan, K., Azevedo, S. G., & Cruz-Machado, V. (2017). Modelling green and lean supply chains: An eco-efficiency perspective. *Resources, Conservation and Recycling, 120*, 75–87. https://doi.org/10.1016/j.resconrec.2016.09.025.

Chatterjee, K., & Kar, S. (2016). Multi-criteria analysis of supply chain risk management using interval valued fuzzy TOPSIS. *Journal of the Operational Research Society of India, 53*, 474–499.

Chopra, S., & Sodhi, M. S. (2014). Reducing the risk of supply chain disruptions. *MIT Sloan Management Review, 73*, 73–80.

Chopra, S., & Sodhi, M. M. S. (2004). Managing risk to avoid supply-chain breakdown. *MIT Sloan Management Review, 46*(1), 52–62.

Corbett, C. J., & Klassen, R. D. (2006). Extending the horizons: Environmental excellence as key to improving operations. *Manufacturing & Service Operations Management, 8*(1), 5–22. https://doi.org/10.1287/msom.1060.0095.

Cousins, P. D., Lamming, R. C., & Bowen, F. (2013). The role of risk in environment-related supplier initiatives. *International Journal of Operations & Production Management, 24*(6), 554–565.

Dhingra, R., Kress, R., & Upreti, G. (2014). Does lean mean green? *Journal of Cleaner Production, 85*, 1–7. https://doi.org/10.1016/j.jclepro.2014.10.032.

Dong, Q., & Cooper, O. (2016). An orders-of-magnitude AHP supply chain risk assessment framework. *International Journal of Production Economics, 182*, 144–156. https://doi.org/10.1016/j.ijpe.2016.08.021.

EPA. (2007). The lean and environment toolkit (pp. 1–96). Retrieved September 22, 2017, from https://www.epa.gov/sites/production/files/2013-10/documents/leanenvirotoolkit.pdf; http://www.epa.gov/lean

Foerstl, K., Reuter, C., Hartmann, E., & Blome, C. (2010). Managing supplier sustainability risks in a dynamically changing environment-sustainable supplier management in the chemical industry. *Journal of Purchasing & Supply Management, 16*(2), 118–130. https://doi.org/10.1016/j.pursup.2010.03.011.

Ganguly, K. K., & Guin, K. K. (2013). A fuzzy AHP approach for inbound supply risk assessment. *Benchmarking: An International Journal, 20*(1), 129–146.

Ho, W., Zheng, T., Yildiz, H., & Talluri, S. (2015). Supply chain risk management: A literature review. *International Journal of Production Research, 53*(16), 5031–5069. https://doi.org/10.1080/00207543.2015.1030467.

Hofmann, H., Busse, C., Bode, C., & Henke, M. (2014). Sustainability-related supply chain risks: Conceptualization and management. *Business Strategy and the Environment, 23*, 160–172.

King, A. A., & Lenox, M. J. (2001). Lean and green? An empirical examination of the relationship between lean production and environmental performance. *Journal Production and Operations Management, 10*(3), 244–256.

Kumar, S., Kumar, P., & Kumar, M. (2015). Risk analysis in green supply chain using fuzzy AHP approach: A case study. *Resources, Conservation and Recycling, 104*, 375–390.

Mital, M., Del Guidice, M., & Papa, A. (2017, February). Comparing supply chain risks for multiple product categories with cognitive mapping and analytic hierarchy process. *Technological Forecasting & Social Change,* 1–12. https://doi.org/10.1016/j.techfore.2017.05.036.

Mollenkopf, D., Stolze, H., Tate, W. L., & Ueltschy, M. (2010). Green, lean, and global supply chains. *International Journal of Physical Distribution & Logistics Management, 40*(1/2), 14–41.

Nazam, M., et al. (2015). A fuzzy AHP-TOPSIS framework for the risk assessment of green supply chain implementation in the textile industry. *International Journal of Supply and Operations Management, 2*(1), 548–568.

Nooraie, S. V., & Mellat Parast, M. (2015). A multi-objective approach to supply chain risk management: Integrating visibility with supply and demand risk. *International Journal of Production Economics, 161*, 192–200. https://doi.org/10.1016/j.ijpe.2014.12.024.

Pettit, T. J., Croxton, K. L., & Fiksel, J. (2013). Ensuring supply chain resilience: Development and implementation of an assessment tool. *Journal of Business Logistics, 34*(1), 46–76.

Radivojević, G., & Gajović, V. (2017). Supply chain risk modeling by AHP and fuzzy AHP methods. *Journal of Risk Research, 17*(3), 337–352.

Prasanna Venkatesan, S., & Kumanan, S. (2012). Supply chain risk prioritisation using a hybrid AHP and PROMETHEE approach. *International Journal of Services and Operations Management, 13*(1), 19–41.

Seuring, S., & Müller, M. (2008). From a literature review to a conceptual framework for sustainable supply chain management. *Journal of Cleaner Production, 16*, 1699–1710.

Simpson, D. F., & Power, D. J. (2005). Use the supply relationship to develop lean and green suppliers. *Supply Chain Management: An International Journal, 10*(1), 60–68. https://doi.org/10.1108/13598540510578388.

Sodhi, M. M. S. (2005). Managing demand risk in tactical supply chain planning for a global consumer electronics company. *Production and Operations Management, 14*(1), 69–79. https://doi.org/10.1111/j.1937-5956.2005.tb00010.x.

Tang, O., & Nurmaya Musa, S. (2011). Identifying risk issues and research advancements in supply chain risk management. *International Journal of Production Economics, 133*, 25–34. https://doi.org/10.1016/j.ijpe.2010.06.013.

Tao, L. (2012). Risk assessment in the supply chain management based on fuzzy AHP model. *Progress in Applied Mathematics, 4*(1), 9–13.

Teuscher, P., Grüninger, B., & Ferdinand, N. (2006). Risk management in sustainable lessons learnt from the case of GMO-free soybeans. *Corporate Social Responsibility and Environmental Management, 13*, 1–10.

Wang, X., Chan, H. K., Yee, R. W. Y., & Diaz-Rainey, I. (2012). A two-stage fuzzy-AHP model for risk assessment of implementing green initiatives in the fashion supply chain. *International Journal of Production Economics, 135*, 595–606. https://doi.org/10.1016/j.ijpe.2011.03.021.

Xiaoping, W. (2016). Food supply chain safety risk evaluation based on AHP fuzzy integrated evaluation method. *International Journal of Security and Its Applications, 10*(3), 233–244.

Zsidisin, G. A. (2003). A grounded definition of supply risk. *Journal of Purchasing and Supply Management, 9*(5–6), 217–224.

# A New Multi Objective Linear Programming Model for Lean and Green Supplier Selection with Fuzzy TOPSIS

**Belkız Torğul and Turan Paksoy**

## 1 Introduction

The success of a company is not only dependent on its performance. The performance of other members in the supply chain directly affects this success, too. For this reason, in order for success to be achieved and sustained, companies must give importance and effectively manage their supply chains. Supplier, the first step and an important link in the supply chain, plays an important role in achieving the aims of companies (Akman and Alkan 2006; Yildiz 2013). Within the new global market, managers have recognized the prominence of choosing the most proper suppliers (Galankashi et al. 2016).

The supplier selection can be defined as the process of deciding on how to receive the raw materials, semi products and materials needed for production and from which supplier, what quantities and at what time (Yildiz 2013). In other words, supplier selection is the choosing the best supplier that will provide competitive advantage among alternative suppliers considering the short and long term plans of the company (Dağdeviren and Erarslan 2008). When a supplier is selected, full compliance with the supplier may not always be possible, in such cases, the supplier that is closest to the ideal supplier is selected and if possible this supplier can be developed later (Şenocak and Gören 2016).

For business, it has been needed to work with suppliers who adopt lean thinking since the suppliers' performance on product cost, quality and delivery is of great importance. As an operational strategy, lean suppliers allow a firm to meet customer expectations at faster delivery, better quality and lower cost than its competitors. However, in order to have a long-term success and presence in

B. Torğul (✉) · T. Paksoy
Department of Industrial Engineering, Konya Technical University, Konya, Turkey



T. Paksoy et al. (eds.), *Lean and Green Supply Chain Management*, International Series in Operations Research & Management Science 273,
https://doi.org/10.1007/978-3-319-97511-5_4

the global market, organizations should not only focus on financial situation but also should take environmental and social concerns into account when evaluating suppliers. As environmental awareness increases by stakeholders, governments, customers, employees, and competitors, organizations are learning to buy goods and services from lean suppliers that can provide low cost, high quality, fast delivery and from green suppliers that can simultaneously provide environmental and social responsibility (Shaik and Abdul-Kader 2011).

Green suppliers collaborate with organizations on environment friendly approaches through green materials, recycled materials, environmental management systems, life cycle assessments, waste prevention and control practices (Shen et al. 2013). In this context, lean and green supplier selection has attracted ever-growing attention recently, which is a multi-criteria decision-making problem that requires consideration of many qualitative and quantitative factors.

Attributes for supplier selection are well documented in the literature, but there is a lack of relevant attributes in the field of lean and green supplier selection. This has led to the development of criteria and general framework for lean and green supplier selection (Shaik and Abdul-Kader 2011). The most popular approaches accepted in the supplier evaluation and selection literature are AHP followed by ANP, Data Envelopment Analysis (DEA), Case-Based Reasoning (CBR), Genetic Algorithm (GA), Artificial Neural Networks, Gray Relational Analysis, Mathematical Programming, other methods and combinations thereof (Junior et al. 2014; Govindan et al. 2015). Along with Multi-Criteria Decision Making (MCDM) methods, Fuzzy Set Theory has been widely used to deal with uncertainty in the supplier selection decision process because it provides a suitable language for processing uncertain criteria, which can integrate the analysis of qualitative and quantitative factors (Junior et al. 2014). The integrated fuzzy methods used in the studies are Fuzzy TOPSIS (FTOPSIS), Fuzzy AHP (FAHP), Fuzzy ANP, Fuzzy SMART (simple multi-attribute rating technique), Fuzzy QFD (Quality Function Deployment), Fuzzy ART (Fuzzy Adaptive Resonance Theory) and Fuzzy multi-objective programming (Yildiz 2013).

## *1.1 Brief Review of the Literature*

Traditionally, the focus of supplier selection is usually on business performance evaluating the lean criteria in a deterministic model with single objective or multi-objective in supply chain management. For example; Talluri and Narasimhan (2003) proposed max–min approach for vendor selection by incorporating performance variability into the evaluation process. They considered product price, quality, and delivery as the most important (lean) factors in evaluating alternative vendors. Birgün Barla (2003) studied the supplier selection and evaluation for a manufacturing company under lean philosophy. She conducted by multi-attribute selection model (MSM) in five basic steps and determined important attributes for the company under lean supply principles, which are reliability, capability, quality organization, geographic location, financial condition, service level and price.

Amid et al. (2006) developed a fuzzy multi objective linear model by applying an asymmetric fuzzy-decision making technique to assign different weights to various lean criteria (price, quality and service). Kumar et al. (2006) treated a fuzzy multi-objective integer programming vendor selection problem formulation that facilitates the vendor selection and their quota allocation under different degrees of information vagueness in the decision parameters that are Net cost, Rejection and Late deliveries objectives, Capacity and Budget constraints.

Ng (2008) proposed a weighted linear program for the multi-criteria supplier selection problem and studied a transformation technique for the proposed model to be solved without an optimizer. He considered five lean criteria, including supply variety, quality, distance, delivery and price for a firm manufacturing agricultural and construction equipment. Jadidi et al. (2008) proposed a single objective function and used TOPSIS to solve the fuzzy multi-item multi-objective model in order to calculate the optimum order quantities to each supplier. They evaluated three alternative suppliers against four lean attributes (e.g., special factor, on-time delivery, performance history and technical capability).

Wang and Yang (2009) introduced the AHP and fuzzy compromise programming for allocating order quantities among suppliers with their quantity discount rate offered. They considered three important lean criteria (Net price, Delivery, Quality) for formulating three objective functions. Amid et al. (2011) developed a weighted max–min fuzzy model which help the decision maker to find out the appropriate order to each supplier, and manage supply chain performance on cost, quality and service. They used AHP to determine the weights of lean criteria that are net price, quality and service. Lin (2012) used the fuzzy ANP (FANP) approach first to identify top suppliers and then integrated FANP with fuzzy multi-objective linear programming in selecting the best suppliers for achieving optimal order allocation under fuzzy conditions. They evaluated four alternative suppliers influenced by four lean criteria of Price, Quality, Delivery, and Technique.

Erginel and Gecer (2016) presented a systematic approach for a calibration supplier selection problem and represented the criteria with a questionnaire and proposed the fuzzy multi-objective linear programming model for selecting the calibration supplier. Warranties and complaint policy, Communication, Service features, Quality and Performance history are the important criteria they deal with. Galankashi et al. (2016) developed an integrated lean-based Data Envelopment Analysis (DEA) model for assessing and selecting suppliers. They designated performance perspectives (cost, quality, lead time and service level) in addition to 16 related metrics as substitutes for leanness.

The decision models have become more complex because of many new dimensions brought in by green paradigms, where the tradeoffs become more apparent and numerous. These new criteria required rethinking some of the more fixed approaches and models. The supplier selection process has only recently started to integrate various environmental dimensions by researchers (Govindan et al. 2015). For example; Lee et al. (2009) proposed a model for evaluating green suppliers with environmental performance criteria including Green image, Pollution control, Environment management, Green product and Green competencies. They applied Delphi method to differentiate the criteria for evaluating traditional suppliers and green suppliers and used FAHP to consider the vagueness of experts' opinions. Kuo et al. (2010) developed a green supplier selection model which integrates Artificial Neural Network (ANN) and two Multi-Attribute Decision Analysis (MADA) methods: Data Envelopment Analysis (DEA) and Analytic Network Process (ANP). This hybrid method considers both traditional and environmental supplier selection criteria. Shaik and Abdul-Kader (2011) proposed a framework consisting of environmental, green and organizational factors that are required for the green supplier selection process and applied the multiple attribute utility theory.

Shaw et al. (2012) presented an integrated approach for selecting the appropriate supplier in the supply chain, addressing the carbon emission issue, using FAHP and fuzzy multi-objective linear programming. Büyüközkan and Çifçi (2012) proposed a case study for green supplier evaluation in Ford Otosan. They formulated a new model defining possible green supplier evaluation criteria and used a combination of DEMATEL, ANP, and TOPSIS in fuzzy environment for the solution of the problem. Kannan et al. (2014) proposed a framework using three types of FTOPSIS to select green suppliers for a Brazilian electronics company. They collected the data from a set of 12 available suppliers according to 17 identified green criteria. Dobos and Vörösmarty (2014) used composite indicators (CI) method to study the extension of traditional supplier selection methods with green factors. In this method, they handled the managerial and environmental criteria separately.

Kannan et al. (2015) proposed a Multi-Criteria Decision-Making (MCDM) approach called Fuzzy Axiomatic Design (FAD) to select the best green supplier for a Singapore-based plastic manufacturing company. They developed environmental criteria which included Environment protection/Environment Management, Corporate social responsibility, Pollution control, Green Product, Green Image, Green Innovation, Environmental Performance, Hazardous Substance Management and their Sub-Criteria with the traditional criteria. Banaeian et al. (2015) formulated an integrated framework to decide on the green criteria for food supply chain using the Analytic Hierarchy Process and Delphi method and also proposed different methods -Fuzzy Grey Relational Analysis and linear programming- for single and multiple supplier selection.

Wu and Barnes (2016) presented a model for green partner selection by combining ANP and multi-objective programming methodologies and developed an additional decision-making tool in the form of the environmental difference, the business difference and the eco-efficiency ratio which quantify the trade-offs between environmental and business performances. Liao et al. (2016) proposed new integrated fuzzy techniques for FAHP, Fuzzy Additive Ratio Assessment (ARASF) and Multi-Segment Goal Programming (MSGP) approach to evaluate and select the best (green) supplier. Hamdan and Cheaitou (2017) proposed a decision-making tool to solve a multi-period green supplier selection and order allocation problem. They used FTOPSIS to rank potential suppliers on the basis of traditional and green criteria separately and used AHP to assign importance weights of criteria. Then, they compared proposed ranking approach with the classic approach adopted in which all the criteria are considered in a single set. They solved mathematical models using the weighted comprehensive criterion method and the branch-and-cut algorithm.

### *1.2 Contributions of the Study*

Although the supplier assessment and selection have been deliberated by numerous scholars, both lean and green supplier selection frameworks and integrating them into mathematical models is less widely investigated (Galankashi et al. 2016). Hence, this study presents an integrated approach for the selection of the appropriate suppliers in a supply chain, aiming to maximize lean and green value of purchasing simultaneously by using integration of FAHP-FTOPSIS and fuzzy multi-objective linear programming.

The rest of the paper is organized as follows. Section 2 provides the problem definition and the main steps of the solution procedure. Section 3 presents Fuzzy set theory, FAHP, FTOPSIS and fuzzy linear programming methodologies. In Sect. 4, developed mathematical model is explained. Then related model is tested with a case study in GAMS package program and obtained results are interpreted in Sect. 2. Section 3 presents the conclusions of the study with suggestions for future work.

## 2 Problem Definition

A supplier selection model will be designed for the supply of materials that a firm will use in the production process. The firm wants to work with suppliers who have lean and green management practices to improve its environmental and economic performance. Besides the complementarities that exist between lean and green, the relationships between them can also reflect some trade-offs (Carvalho et al. 2017). In fact, firms may have to compromise some of their lean practices in order to be environment friendly. Nevertheless, while companies can deal with the environment on the one hand, they try to improve their profits on the other hand. This balance varies according to the lean and green tools of their suppliers. Therefore, in this study, potential suppliers will be evaluated separately in terms of lean and green tools that are important for the firm and will be selected according to the assumptions of lean-green balance determined by the firm.

This study proposes a two-stage mathematical model to evaluate suppliers, given a number of lean and green tools. In the first stage, FAHP is used to obtain the relative weights for lean and green criteria and FTOPSIS is used for ranking (weights) of suppliers. Then, formulation of multi objective linear programming of supplier selection is constructed using fuzzy linear programming. In the second stage, the weights of suppliers are incorporated into the multi-objective linear programming model to determine the optimal order quantity from each supplier with some resource constraints. The weights of suppliers are used as coefficients of objective functions to increase the lean or green or both of their values for the company so that the sum of the purchasing value is maximized and order quantities are allocated between suppliers. The main steps in the solution procedure adopted for this study are presented in Fig. 1.

All cost, demand, capacity, $CO_2$ emission, lead time and budget information were assumed fixed and already known. Also, it was assumed that only one type of product can be purchased from each supplier and there is no provision for quantity reduction in this model.

## 3 Methodologies

Fuzzy set theory, FAHP, FTOPSIS and fuzzy linear programming that were utilized in the current study are elaborated in this section.

Fuzzy set theory: Fuzzy sets were interpreted by Lotfi A. Zadeh who noted in 1965 that human thinking is mostly fuzzy. Fuzzy set theory has been used for modeling decision making processes based on uncertain and vague information such as judgment of decision makers. Fuzzy set theory removes uncertainty in decision making through quantification to qualitative values (Junior et al. 2014; Kumar et al. 2017).

A fuzzy set is defined with a membership function and all its elements have membership degrees that vary between zero and one (Zadeh 1965). In this study, triangular fuzzy numbers have been used as well as the possibility of different fuzzy numbers. A triangular fuzzy number is shown in Fig. 2. A triangular fuzzy number is shown as (l, m, u) with l < m < u (Junior et al. 2014; Kargi 2016).

A triangular membership function and its elements are represented as follows:

$$\mu_A(x) = \begin{cases} 0 & for\, x < l, \\ \frac{x-l}{m-l} & for\, l \le x \le m, \\ \frac{u-x}{u-m} & for\, m \le x \le u, \\ 0 & for\, x > u, \end{cases} \tag{1}$$

*Fuzzy AHP*: In this study, Buckley's method was used to determine the relative importance weights for the lean and green tools. Buckley (1985) used the geometric mean method to obtain fuzzy weights and performance scores. This method can be easily generalized to fuzzy situations and it obtains a single solution from

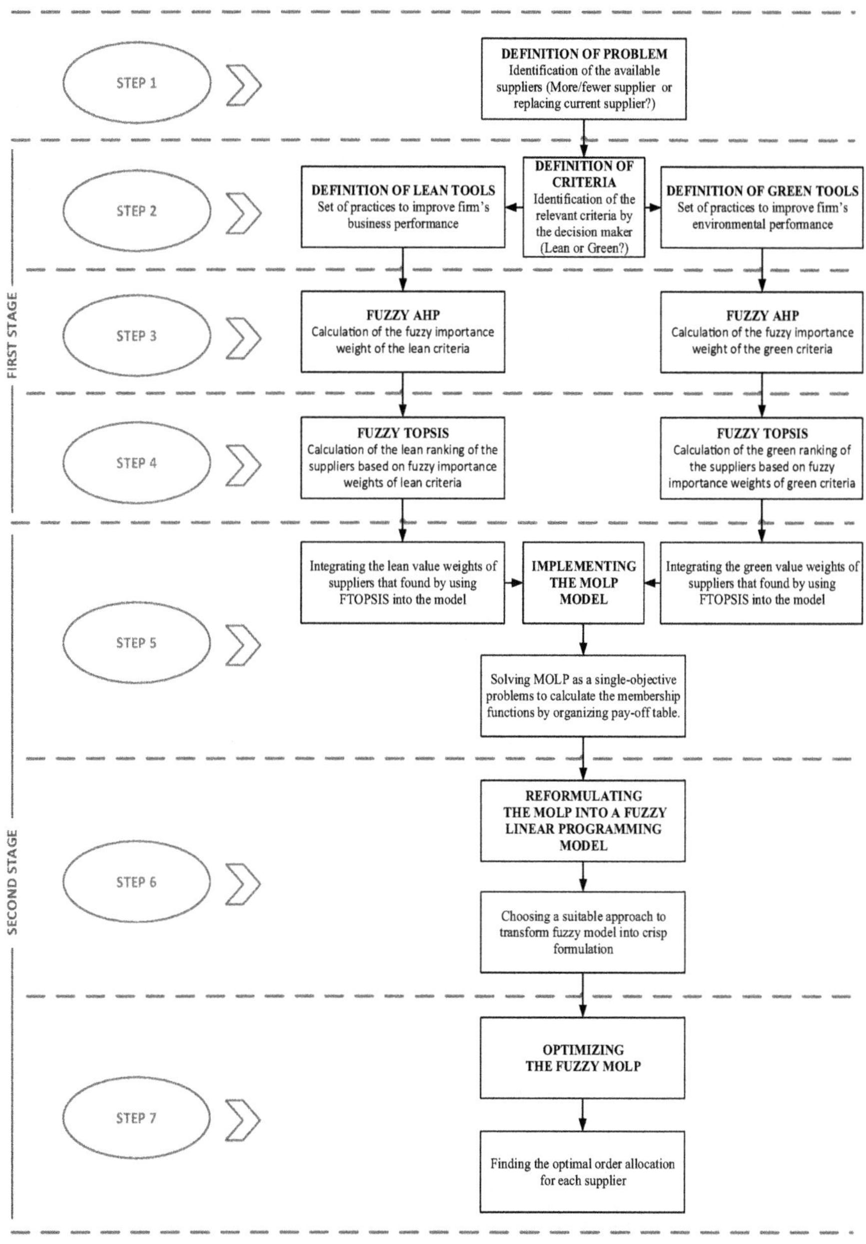


**Fig. 1** The lean and green supplier selection framework

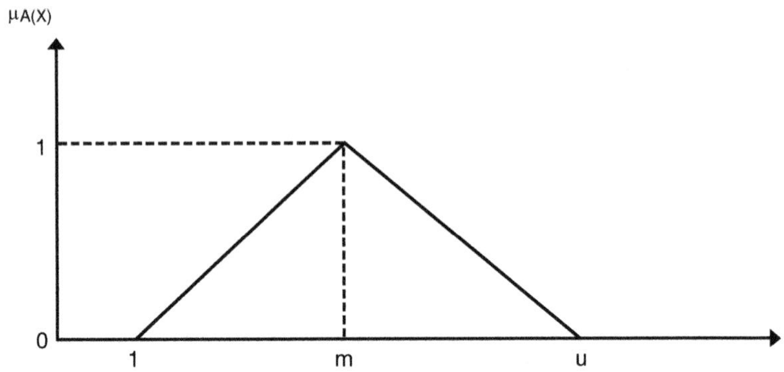


**Fig. 2** Triangular membership function

comparison matrices. The method consists of the following steps (Hsieh et al. 2004);

1. A matrix of comparison to the opinion of the decision maker is obtained. If the fuzzy positive comparison matrix is given as follows;

$$A = \begin{bmatrix} a_{11} & \dots & a_{1n} \\ \vdots & \ddots & \vdots \\ a_{n1} & \dots & a_{nn} \end{bmatrix} \tag{2}$$

2. In order to calculate $w_i$ fuzzy weights, geometric averages are calculated for each row first.

$$Z_i = \left[ \prod_{j=1}^{n} a_{ij} \right]^{1/n}, \forall i. \tag{3}$$

3. After finding the geometric mean of each row, $w_i$ is calculated using (4).

$$w_i = \frac{z_i}{z_i + \cdots + z_n}, \forall i. \tag{4}$$

   $w_i$, can be shown as $w_i = (lw_i, \text{m}w_i, \text{u}w_i,)$ where $lw_i$, $mw_i$ and $uw_i$ are the lower, middle and upper fuzzy values.
4. The defuzzification value of the fuzzy number $w_i$ can be found by the following equation.

$$F_i = \frac{[(\text{u}w_i - lw_i) + (\text{u}w_i - lw_i)]}{(3 + lw_i)}, \forall i. \tag{5}$$

According to this calculated $F_i$ value, the ranking of each alternative is made.

*Fuzzy TOPSIS*: The FTOPSIS proposed by Chen (2000) was developed to solve group decision problems under fuzzy environment. In this method, linguistic expressions are used instead of numerical values in order to reflect the environment more realistically. Linguistic variables are used by the decision makers, $D_r$ (r = 1, ..., k), to assess the weights of the criteria and the ratings of the alternatives. Thus, $\tilde{W}_r^j$ describes the weight of the jth criterion, $C_j$ (j = 1, ..., m), given by the rth decision maker. Similarly, $\tilde{X}_{ij}^r$ describes the rating of the ith alternative, $A_i$ (i = 1, ..., n), with respect to criterion j, given by the rth decision maker. Given that, the method comprises the following steps (Junior et al. 2014; Kargi 2016; Kannan et al. 2013; Chen et al. 2006):

1. Aggregate the weights of criteria and ratings of alternatives given by k decision makers, as expressed in (6) and (7) respectively:

$$\tilde{w}_j = 1/k\left[\tilde{w}_j^1 + \tilde{w}_j^2 + \cdots + \tilde{w}_j^k\right] \tag{6}$$

$$\tilde{x}_{ij} = 1/k\left[\tilde{x}_{ij}^1 + \tilde{x}_{ij}^2 + \cdots + \tilde{x}_{ij}^k\right] \tag{7}$$

2. Assemble the fuzzy decision matrix of the alternatives ($\tilde{D}$) and the criteria ($\tilde{W}$), according to (8) and (9):

$$\begin{array}{c} \quad\quad\quad\quad C_1 \quad C_2 \quad C_j \quad C_m \\ \tilde{D} = \begin{array}{c} A_1 \\ A_i \\ A_n \end{array} \begin{bmatrix} \tilde{x}_{11} & \tilde{x}_{12} & \tilde{x}_{1j} & \tilde{x}_{1m} \\ \vdots & \vdots & \vdots & \vdots \\ \tilde{x}_{n1} & \tilde{x}_{n2} & \tilde{x}_{nj} & \tilde{x}_{nm} \end{bmatrix} \end{array} \tag{8}$$

$$\tilde{W} = \left[\tilde{w}_1 + \tilde{w}_2 + \cdots + \tilde{w}_m\right] \tag{9}$$

3. Normalize the fuzzy decision matrix of the alternatives ($\tilde{D}$) using linear scale transformation. The normalized fuzzy decision matrix $\tilde{R}$ is given by:

$$\tilde{R} = \left[\tilde{r}_{ij}\right]_{m\times n} \tag{10}$$

$$\tilde{r}_{ij} = \left(\frac{l_{ij}}{u_j^+}, \frac{m_{ij}}{u_j^+}, \frac{u_{ij}}{u_j^+}\right) \text{ and } u_j^+ = \max_i u_{ij} \text{ (benefit criteria)} \quad (11)$$

$$\tilde{r}_{ij} = \left(\frac{l_j^-}{u_{ij}}, \frac{l_j^-}{m_{ij}}, \frac{l_j^-}{l_{ij}}\right) \text{ and } l_j^- = \min_i l_{ij} \text{ (cost criteria)} \quad (12)$$

4. Compute the weighted normalized decision matrix, $\tilde{V}$ by multiplying the weights of the evaluation criteria, $\tilde{w}_j$, by the elements $\tilde{r}_{ij}$ of the normalized fuzzy decision matrix.

$$\tilde{V} = \left[\tilde{v}_{ij}\right]_{m\times n} \quad (13)$$

where $\tilde{v}_{ij}$ is given by (14)

$$\tilde{v}_{ij} = \tilde{x}_{ij} \times \tilde{w}_j \quad (14)$$

5. Define the Fuzzy Positive Ideal Solution (FPIS, $A^+$) and the Fuzzy Negative Ideal Solution (FNIS, $A^-$), according to (15) and (16).

$$A^+ = \left\{\tilde{v}_1^+, \tilde{v}_j^+, \ldots, \tilde{v}_m^+\right\} \quad (15)$$

$$A^- = \left\{\tilde{v}_1^-, \tilde{v}_j^-, \ldots, \tilde{v}_m^-\right\} \quad (16)$$

where $\tilde{v}_j^+ = \max\{\tilde{v}_{ij\,3}\}$ and $\tilde{v}_j^- = \min\{\tilde{v}_{ij\,1}\}$

6. Compute the distances $d_i^+$ and $d_i^-$ of each alternative from respectively $\tilde{v}_j^+$ and $\tilde{v}_j^-$ according to (17) and (18).

$$d_i^+ = \sum\nolimits_{j=1}^{n} d_v\left(\tilde{v}_{ij}, \tilde{v}_j^+\right) \quad (17)$$

$$d_i^- = \sum\nolimits_{j=1}^{n} d_v\left(\tilde{v}_{ij}, \tilde{v}_j^-\right) \quad (18)$$

where $d(.,.)$ represents the distance between two fuzzy numbers according to the vertex method. For triangular fuzzy numbers, this is expressed as in (19).

$$d\left(\tilde{x}, \tilde{z}\right) = \sqrt{\frac{1}{3}\left[(l_x - l_z)^2 + (m_x - m_z)^2 + (u_x - u_z)^2\right]} \quad (19)$$

7. Compute the closeness coefficient, $CC_i$, according to (20).

$$CC_i = \frac{d_i^-}{d_i^+ + d_i^-} \tag{20}$$

8. Define the ranking of the alternatives according to the closeness coefficient, $CC_i$, in decreasing order. The best alternative is closest to the FPIS and farthest to the FNIS.

*Fuzzy linear programming*: Zimmermann (1978) proposed fuzzy linear programming by using Bellman and Zadeh (1970) approach. After fuzzification, the fuzzy linear program that included fuzzy goals and fuzzy constraints can be solved as a linear programming problem. Zimmermann (1978) proposed a conventional linear programming problem as fallows (21, 22 and 23):

$$\text{Minimize Z} = Cx \tag{21}$$

Subject to,

$$Ax \leq \text{b} \tag{22}$$

$$x \geq 0 \tag{23}$$

After fuzzification the equation can be represented as (24, 25 and 26),

$$\tilde{C}x \precsim \text{Z} \tag{24}$$

$$\tilde{A}x \precsim \text{b} \tag{25}$$

$$x \geq 0 \tag{26}$$

The symbol $\precsim$ in the constraint set denotes 'smaller than or equal to' and allows one to reach some aspiration level where $\tilde{C}$ and $\tilde{A}$ represent the fuzzy values. According to Bellman and Zadeh (1970) approach, the fuzzy set $A$ in X is defined as: $A = \{x, \mu_A(x)/x \in X\}$. Where $\mu_A(x) : x \rightarrow [0, 1]$ is called the membership function of A and $\mu_A(x)$ is the degree of membership to which x belongs to A.

A fuzzy objective $\tilde{\text{Z}} \in \text{X}$ is a fuzzy subset of X characterized by its membership function $\mu_Z(x) : x \rightarrow [0, 1]$. In this approach, Zimmermann formulated the membership function of objectives by separating every objective function into its maximum and minimum values. The linear membership functions for minimization and maximization objectives are given as:

$$\mu_{Zj}(x) = \begin{cases} 1 & \mathit{if}\ Z_j(x) \le Z_j^{min} \\ \frac{Z_j^{max} - Z_j(x)}{Z_j^{max} - Z_j^{min}} & \mathit{if}\ Z_j^{min} \le Z_j(x) \le Z_j^{max}, \mathit{where}\ j = 1, 2, \ldots, j \\ 0 & \mathit{if}\ Z_j(x) \ge Z_j^{max} \end{cases} \text{ (for minimization)} \tag{27}$$

$$\mu_{Zj}(x) = \begin{cases} 0 & \mathit{if}\ Z_j(x) \le Z_j^{min} \\ \frac{Z_j(x) - Z_j^{min}}{Z_j^{max} - Z_j^{min}} & \mathit{if}\ Z_j^{min} \le Z_j(x) \le Z_j^{max}, \mathit{where}\ j = 1, 2, \ldots, j \\ 1 & \mathit{if}\ Z_j(x) \ge Z_j^{max} \end{cases} \text{ (for maximization)} \tag{28}$$

In Eqs. (27) and (28), $Z_j^{min}$ is $min_j\ Z_j$ (x*) and $Z_j^{max}$ is $max_j\ Z_j$ (x*) and x* is the optimum solution.

A fuzzy constraint $\tilde{C} \in X$ is a fuzzy subset of X characterized by its membership function $\mu_C(x) : x \rightarrow [0, 1]$. The linear membership function for the fuzzy constraints is given by (29):

$$\mu_{ck}(x) = \begin{cases} 1 & \mathit{if}\ g_k(x) \le b_k \\ [1 - \{g_k(x) - b_k\} / d_k] & \mathit{if}\ b_k \le g_k(x) \le b_k + d_k \\ 0 & \mathit{if}\ b_k + d_k \ge g_k(x) \end{cases} \tag{29}$$

According to Zimmermann (1978) each objective function should be solved as a single objective function by using each time only one of the objectives and ignoring all others. This process is repeated j times for j different objective functions. All found solutions are used to construct a pay-off matrix of size j by j. Then the lower bound ($Z_j^{min}$) and the upper bound ($Z_j^{max}$) are estimated from the pay-off matrix for the jth objective function $Z_j$ as (Behera and Nayak 2011):

$$Z_j^{min} \le Z_j \le Z_j^{max}\ \mathit{for\ all}\ j,\ j = 1, 2, \ldots, J \tag{30}$$

*Crisp formulation*: With Zimmermann's approach, the fuzzy programming model for j objectives and k constraints are transformed into the crisp formulation as fallows (Kumar et al. 2006, 2017; Shaw et al. 2012).

$$\text{Maximize } \lambda \tag{31}$$

Subject to,

$$\lambda \left(Z_j^{max} - Z_j^{min}\right) + Z_j(x) \le Z_j^{max}\ \mathit{for\ all}\ j,\ j = 1, 2, \ldots, J, \text{ (for minimization)} \tag{32}$$

$$\lambda\left(Z_j^{max} - Z_j^{min}\right) + Z_j^{min} \leq Z_j(x)\, for\ all\ j,\ j = 1, 2, \ldots, J, \text{ (for maximization)} \tag{33}$$

$$\lambda\,(\mathrm{d}_x) + g_k(x) \leq b_k + d_k\, for\ all\ k,\ k = 1, 2, \ldots, K, \tag{34}$$

$$Ax \leq \mathrm{b}\, for\ all\ the\ deterministic\ constant, \tag{35}$$

$$x \geq 0 \text{ and integer,} \tag{36}$$

$$0 \leq \lambda \leq 1. \tag{37}$$

In Zimmermann approach, all factors are assigned the same weight. However, the weights of all objectives may not be same in a real-life supplier selection problem. For example, if the work of the study is considered, the lean value may be more important than the green value for an organization. Thus, the organization should assign a higher weight for the objective that maximizes lean value than for the other. To deal with such situations, many researchers have used a weighted additive model for multi-objective optimization problems. The weighted additive model proposed by Tiwari et al. (1987) is shown in (38, 39, 40, 41, 42 and 43).

$$\text{Maximize} \sum_{j}^{J} w_j\, \lambda_j \tag{38}$$

Subject to,

$$\lambda_j \leq \mu_{Zj}(x)\, for\ all\ j,\ j = 1, 2, \ldots, J, \tag{39}$$

$$Ax \leq \mathrm{b}\, for\ all\ the\ deterministic\ constant, \tag{40}$$

$$\sum_{j}^{J} w_j = 1 \tag{41}$$

$$w_j,\ x\ \geq 0 \text{ and integer, } for\ all\ j,\ j = 1, 2, \ldots, J, \tag{42}$$

$$0 \leq \lambda \leq 1. \tag{43}$$

Where $w_j$ is the coefficient weighting that presents the relative importance among the fuzzy goals.

## 4 Mathematical Model

**Indices**
$\boldsymbol{i}$: *Set of suppliers* ($i = 1, 2, \ldots, I$)
$\boldsymbol{j}$: *Set of objectives* ($j = 1, 2, \ldots, J$)

**Parameters**
$\boldsymbol{D}$: *Total demand of the product*
$\boldsymbol{N}$: *Number of suppliers*
$\boldsymbol{B}$: *Total budget allocated to suppliers*
$\boldsymbol{O}$: *Total carbon emission limit for supply*
$\boldsymbol{R}$: *Maximum acceptable ratio (percent) of the late delivered products*
$\boldsymbol{P_i}$: *Unit purchasing cost of product from supplier i*
$\boldsymbol{L_i}$ : *The ratio (percent) of the late delivered products by the supplier i*
$\boldsymbol{C_i}$: *The capacity of the supplier i*
$\boldsymbol{G_i}$ : *The carbon emission for product supplied by supplier i*
$\boldsymbol{Wl_i}$: *The lean weight (lean priority value) of the supplier i*
$\boldsymbol{Wg_i}$: *The green weight (green priority value) of the supplier I*

**Decision Variables**
$\boldsymbol{X_i}$: *The amount of product ordered from supplier i*

**Objective Functions**

$$\text{Max } Z_1 = \sum_{i}^{N} Wl_i \; X_i \tag{44}$$

$$\text{Max } Z_2 = \sum_{i}^{N} Wg_i \; X_i \tag{45}$$

**Constraints**

$$\sum_{i}^{N} X_i \geq D \tag{46}$$

$$X_i \leq C_i \quad \forall i \tag{47}$$

$$\sum_{i}^{N} L_i \; X_i \leq RD \tag{48}$$

$$\sum_{i}^{N} G_i \, X_i \leq O \tag{49}$$

$$\sum_{i}^{N} P_i \, X_i \leq B \tag{50}$$

$$X_i \geq 0 \; and \; integer \quad \forall i \tag{51}$$

Objective function (44) maximizes the total lean value of purchasing. Objective function (45) maximizes the total green value of purchasing. Constraint (46) is the demand constraint which ensures that the total demand is satisfied. Constraint (47) is the capacity constraint of suppliers, meaning that the amount of product ordered from each supplier cannot exceed its capacity. Constraint (48) is the lead time constraint, meaning that the total late delivered products cannot exceed maximum total acceptable for late delivered products. Constraint (49) is the carbon footprint constraint, meaning that the amount of carbon emission released by each supplier cannot exceed maximum total carbon emission limit. Constraint (50) is the budget constraint which means that total purchasing cost of products cannot exceed budget amount allocated to the suppliers. Constraint (51) is the non-negativity and integrity constraint.

## *4.1 Application of Fuzzy Linear Programming*

A fuzzy multi objective linear programming model is proposed for lean and green supplier selection. Fuzzy set theory is used for modelling and solving the multi objective linear programming due to the vagueness of the model parameters. The lean and green value of purchasing are considered fuzzy numbers because of their inherent uncertainty and vague structure. After fuzzification, the equations can be represented as follows (52, 53, 54, 55, 56, 57, 58 and 59).

$$\sum_{i}^{N} Wl_i \, X_i \precsim \tilde{Z}_1 \tag{52}$$

$$\sum_{i}^{N} Wg_i \, X_i \precsim \tilde{Z}_2 \tag{53}$$

$$\sum_{i}^{N} X_i \geq D \tag{54}$$

$$X_i \leq C_i \quad \forall i \tag{55}$$

$$\sum_{i}^{N} L_i \, X_i \leq RD \tag{56}$$

$$\sum_{i}^{N} G_i \, X_i \leq O \tag{57}$$

$$\sum_{i}^{N} P_i \, X_i \leq B \tag{58}$$

$$X_i \geq 0 \; and \; integer \quad \forall i \tag{59}$$

where $\sim$ denotes the fuzzy number.

## 5 A Case Study

**STEP 1**: *Definition of Problem*; Supplier selection problem can be defined as searching for new suppliers or choosing suppliers from the pool of suppliers.

The case study is about an automotive supply chain which consists of one buyer (OEM) and suppliers (OES) in Turkey (Process illustration of the case study is given in Fig. 3). But the real-life data is distorted because of firm's privacy policy. In the case study, 10 potential suppliers will be evaluated separately according to 12 lean and 12 green tools. The lean and green tools identified by the organization are given in Table 1. The importance weights of decision criteria and alternatives are defined with linguistic variables by the decision maker. The triangular fuzzy numbers and linguistic expressions indicating the importance ratings of the criteria and alternatives are given in Table 2 (Chen 2000; Kannan et al. 2013). The fuzzy pair wise comparisons among the lean and green criteria are given in Tables 3 and 4 respectively. Weights of the lean and green criteria are found separately by the FAHP method by using the pair wise comparison matrixes. The evaluation of alternative suppliers with linguistic variables by the decision maker on lean and green criteria are given in Tables 5 and 6 respectively. Tables 7 and 8 represent the conversion of these linguistic evaluations made by the decision maker into triangular fuzzy numbers as shown in Table 2.

Lean and green weights of the suppliers are found by the FTOPSIS method by using fuzzy decision matrixes and fuzzy weights of criteria (as shown in Tables 7 and 8). To perform optimal order allocation, the lean and green weights of the

**Fig. 3** The depiction of the case study

suppliers based on the results of FTOPSIS are used as coefficients in the objective functions of the fuzzy multi objective linear programing model.

The normalized fuzzy decision matrixes for lean and green tools are given in Tables 16 and 17 respectively. The weighted normalized fuzzy decision matrixes for lean and green tools are given in Tables 18 and 19 respectively. The distances of each alternative from $A^+$ and $A^-$ with respect to each criterion for lean tools are given in Tables 20 and 21. The distances of each alternative from $A^+$ and $A^-$ with respect to each criterion for green tools are given in Tables 22 and 23. The outranking of alternative suppliers for lean and green tools are given in Tables 9 and 10 respectively.

**STEP 2**: *Definition of Criteria*; Converting the requirements of the decision maker into decision criteria and defining them.

**STEP 3**: FAHP

**STEP 4**: FTOPSIS

See Appendix for Tables 16, 17, 18, 19, 20, 21, 22 and 23.

**STEP 5**: Implementing the MOLP model

Supplier quantitative information is given in Table 11. Additionally, the total budget (B) is taken 180,000 and the total demand value (D) is taken 20,400, maximum acceptable ratio of the late delivered products value (R) is taken 0.08 and the total carbon emission limit value (O) is taken 36,000 in this model. The

**Table 1** Lean and green tools (Shaik and Abdul-Kader 2011; Shen et al. 2013; Kuo et al. 2010; Kannan et al. 2015; Hamdan and Cheaitou 2017; Carvalho et al. 2017; Anvari et al. 2014; Pandya et al. 2017)

| No | Criteria name | Code | Definition |
|---|---|---|---|
| *Lean tools* | | | |
| 1 | JIT (Pull system)[a] and Kanban[b] | C1 | [a] The delivering of raw materials, work in process, and finished goods, at the exact time they are necessary<br>[b] The method of production control in which downstream activities signal their needs to upstream activities |
| 2 | One-piece-flow/continuous flow | C2 | The production and moving of one item/items at a time through a series of processing steps as continuously as possible, with each step making just what is requested by the next step |
| 3 | TPM[c] & OEE[d] | C3 | [c] The set of techniques to ensure that every machine in a production process is always able to perform its required tasks<br>[d] The metric that identifies the percentage of planned production time that is truly productive |
| 4 | Jidoka (Autonomation) | C4 | The design equipment to partially automate the manufacturing process- stopping a line automatically when a defective part is detected |
| 5 | Kaizen & Continues Improvement Techniques | C5 | The practice of continuous improvement in quality, productivity, safety and workplace culture |
| 6 | 5S | C6 | Eliminating the waste that occurs because of poorly designed workplace- the term 5S refers to five steps (sort, set in order, shine, standardize, and sustain) |
| 7 | Value stream mapping | C7 | Visually mapping the flow of production- the current and future state of processes in a way that highlights opportunities for improvement |
| 8 | Poka-Yoke | C8 | The mistake-proofing device or procedure used to prevent a defect during the production process |
| 9 | Visual management & Andons | C9 | The visual feedback system for the plant floor that indicates production status |

(continued)

**Table 1** (continued)

| No | Criteria name | Code | Definition |
|---|---|---|---|
| 10 | SMED | C10 | The reduction of set up time to less than 10 min- the quick and efficient way of converting a manufacturing process from running the current product to running the following product |
| 11 | Employee engagement | C11 | The workplace approach resulting in the right conditions for all members of an organization to give of their best each day, committed to their organization's goals and values, motivated to contribute to organizational success, with an enhanced sense of their own well-being |
| 12 | Heijunka | C12 | The form of production scheduling that purposely manufactures in much smaller batches by sequencing (mixing) product variants within the same process |
| *Green tools* | | | |
| 13 | $CO_2$ emission | C13 | The quantity control and treatment of hazardous emission, such as $SO_2$, $NH_3$, CO and HCl |
| 14 | Energy consumption | C14 | The efforts and controls to reduce energy consumption- use of efficient energy, low energy consumption and energy saving |
| 15 | Hazardous substance management (RoHS) | C15 | The quantity control and treatment of chemical waste, restriction of hazardous substance |
| 16 | Environment management system | C16 | The environmental certifications such as ISO 14000, environmental policies, planning of environmental objectives, practice, checking and control of environmental activities |
| 17 | Green image | C17 | The ratio of green customers to total customers- the measuring effort and success in protecting the environment, implementing environmental friendly products and green programs indicating competitive advantage |
| 18 | Environment friendly transportation | C18 | The greener transport system-more efficient use of traditional fuels, the utilization of electric vehicle technologies and using bio-gas |

(continued)

**Table 1** (continued)

| No | Criteria name | Code | Definition |
|---|---|---|---|
| 19 | Green CSR projects | C19 | The business approach, culture and voluntary role that contributes to sustainable development by delivering economic, social and environmental benefits for all stakeholders |
| 20 | Recovery of company's end of life products | C20 | The level of recycling, reusing, refurbishing, remanufacturing, disassembly and disposal of the products |
| 21 | Environment friendly packaging | C21 | The level of green recyclable, reusable, returnable, packing materials (boxes, bags, pallets, racks, containers, among others) used in packaging of goods |
| 22 | Green design and innovation | C22 | The design for environment which includes checking the supplier's design for environment capability-the level of implementation of green and clean technologies |
| 23 | Waste management | C23 | The control and treatment of the amount, collection-accumulation, transport-transfer, processing-recovery, disposal-final away and storage of waste |
| 24 | Environment friendly raw materials | C24 | The use of materials in the products which have a lower impact on natural resources- the level of green recyclable material used in manufacturing of goods |

**Table 2** Linguistic variables and fuzzy numbers used in the evaluation of criteria and alternatives

| Linguistic variables for the importance weight of each criterion | | Linguistic variables for the ratings of alternatives | |
|---|---|---|---|
| Linguistic variables | Triangular fuzzy numbers | Linguistic variables | Triangular fuzzy numbers |
| Equally strong | (1, 1, 1) | Very poor (VP) | (0, 0, 1) |
| Intermediate | (1, 2, 3) | Poor (P) | (0, 1, 3) |
| Moderately strong | (2, 3, 4) | Medium poor (MP) | (1, 3, 5) |
| Intermediate | (3, 4, 5) | Fair (F) | (3, 5, 7) |
| Strong | (4, 5, 6) | Medium good (MG) | (5, 7, 9) |
| Intermediate | (5, 6, 7) | Good (G) | (7, 9, 10) |
| Very strong | (6, 7, 8) | Very good (VG) | (9, 10, 10) |
| Intermediate | (7, 8, 9) | | |
| Extremely strong | (9, 9, 9) | | |

**Table 3** The fuzzy pair wise comparisons among the lean criteria

| | C1 | C2 | C3 | C4 | C5 | C6 | C7 | C8 | C9 | C10 | C11 | C12 |
|---|---|---|---|---|---|---|---|---|---|---|---|---|
| C1 | (1.00,1.00,1.00) | (2.00,3.00,4.00) | (5.00,6.00,7.00) | (3.00,4.00,5.00) | (4.00,5.00,6.00) | (7.00,8.00,9.00) | (9.00,9.00,9.00) | (7.00,8.00,9.00) | (9.00,9.00,9.00) | (5.00,6.00,7.00) | (3.00,4.00,5.00) | (5.00,6.00,7.00) |
| C2 | (0.25,0.33,0.50) | (1.00,1.00,1.00) | (5.00,6.00,7.00) | (3.00,4.00,5.00) | (4.00,5.00,6.00) | (7.00,8.00,9.00) | (9.00,9.00,9.00) | (7.00,8.00,9.00) | (6.00,7.00,8.00) | (3.00,4.00,5.00) | (2.00,3.00,4.00) | (3.00,4.00,5.00) |
| C3 | (0.14,0.17,0.20) | (0.14,0.17,0.20) | (1.00,1.00,1.00) | (0.17,0.20,0.25) | (0.20,0.25,0.33) | (5.00,6.00,7.00) | (6.00,7.00,8.00) | (4.00,5.00,6.00) | (9.00,9.00,9.00) | (2.00,3.00,4.00) | (1.00,2.00,3.00) | (3.00,4.00,5.00) |
| C4 | (0.20,0.25,0.33) | (0.20,0.25,0.33) | (4.00,5.00,6.00) | (1.00,1.00,1.00) | (2.00,3.00,4.00) | (9.00,9.00,9.00) | (9.00,9.00,9.00) | (7.00,8.00,9.00) | (9.00,9.00,9.00) | (3.00,4.00,5.00) | (2.00,3.00,4.00) | (4.00,5.00,6.00) |
| C5 | (0.17,0.20,0.25) | (0.17,0.20,0.25) | (3.00,4.00,5.00) | (0.25,0.33,0.50) | (1.00,1.00,1.00) | (7.00,8.00,9.00) | (9.00,9.00,9.00) | (6.00,7.00,8.00) | (1.00,2.00,3.00) | (2.00,3.00,4.00) | (1.00,2.00,3.00) | (3.00,4.00,5.00) |
| C6 | (0.11,0.13,0.14) | (0.11,0.13,0.14) | (0.14,0.17,0.20) | (0.11,0.11,0.11) | (0.11,0.13,0.14) | (1.00,1.00,1.00) | (1.00,2.00,3.00) | (0.33,0.50,1.00) | (0.33,0.50,1.00) | (0.17,0.20,0.25) | (0.13,0.14,0.17) | (0.20,0.25,0.33) |
| C7 | (0.11,0.11,0.11) | (0.11,0.11,0.11) | (0.13,0.14,0.17) | (0.11,0.11,0.11) | (0.11,0.11,0.11) | (0.33,0.50,1.00) | (1.00,1.00,1.00) | (0.25,0.33,0.50) | (2.00,3.00,4.00) | (0.17,0.20,0.25) | (0.13,0.14,0.17) | (0.20,0.25,0.33) |
| C8 | (0.11,0.13,0.14) | (0.11,0.13,0.14) | (0.17,0.20,0.25) | (0.11,0.13,0.14) | (0.13,0.14,0.17) | (1.00,2.00,3.00) | (2.00,3.00,4.00) | (1.00,1.00,1.00) | (1.00,1.00,1.00) | (0.20,0.25,0.33) | (0.14,0.17,0.20) | (0.25,0.33,0.50) |
| C9 | (0.11,0.11,0.11) | (0.11,0.11,0.11) | (0.13,0.14,0.17) | (0.11,0.11,0.11) | (0.11,0.11,0.11) | (0.33,0.50,1.00) | (1.00,2.00,3.00) | (0.25,0.33,0.50) | (6.00,7.00,8.00) | (0.13,0.14,0.17) | (0.11,0.13,0.14) | (0.14,0.17,0.20) |
| C10 | (0.14,0.17,0.20) | (0.20,0.25,0.33) | (0.25,0.33,0.50) | (0.20,0.25,0.33) | (0.25,0.33,0.50) | (4.00,5.00,6.00) | (4.00,5.00,6.00) | (3.00,4.00,5.00) | (7.00,8.00,9.00) | (1.00,1.00,1.00) | (0.25,0.33,0.50) | (1.00,2.00,3.00) |
| C11 | (0.20,0.25,0.33) | (0.25,0.33,0.50) | (0.33,0.50,1.00) | (0.25,0.33,0.50) | (0.33,0.50,1.00) | (6.00,7.00,8.00) | (6.00,7.00,8.00) | (5.00,6.00,7.00) | (5.00,6.00,7.00) | (2.00,3.00,4.00) | (1.00,1.00,1.00) | (2.00,3.00,4.00) |
| C12 | (0.14,0.17,0.20) | (0.20,0.25,0.33) | (0.20,0.25,0.33) | (0.17,0.20,0.25) | (0.20,0.25,0.33) | (3.00,4.00,5.00) | (3.00,4.00,5.00) | (2.00,3.00,4.00) | (9.00,9.00,9.00) | (0.33,0.50,1.00) | (0.25,0.33,0.50) | (1.00,1.00,1.00) |

The final FAHP importance lean criteria weights are calculated as below

$Wl_1 = (0.182,0.258,0.360)$, $Wl_2 = (0.136,0.196,0.281)$, $Wl_3 = (0.047,0.071,0.105)$, $Wl_4 = (0.103,0.147,0.210)$, $Wl_5 = (0.071,0.105,0.155)$, $Wl_6 = (0.010,0.015,0.023)$, $Wl_7 = (0.008,0.011,0.017)$, $Wl_8 = (0.013,0.020,0.031)$, $Wl_9 = (0.008,0.012,0.017)$, $Wl_{10} = (0.031,0.049,0.077)$, $Wl_{11} = (0.050,0.078,0.128)$, $Wl_{12} = (0.025,0.038,0.060)$

**Table 4** The fuzzy pair wise comparisons among the green criteria

| | C13 | C14 | C15 | C16 | C17 | C18 | C19 | C20 | C21 | C22 | C23 | C24 |
|---|---|---|---|---|---|---|---|---|---|---|---|---|
| C13 | (1.00,1.00,1.00) | (1.00,2.00,3.00) | (3.00,4.00,5.00) | (1.00,2.00,3.00) | (2.00,3.00,4.00) | (4.00,5.00,6.00) | (5.00,6.00,7.00) | (4.00,5.00,6.00) | (5.00,6.00,7.00) | (2.00,3.00,4.00) | (3.00,4.00,5.00) | (2.00,3.00,4.00) |
| C14 | (0.33,0.50,1.00) | (1.00,1.00,1.00) | (2.00,3.00,4.00) | (1.00,2.00,3.00) | (2.00,3.00,4.00) | (4.00,5.00,6.00) | (6.00,7.00,8.00) | (5.00,6.00,7.00) | (4.00,5.00,6.00) | (3.00,4.00,5.00) | (3.00,4.00,5.00) | (4.00,5.00,6.00) |
| C15 | (0.20,0.25,0.33) | (0.25,0.33,0.50) | (1.00,1.00,1.00) | (0.17,0.20,0.25) | (0.20,0.25,0.33) | (5.00,6.00,7.00) | (6.00,7.00,8.00) | (4.00,5.00,6.00) | (4.00,5.00,6.00) | (2.00,3.00,4.00) | (1.00,2.00,3.00) | (1.00,2.00,3.00) |
| C16 | (0.33,0.50,1.00) | (0.33,0.50,1.00) | (3.00,4.00,5.00) | (1.00,1.00,1.00) | (2.00,3.00,4.00) | (4.00,5.00,6.00) | (6.00,7.00,8.00) | (5.00,6.00,7.00) | (4.00,5.00,6.00) | (3.00,4.00,5.00) | (2.00,3.00,4.00) | (3.00,4.00,5.00) |
| C17 | (0.25,0.33,0.50) | (0.25,0.33,0.50) | (3.00,4.00,5.00) | (0.25,0.33,0.50) | (1.00,1.00,1.00) | (6.00,7.00,8.00) | (5.00,6.00,7.00) | (4.00,5.00,6.00) | (3.00,4.00,5.00) | (1.00,2.00,3.00) | (1.00,2.00,3.00) | (2.00,3.00,4.00) |
| C18 | (0.17,0.20,0.25) | (0.17,0.20,0.25) | (0.14,0.17,0.20) | (0.17,0.20,0.25) | (0.13,0.14,0.17) | (1.00,1.00,1.00) | (1.00,2.00,3.00) | (0.33,0.50,1.00) | (1.00,2.00,3.00) | (0.17,0.20,0.25) | (0.13,0.14,0.17) | (0.20,0.25,0.33) |
| C19 | (0.14,0.17,0.20) | (0.13,0.14,0.17) | (0.17,0.20,0.25) | (0.13,0.14,0.17) | (0.14,0.17,0.20) | (0.33,0.50,1.00) | (1.00,1.00,1.00) | (0.25,0.33,0.50) | (0.33,0.50,1.00) | (0.17,0.20,0.25) | (0.13,0.14,0.17) | (0.20,0.25,0.33) |
| C20 | (0.17,0.20,0.25) | (0.14,0.17,0.20) | (0.17,0.20,0.25) | (0.14,0.17,0.20) | (0.17,0.20,0.25) | (1.00,2.00,3.00) | (2.00,3.00,4.00) | (1.00,1.00,1.00) | (2.00,3.00,4.00) | (0.20,0.25,0.33) | (0.14,0.17,0.20) | (0.25,0.33,0.50) |
| C21 | (0.14,0.17,0.20) | (0.17,0.20,0.25) | (0.17,0.20,0.25) | (0.17,0.20,0.25) | (0.20,0.25,0.33) | (0.33,0.50,1.00) | (1.00,2.00,3.00) | (0.25,0.33,0.50) | (1.00,1.00,1.00) | (0.13,0.14,0.17) | (0.11,0.13,0.14) | (0.14,0.17,0.20) |
| C22 | (0.25,0.33,0.50) | (0.20,0.25,0.33) | (0.25,0.33,0.50) | (0.20,0.25,0.33) | (0.33,0.50,1.00) | (3.00,4.00,5.00) | (3.00,4.00,5.00) | (2.00,3.00,4.00) | (4.00,5.00,6.00) | (1.00,1.00,1.00) | (0.25,0.33,0.50) | (1.00,2.00,3.00) |
| C23 | (0.20,0.25,0.33) | (0.20,0.25,0.33) | (0.33,0.50,1.00) | (0.25,0.33,0.50) | (0.33,0.50,1.00) | (4.00,5.00,6.00) | (4.00,5.00,6.00) | (3.00,4.00,5.00) | (5.00,6.00,7.00) | (1.00,2.00,3.00) | (1.00,1.00,1.00) | (2.00,3.00,4.00) |
| C24 | (0.25,0.33,0.50) | (0.17,0.20,0.25) | (0.33,0.50,1.00) | (0.20,0.25,0.33) | (0.25,0.33,0.50) | (3.00,4.00,5.00) | (3.00,4.00,5.00) | (2.00,3.00,4.00) | (5.00,6.00,7.00) | (0.33,0.50,1.00) | (0.25,0.33,0.50) | (1.00,1.00,1.00) |

The final FAHP importance green criteria weights are calculated as below
$Wg_1 = (0.108,0.196,0.333)$, $Wg_2 = (0.106,0.185,0.318)$, $Wg_3 = (0.048,0.083,0.143)$, $Wg_4 = (0.094,0.162,0.285)$, $Wg_5 = (0.062,0.110,0.193)$, $Wg_6 = (0.012,0.021,0.036)$, $Wg_7 = (0.009,0.015,0.027)$, $Wg_8 = (0.016,0.026,0.045)$, $Wg_9 = (0.011,0.017,0.030)$, $Wg_{10} = (0.032,0.057,0.105)$, $Wg_{11} = (0.044,0.076,0.141)$, $Wg_{12} = (0.030,0.051,0.095)$

**Table 5** Linguistic ratings of the alternative suppliers by decision maker on lean criteria

| Suppliers | C1 | C2 | C3 | C4 | C5 | C6 | C7 | C8 | C9 | C10 | C11 | C12 |
|---|---|---|---|---|---|---|---|---|---|---|---|---|
| 1 | F | G | F | F | MP | VP | MP | F | F | VG | G | MG |
| 2 | VG | VG | MG | G | MG | P | F | MP | P | F | MG | MP |
| 3 | P | MP | VG | MP | F | MP | P | F | MG | VG | VG | P |
| 4 | VG | G | F | MG | F | P | VP | P | VP | F | MG | F |
| 5 | MG | F | G | P | G | MP | P | F | F | G | VG | G |
| 6 | MP | P | F | F | MP | G | MG | P | VG | MG | F | F |
| 7 | F | G | VP | MG | P | F | F | MG | VG | VP | G | G |
| 8 | G | VG | MP | F | P | VP | F | MG | G | P | P | MG |
| 9 | VP | P | F | MP | F | G | VG | G | VG | F | MP | F |
| 10 | G | F | G | F | MP | P | F | F | MG | G | G | F |

**Table 6** Linguistic ratings of the alternative suppliers by decision maker on green criteria

| Suppliers | C13 | C14 | C15 | C16 | C17 | C18 | C19 | C20 | C21 | C22 | C23 | C24 |
|---|---|---|---|---|---|---|---|---|---|---|---|---|
| 1 | G | MG | F | MG | F | P | P | VP | P | G | G | F |
| 2 | G | MG | F | P | MP | F | P | MG | F | G | VP | MG |
| 3 | F | F | G | F | MG | F | G | G | MP | MP | F | F |
| 4 | P | MP | F | G | VG | VG | F | F | P | MP | MP | VP |
| 5 | F | MG | MP | G | G | MP | P | F | VP | MG | VG | P |
| 6 | VP | P | P | F | F | MG | VG | MG | MG | F | G | G |
| 7 | VG | G | MG | G | G | MP | P | MP | VP | F | F | MG |
| 8 | MG | MG | G | P | F | MP | MG | MG | F | F | MP | VP |
| 9 | MP | G | MG | F | F | F | VG | P | MP | MP | G | G |
| 10 | G | VG | P | P | F | VP | MG | MG | MG | MP | F | G |

multi objective linear formulation of case study is presented as the following.

$$Z_1 = 0.109x_1 + 0.143x_2 + 0.075x_3 + 0.129x_4 + 0.113x_5 + 0.064x_6 + 0.101x_7 + 0.107x_8 + 0.043x_9 + 0.116x_{10}$$

$$Z_2 = 0.121x_1 + 0.092x_2 + 0.101x_3 + 0.085x_4 + 0.116x_5 + 0.051x_6 + 0.138x_7 + 0.094x_8 + 0.104x_9 + 0.098x_{10}$$

Subject to;

$$x_1 + x_2 + x_3 + x_4 + x_5 + x_6 + x_7 + x_8 + x_9 + x_{10} \geq 20400$$

$$x_1 \leq 7500$$

**Table 7** Fuzzy numbers of the ratings of the alternative suppliers and lean criteria

| Suppliers | C1 | C2 | C3 | C4 | C5 | C6 | C7 | C8 | C9 | C10 | C11 | C12 |
|---|---|---|---|---|---|---|---|---|---|---|---|---|
| 1 | (3,5,7) | (7,9,10) | (3,5,7) | (3,5,7) | (1,3,5) | (0,0,1) | (1,3,5) | (3,5,7) | (3,5,7) | (9,10,10) | (7,9,10) | (5,7,9) |
| 2 | (9,10,10) | (9,10,10) | (5,7,9) | (7,9,10) | (5,7,9) | (0,1,3) | (3,5,7) | (1,3,5) | (0,1,3) | (3,5,7) | (5,7,9) | (1,3,5) |
| 3 | (0,1,3) | (1,3,5) | (9,10,10) | (1,3,5) | (3,5,7) | (1,3,5) | (0,1,3) | (3,5,7) | (5,7,9) | (9,10,10) | (9,10,10) | (0,1,3) |
| 4 | (9,10,10) | (7,9,10) | (3,5,7) | (5,7,9) | (3,5,7) | (0,1,3) | (0,0,1) | (0,1,3) | (0,0,1) | (3,5,7) | (5,7,9) | (3,5,7) |
| 5 | (5,7,9) | (3,5,7) | (7,9,10) | (0,1,3) | (7,9,10) | (1,3,5) | (0,1,3) | (3,5,7) | (3,5,7) | (7,9,10) | (9,10,10) | (7,9,10) |
| 6 | (1,3,5) | (0,1,3) | (3,5,7) | (3,5,7) | (1,3,5) | (7,9,10) | (5,7,9) | (0,1,3) | (9,10,10) | (5,7,9) | (3,5,7) | (3,5,7) |
| 7 | (3,5,7) | (7,9,10) | (0,0,1) | (5,7,9) | (0,1,3) | (3,5,7) | (3,5,7) | (5,7,9) | (9,10,10) | (0,0,1) | (7,9,10) | (7,9,10) |
| 8 | (7,9,10) | (9,10,10) | (1,3,5) | (3,5,7) | (0,1,3) | (0,0,1) | (3,5,7) | (5,7,9) | (7,9,10) | (0,1,3) | (0,1,3) | (5,7,9) |
| 9 | (0,0,1) | (0,1,3) | (3,5,7) | (1,3,5) | (3,5,7) | (7,9,10) | (9,10,10) | (7,9,10) | (9,10,10) | (3,5,7) | (1,3,5) | (3,5,7) |
| 10 | (7,9,10) | (3,5,7) | (7,9,10) | (3,5,7) | (1,3,5) | (0,1,3) | (3,5,7) | (3,5,7) | (5,7,9) | (7,9,10) | (7,9,10) | (3,5,7) |
| Weights of criteria | (0.182, 0.258, 0.360) | (0.136, 0.196, 0.281) | (0.047, 0.071, 0.105) | (0.103, 0.147, 0.210) | (0.071, 0.105, 0.155) | (0.010, 0.015, 0.023) | (0.008, 0.011, 0.017) | (0.013, 0.020, 0.031) | (0.008, 0.012, 0.017) | (0.031, 0.049, 0.077) | (0.050, 0.078, 0.128) | (0.025, 0.038, 0.060) |

**Table 8** Fuzzy numbers of the ratings of the alternative suppliers and green criteria

| Suppliers | C13 | C14 | C15 | C16 | C17 | C18 | C19 | C20 | C21 | C22 | C23 | C24 |
|---|---|---|---|---|---|---|---|---|---|---|---|---|
| 1 | (7,9,10) | (5,7,9) | (3,5,7) | (5,7,9) | (3,5,7) | (0,1,3) | (0,1,3) | (0,0,1) | (0,1,3) | (7,9,10) | (7,9,10) | (3,5,7) |
| 2 | (7,9,10) | (5,7,9) | (3,5,7) | (0,1,3) | (1,3,5) | (3,5,7) | (0,1,3) | (5,7,9) | (3,5,7) | (7,9,10) | (0,0,1) | (5,7,9) |
| 3 | (3,5,7) | (3,5,7) | (7,9,10) | (3,5,7) | (5,7,9) | (3,5,7) | (7,9,10) | (7,9,10) | (1,3,5) | (1,3,5) | (3,5,7) | (3,5,7) |
| 4 | (0,1,3) | (1,3,5) | (3,5,7) | (7,9,10) | (9,10,10) | (9,10,10) | (3,5,7) | (3,5,7) | (0,1,3) | (1,3,5) | (1,3,5) | (0,0,1) |
| 5 | (3,5,7) | (5,7,9) | (1,3,5) | (7,9,10) | (7,9,10) | (1,3,5) | (0,1,3) | (3,5,7) | (0,0,1) | (5,7,9) | (9,10,10) | (0,1,3) |
| 6 | (0,0,1) | (0,1,3) | (0,1,3) | (3,5,7) | (3,5,7) | (5,7,9) | (9,10,10) | (5,7,9) | (5,7,9) | (3,5,7) | (7,9,10) | (7,9,10) |
| 7 | (9,10,10) | (7,9,10) | (5,7,9) | (7,9,10) | (7,9,10) | (1,3,5) | (0,1,3) | (1,3,5) | (0,0,1) | (3,5,7) | (3,5,7) | (5,7,9) |
| 8 | (5,7,9) | (5,7,9) | (7,9,10) | (0,1,3) | (3,5,7) | (1,3,5) | (5,7,9) | (5,7,9) | (3,5,7) | (3,5,7) | (1,3,5) | (0,0,1) |
| 9 | (1,3,5) | (7,9,10) | (5,7,9) | (3,5,7) | (3,5,7) | (3,5,7) | (9,10,10) | (0,1,3) | (1,3,5) | (1,3,5) | (7,9,10) | (7,9,10) |
| 10 | (7,9,10) | (9,10,10) | (0,1,3) | (0,1,3) | (3,5,7) | (0,0,1) | (5,7,9) | (5,7,9) | (5,7,9) | (1,3,5) | (3,5,7) | (7,9,10) |
| Weights of criteria | (0.108, 0.196, 0.333) | (0.106, 0.185, 0.318) | (0.048, 0.083, 0.143) | (0.094, 0.162, 0.285) | (0.062, 0.110, 0.193) | (0.012, 0.021, 0.036) | (0.009, 0.015, 0.027) | (0.016, 0.026, 0.045) | (0.011, 0.017, 0.030) | (0.032, 0.057, 0.105) | (0.044, 0.076, 0.141) | (0.030, 0.051, 0.095) |

**Table 9** Outranking of alternative suppliers for lean tools

| Suppliers | $d_j^+$ | $d_j^-$ | $CC_j$ | Normal weights | Rank |
|---|---|---|---|---|---|
| 1 | 0.33 | 0.63 | 0.654 | 0.109 | 5 |
| 2 | 0.14 | 0.83 | 0.856 | 0.143 | 1 |
| 3 | 0.54 | 0.44 | 0.449 | 0.075 | 8 |
| 4 | 0.22 | 0.75 | 0.771 | 0.129 | 2 |
| 5 | 0.31 | 0.64 | 0.673 | 0.113 | 4 |
| 6 | 0.57 | 0.35 | 0.382 | 0.064 | 9 |
| 7 | 0.39 | 0.59 | 0.602 | 0.101 | 7 |
| 8 | 0.36 | 0.64 | 0.643 | 0.107 | 6 |
| 9 | 0.67 | 0.24 | 0.259 | 0.043 | 10 |
| 10 | 0.29 | 0.66 | 0.693 | 0.116 | 3 |

**Table 10** Outranking of alternative suppliers for green tools

| Suppliers | $d_j^+$ | $d_j^-$ | $CC_j$ | Normal weights | Rank |
|---|---|---|---|---|---|
| 1 | 0.27 | 0.73 | 0.727 | 0.121 | 2 |
| 2 | 0.44 | 0.54 | 0.551 | 0.092 | 8 |
| 3 | 0.38 | 0.58 | 0.606 | 0.101 | 5 |
| 4 | 0.52 | 0.54 | 0.511 | 0.085 | 9 |
| 5 | 0.31 | 0.71 | 0.698 | 0.116 | 3 |
| 6 | 0.61 | 0.27 | 0.303 | 0.051 | 10 |
| 7 | 0.17 | 0.81 | 0.825 | 0.138 | 1 |
| 8 | 0.45 | 0.58 | 0.564 | 0.094 | 7 |
| 9 | 0.37 | 0.60 | 0.621 | 0.104 | 4 |
| 10 | 0.39 | 0.56 | 0.590 | 0.098 | 6 |

**Table 11** Supplier's quantitative information

| Suppliers | $P_i$ (\$/unit) | $C_i$ (kg) | $G_i$ (kg/unit) | $L_i$ (%) |
|---|---|---|---|---|
| 1 | 8 | 7500 | 1.3 | 0.08 |
| 2 | 7.5 | 7200 | 2.1 | 0.03 |
| 3 | 6.8 | 7000 | 1.7 | 0.15 |
| 4 | 8.5 | 9950 | 2.3 | 0.04 |
| 5 | 9.5 | 11500 | 1.4 | 0.07 |
| 6 | 7 | 7000 | 2.5 | 0.17 |
| 7 | 8 | 8000 | 1.2 | 0.13 |
| 8 | 9 | 13500 | 2.0 | 0.11 |
| 9 | 10 | 15000 | 1.5 | 0.18 |
| 10 | 7.8 | 9000 | 1.8 | 0.50 |

$$x_2 \leq 7200$$

$$x_3 \leq 7000$$

$$x_4 \leq 9950$$

$$x_5 \leq 11500$$

$$x_6 \leq 7000$$

$$x_7 \leq 8000$$

$$x_8 \leq 13500$$

$$x_9 \leq 15000$$

$$x_{10} \leq 9000$$

$$0.09x_1 + 0.03x_2 + 0.15x_3 + 0.04x_4 + 0.08x_5 + 0.17x_6 + 0.13x_7 + 0.11x_8 + 0.18x_9 + 0.50x_{10} \leq 0.08 * 20400$$

$$1.2x_1 + 2.1x_2 + 1.7x_3 + 2.3x_4 + 1.4x_5 + 2.5x_6 + 1.2x_7 + 2x_8 + 1.5x_9 + 1.8x_{10} \leq 36000$$

$$8x_1 + 7.5x_2 + 6.8x_3 + 8.5x_4 + 9.5x_5 + 7x_6 + 8x_7 + 9x_8 + 10x_9 + 7.8x_{10} \leq 180000$$

$$x_1 \geq 0, x_2 \geq 0, x_3 \geq 0, x_4 \geq 0, x_5 \geq 0, x_6 \geq 0, x_7 \geq 0, x_8 \geq 0, x_9 \geq 0, x_{10} \geq 0$$

$$x_1, x_2, x_3, x_4, x_5, x_6, x_7, x_8, x_9, x_{10} \text{ are integer.}$$

The multi objective supplier selection problem is solved as a single-objective supplier selection problem by using only one objective each time. Firstly, the objective function $Z_1$ is maximized by using all constraints for getting the upper-bound of the objective function—this value is the best value for this objective as other objectives are absent—and $Z_2$ gets one of its lower bounds on the other hand. This procedure is repeated for objective functions $Z_2$ to get its upper bound and $Z_1$'s lower bound. The data set for pay-offs matrix and the lower bounds and upper bounds values of the objective functions are given in Tables 12 and 13 respectively.

**Table 12** The pay-offs matrix

| Objective functions | $Z_1$ | $Z_2$ | $x_1$ | $x_2$ | $x_3$ | $x_4$ | $x_5$ | $x_6$ | $x_7$ | $x_8$ | $x_9$ | $x_{10}$ |
|---|---|---|---|---|---|---|---|---|---|---|---|---|
| $Z_1$(max) | **2563.87** | 2110.93 | 0 | 7200.00 | 0 | 2666.67 | 10533.33 | 0 | 0 | 0 | 0 | 0 |
| $Z_2$(max) | 2249.40 | **2451.20** | 7500.00 | 0 | 0 | 0 | 10750.00 | 0 | 2150.00 | 0 | 0 | 0 |

**Table 13** The data set for membership functions

| Objective functions | *Lower bounds* $\mu = 1$ | *Upper bounds* $\mu = 0$ |
|---|---|---|
| $Z_1$ | 2249.40 | 2563.87 |
| $Z_2$ | 2110.93 | 2451.20 |

**Table 14** The optimal solutions according to each weight pairs

| Number of experiment | $w_1$ | $w_2$ | $Z_1$ | $Z_2$ | Total purchasing cost | Total $CO_2$ emission | Total late delivery |
|---|---|---|---|---|---|---|---|
| 1 | 0 | 1 | 2249.40 | 2451.20 | 179325 | 27380 | 1632 |
| 2 | 0.1 | 0.9 | 2249.40 | 2451.20 | 179325 | 27380 | 1632 |
| 3 | 0.2 | 0.8 | 2249.40 | 2451.20 | 179325 | 27380 | 1632 |
| 4 | 0.3 | 0.7 | 2391.30 | 2391.00 | 159975 | 31035 | 1632 |
| 5 | 0.4 | 0.6 | 2410.80 | 2382.00 | 159600 | 31560 | 1632 |
| 6 | 0.5 | 0.5 | 2425.20 | 2369.60 | 167400 | 32000 | 1620 |
| 7 | 0.6 | 0.4 | 2425.20 | 2369.60 | 167400 | 32000 | 1620 |
| 8 | 0.7 | 0.3 | 2563.86 | 2110.95 | 176734 | 35999.40 | 1060.02 |
| 9 | 0.8 | 0.2 | 2563.86 | 2110.95 | 176734 | 35999.40 | 1060.02 |
| 10 | 0.9 | 0.1 | 2563.86 | 2110.94 | 176728.50 | 36000 | 1060.02 |
| 11 | 1 | 0 | 2563.86 | 2110.94 | 176728.50 | 36000 | 1060.02 |

**STEP 6:** Reformulating the MOLP into a fuzzy linear programming model

Setting up an appropriate weight for objective functions depends on the proportion of each objective's relative importance in the problem. According to the expectations of the organizations, 11 different weight pairs were determined considering different supplier profiles and the results in Table 14 were obtained by the use of Tiwari method. Based on the weighted additive method, the crisp formulation of the case study (for the experiment 3) is presented as follows. The following crisp single objective programming is equivalent to the above fuzzy model.

$$\text{Maximize } 0,2^{*}\lambda_1 + 0,8^{*}\lambda_2$$

Subject to

$$\lambda_1 \leq \frac{(0.109x_1 + 0.143x_2 + 0.075x_3 + 0.129x_4 + 0.113x_5 + 0.064x_6 + 0.101x_7 + 0.107x_8 + 0.043x_9 + 0.116x_{10}) - 2249.40}{314.47}$$

$$\lambda_2 \leq \frac{(0.121x_1 + 0.092x_2 + 0.101x_3 + 0.085x_4 + 0.116x_5 + 0.051x_6 + 0.138x_7 + 0.094x_8 + 0.104x_9 + 0.098x_{10}) - 2110.93}{340.27}$$

$$x_1 + x_2 + x_3 + x_4 + x_5 + x_6 + x_7 + x_8 + x_9 + x_{10} \geq 20400$$

$$x_1 \leq 7500$$

$$x_2 \leq 7200$$

$$x_3 \leq 7000$$

$$x_4 \leq 9950$$

$$x_5 \leq 11500$$

$$x_6 \leq 7000$$

$$x_7 \leq 8000$$

$$x_8 \leq 13500$$

$$x_9 \leq 15000$$

$$x_{10} \leq 9000$$

$$0.09x_1 + 0.03x_2 + 0.15x_3 + 0.04x_4 + 0.08x_5 + 0.17x_6 + 0.13x_7 + 0.11x_8 + 0.18x_9 + 0.50x_{10} \leq 0.08 * 20400$$

$$1.2x_1 + 2.1x_2 + 1.7x_3 + 2.3x_4 + 1.4x_5 + 2.5x_6 + 1.2x_7 + 2x_8 + 1.5x_9 + 1.8x_{10} \leq 36000$$

$$8x_1 + 7.5x_2 + 6.8x_3 + 8.5x_4 + 9.5x_5 + 7x_6 + 8x_7 + 9x_8 + 10x_9 + 7.8x_{10} \leq 180000$$

$$x_1 \geq 0, x_2 \geq 0, x_3 \geq 0, x_4 \geq 0, x_5 \geq 0, x_6 \geq 0, x_7 \geq 0, x_8 \geq 0, x_9 \geq 0, x_{10} \geq 0$$

$$x_1, x_2, x_3, x_4, x_5, x_6, x_7, x_8, x_9, x_{10} \text{ are integer.}$$

**STEP 7**: Optimizing the fuzzy MOLP

Fuzzy multi objective linear programming model developed in accordance with these data which was solved in GAMS/CPLEX 24.0 software package.

The optimal solution for the above formulation is given below (Tables 14 and 15):

In this solution, no products were purchased from suppliers 3, 6, 8, 9 and 10, because the supplier 3 is below average in terms of both gas release-speed performance and lean-green value ranking compared to other suppliers, despite having low procurement cost, and supplier 6 has the worst ranks (9 and 10 respectively) in the lean and green rankings. Suppliers 8 and 9 have the highest procurement costs and as well it can be said that they have bad positions in terms of gas release and speed performance. Also, supplier 9 has the worst ranking for lean value. Supplier 10 was not preferred by the firm for providing all the constraints despite having good (third) ranking for lean value.

The results of experiments 1, 2 and 3 are identical and the firm has worked with suppliers 1 (with maximum capacity), 5 and 7, because they are the best 3 green suppliers. In the experiments 4 and 5, the firm has worked with suppliers 1, 2 and 7. The results of experiments 6 and 7, where the weights are equal or too close (0.5–0.5 and 0.6–0.4), are identical and as this means that the aim of the firm is to work with both lean and green suppliers simultaneously, so it has preferred supplier 2 that is the first in the lean rankings, supplier 7 that is the first in the green rankings, and supplier 5 that is both lean and green. In the experiments 8, 9, 10 and 11, the firm

**Table 15** The optimal order quantities and λ values

| Number of experiment | $\lambda_1$ | $\lambda_2$ | $x_1$ | $x_2$ | $x_3$ | $x_4$ | $x_5$ | $x_6$ | $x_7$ | $x_8$ | $x_9$ | $x_{10}$ |
|---|---|---|---|---|---|---|---|---|---|---|---|---|
| 1 | 0 | 1 | 7500 | 0 | 0 | 0 | 10750 | 0 | 2150 | 0 | 0 | 0 |
| 2 | 0 | 1 | 7500 | 0 | 0 | 0 | 10750 | 0 | 2150 | 0 | 0 | 0 |
| 3 | 0 | 1 | 7500 | 0 | 0 | 0 | 10750 | 0 | 2150 | 0 | 0 | 0 |
| 4 | 0.45 | 0.82 | 7500 | 6450 | 0 | 0 | 0 | 0 | 6450 | 0 | 0 | 0 |
| 5 | 0.51 | 0.80 | 6000 | 7200 | 0 | 0 | 0 | 0 | 7200 | 0 | 0 | 0 |
| 6 | 0.56 | 0.76 | 0 | 7200 | 0 | 0 | 5200 | 0 | 8000 | 0 | 0 | 0 |
| 7 | 0.56 | 0.76 | 0 | 7200 | 0 | 0 | 5200 | 0 | 8000 | 0 | 0 | 0 |
| 8 | 1 | 0 | 0 | 7200 | 0 | 2666 | 10534 | 0 | 0 | 0 | 0 | 0 |
| 9 | 1 | 0 | 0 | 7200 | 0 | 2666 | 10534 | 0 | 0 | 0 | 0 | 0 |
| 10 | 1 | 0 | 3 | 7200 | 0 | 2667 | 10530 | 0 | 0 | 0 | 0 | 0 |
| 11 | 1 | 0 | 3 | 7200 | 0 | 2667 | 10530 | 0 | 0 | 0 | 0 | 0 |

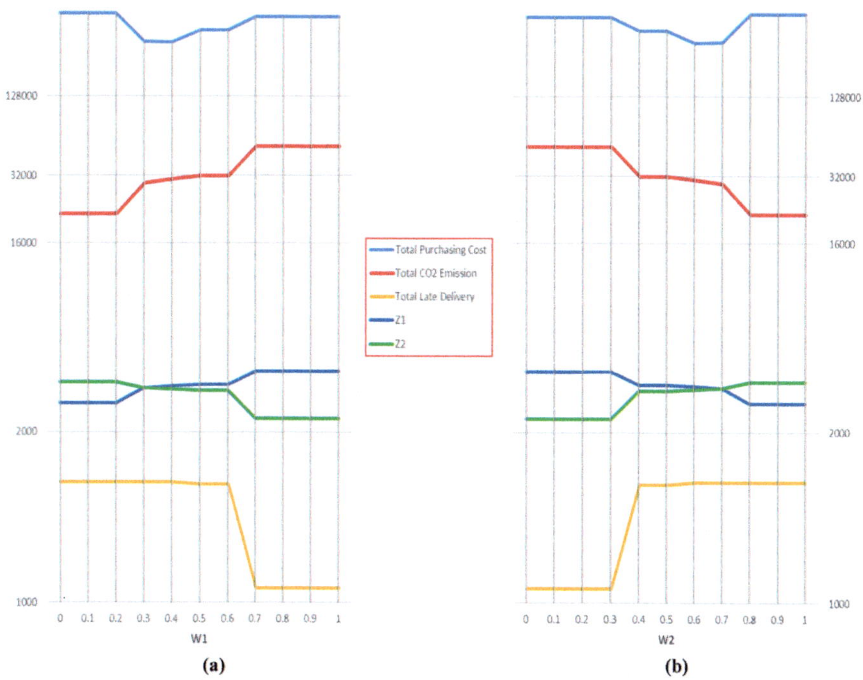


**Fig. 4** The optimal solutions of different weights—for lean value ($Z_1$), green value ($Z_2$), purchasing cost, $CO_2$ emission and late delivery

has preferred suppliers 2, 4 and 5, because they are the first, second and fourth green suppliers respectively.

In Fig. 4, it can be clearly seen how the optimal solutions change according to different weight pairs. As expected, the lean value moved up by increasing lean weight ($w_1$) while decreasing green weight ($w_2$) and green value moved up in the opposite situation. Similarly, it can be observed that the $CO_2$ emission moved up while the late delivery moved down by increasing lean weight ($w_1$) in Fig. 4(a) and vice versa in Fig. 4(b). Other than these, some fluctuations are observed in total purchasing cost. However, in the experiments 4 and 5, the total purchasing cost is low as well as the lean and green values are close to each other. Thus, it can be said that these weight pairs are suitable for firms that want to work with lean and green suppliers. Nevertheless, in the existence of multiple and conflictive objectives, it will be inevitable that multiple solutions exist because no one solution may be the optimal for all changing and conflicting aims.

Figure 5 illustrates the values of total purchasing cost, total $CO_2$ emission and total late delivery according to lean value weights in the form of a smile curve. When looking at the graph, it can be seen that total $CO_2$ emission shows an increasing slope, total late delivery shows a decreasing slope and total purchasing cost first shows a decreasing slope then an increasing slope as the lean value weight increases.

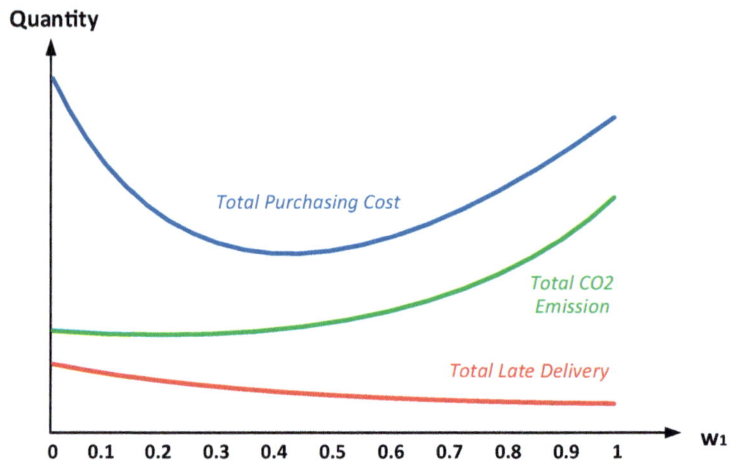


**Fig. 5** Smile curve graph for optimal solution

As a main implication of these experiments, it is possible to say that if decision maker does not give up one of the initiatives, lean and green, and if they give similar priorities to both of them, the total purchasing cost could get better values and the firm would get benefit of this hybrid strategy that includes lean and green approaches simultaneously.

## 6 Conclusions

The multiple supplier selection problems include both selecting suppliers and achieving optimal order allocation among the selected suppliers. According to the literature, few papers have developed practical methods that combine lean and green supplier selection criteria and considered order allocation methods (Kannan et al. 2013). Therefore, the lack of an efficient approach to develop lean and green specific measures handled simultaneously for supplier selection activated us to do this study although there are numerous measures in the literature.

Integration of the lean and green paradigms on supplier operations results in the improvement of total performance of organizations. However, the organizations encounter the challenge of benchmarking on lean and green value because it may bring some trade-offs between them according to the priorities of the organizations. Therefore, we evaluated eleven assumptions of lean and green balance that were determined for different company perspectives.

In this study, we proposed a fuzzy multi objective linear programing model for lean and green supplier selection problem. We summarized the possible 12 lean and 12 green tools and considered them as two separate groups. The problem includes two objective functions: maximizing lean and green value of purchasing individually while satisfying demand, capacity carbon emission limit, on-time delivery performance and budget requirement constraints. At first, we used FAHP to obtain the relative weights for lean criteria and also for green criteria and then we used them in FTOPSIS method to evaluate the selected suppliers according to the lean and green criteria separately. Next, we developed a fuzzy multi objective linear programming model to find out the optimum solution of the problem. Finally, we verified our model with the aid of a case study that had different experiments by solving in GAMS/CPLEX 24.0 software package. The main findings of this study are; (1) lean and green practices have an overall positive effect on each other. (2) If both practices are applied simultaneously, they can bring greater benefits on the total purchasing cost than applied separately.

In practice, decision makers may not have certain and complete information related to sources and constraints, too; thus, for future research, some constraints such as demand and capacity can be considered fuzzy as well as objective functions. Also, other than the FTOPSIS method used in this study, different multi-criteria decision making methods can be used in a fuzzy environment and compared in terms of suitability. In addition, the proposed model approach can be applied to broader real-life situations in order to obtain more complex and realistic models.

## Appendix

**Table 16** Normalized fuzzy decision matrix for lean tools

| Suppliers | C1 | C2 | C3 | C4 | C5 | C6 | C7 | C8 | C9 | C10 | C11 | C12 |
|---|---|---|---|---|---|---|---|---|---|---|---|---|
| 1 | (0.30,0.50,0.70) | (0.70,0.90,1.00) | (0.30,0.50,0.70) | (0.30,0.50,0.70) | (0.10,0.30,0.50) | (0.00,0.00,0.10) | (0.10,0.30,0.50) | (0.30,0.50,0.70) | (0.30,0.50,0.70) | (0.90,1.00,1.00) | (0.70,0.90,1.00) | (0.50,0.70,0.90) |
| 2 | (0.90,1.00,1.00) | (0.90,1.00,1.00) | (0.50,0.70,0.90) | (0.70,0.90,1.00) | (0.50,0.70,0.90) | (0.00,0.10,0.30) | (0.30,0.50,0.70) | (0.10,0.30,0.50) | (0.00,0.10,0.30) | (0.30,0.50,0.70) | (0.50,0.70,0.90) | (0.10,0.30,0.50) |
| 3 | (0.00,0.10,0.30) | (0.10,0.30,0.50) | (0.90,1.00,1.00) | (0.10,0.30,0.50) | (0.30,0.50,0.70) | (0.10,0.30,0.50) | (0.00,0.10,0.30) | (0.30,0.50,0.70) | (0.50,0.70,0.90) | (0.90,1.00,1.00) | (0.90,1.00,1.00) | (0.00,0.10,0.30) |
| 4 | (0.90,1.00,1.00) | (0.70,0.90,1.00) | (0.30,0.50,0.70) | (0.50,0.70,0.90) | (0.30,0.50,0.70) | (0.00,0.10,0.30) | (0.00,0.00,0.10) | (0.00,0.10,0.30) | (0.00,0.00,0.10) | (0.30,0.50,0.70) | (0.50,0.70,0.90) | (0.30,0.50,0.70) |
| 5 | (0.50,0.70,0.90) | (0.30,0.50,0.70) | (0.70,0.90,1.00) | (0.00,0.10,0.30) | (0.70,0.90,1.00) | (0.10,0.30,0.50) | (0.00,0.10,0.30) | (0.30,0.50,0.70) | (0.30,0.50,0.70) | (0.70,0.90,1.00) | (0.90,1.00,1.00) | (0.70,0.90,1.00) |
| 6 | (0.10,0.30,0.50) | (0.00,0.10,0.30) | (0.30,0.50,0.70) | (0.30,0.50,0.70) | (0.10,0.30,0.50) | (0.70,0.90,1.00) | (0.50,0.70,0.90) | (0.00,0.10,0.30) | (0.90,1.00,1.00) | (0.50,0.70,0.90) | (0.30,0.50,0.70) | (0.30,0.50,0.70) |
| 7 | (0.70,0.90,1.00) | (0.70,0.90,1.00) | (0.00,0.00,1.00) | (0.50,0.70,0.90) | (0.00,0.10,0.30) | (0.30,0.50,0.70) | (0.30,0.50,0.70) | (0.50,0.70,0.90) | (0.90,1.00,1.00) | (0.00,0.00,0.10) | (0.70,0.90,1.00) | (0.70,0.90,1.00) |
| 8 | (0.70,0.90,1.00) | (0.90,1.00,1.00) | (0.10,0.30,0.50) | (0.30,0.50,0.70) | (0.00,0.10,0.30) | (0.00,0.00,0.10) | (0.30,0.50,0.70) | (0.50,0.70,0.90) | (0.70,0.90,1.00) | (0.00,0.10,0.30) | (0.00,0.10,0.30) | (0.50,0.70,0.90) |
| 9 | (0.00,0.00,0.10) | (0.00,0.10,0.30) | (0.30,0.50,0.70) | (0.10,0.30,0.50) | (0.30,0.50,0.70) | (0.70,0.90,1.00) | (0.90,1.00,1.00) | (0.70,0.90,1.00) | (0.90,1.00,1.00) | (0.30,0.50,0.70) | (0.10,0.30,0.50) | (0.30,0.50,0.70) |
| 10 | (0.70,0.90,1.00) | (0.30,0.50,0.70) | (0.70,0.90,1.00) | (0.30,0.50,0.70) | (0.10,0.30,0.50) | (0.00,0.10,0.30) | (0.30,0.50,0.70) | (0.30,0.50,0.70) | (0.50,0.70,0.90) | (0.70,0.90,1.00) | (0.70,0.90,1.00) | (0.30,0.50,0.70) |

**Table 17** Normalized fuzzy decision matrix for green tools

| Suppliers | C13 | C14 | C15 | C16 | C17 | C18 | C19 | C20 | C21 | C22 | C23 | C24 |
|---|---|---|---|---|---|---|---|---|---|---|---|---|
| 1 | (0.70,0.90,1.00) | (0.50,0.70,0.90) | (0.33,0.50,0.70) | (0.50,0.70,0.90) | (0.33,0.50,0.70) | (0.00,0.10,0.30) | (0.00,0.10,0.30) | (0.00,0.00,0.10) | (0.00,0.11,0.33) | (0.70,0.90,1.00) | (0.70,0.90,1.00) | (0.33,0.50,0.70) |
| 2 | (0.70,0.90,1.00) | (0.50,0.70,0.90) | (0.33,0.50,0.70) | (0.00,0.10,0.30) | (0.10,0.30,0.50) | (0.33,0.50,0.70) | (0.00,0.10,0.30) | (0.50,0.70,0.90) | (0.37,0.56,0.78) | (0.70,0.90,1.00) | (0.00,0.00,0.10) | (0.50,0.70,0.90) |
| 3 | (0.33,0.50,0.70) | (0.33,0.50,0.70) | (0.70,0.90,1.00) | (0.33,0.50,0.70) | (0.50,0.70,0.90) | (0.33,0.50,0.70) | (0.70,0.90,1.00) | (0.70,0.90,1.00) | (0.11,0.33,0.56) | (0.10,0.30,0.50) | (0.33,0.50,0.70) | (0.33,0.50,0.70) |
| 4 | (0.00,0.10,0.30) | (0.10,0.30,0.50) | (0.33,0.50,0.70) | (0.70,0.90,1.00) | (0.90,1.00,1.00) | (0.90,1.00,1.00) | (0.33,0.50,0.70) | (0.33,0.50,0.70) | (0.00,0.11,0.33) | (0.10,0.30,0.50) | (0.10,0.30,0.50) | (0.00,0.00,0.10) |
| 5 | (0.33,0.50,0.70) | (0.50,0.70,0.90) | (0.10,0.30,0.50) | (0.70,0.90,1.00) | (0.70,0.90,1.00) | (0.10,0.30,0.50) | (0.00,0.10,0.30) | (0.33,0.50,0.70) | (0.00,0.00,0.11) | (0.50,0.70,0.90) | (0.90,1.00,1.00) | (0.00,0.10,0.30) |
| 6 | (0.00,0.00,0.10) | (0.00,0.10,0.30) | (0.00,0.10,0.30) | (0.33,0.50,0.70) | (0.33,0.50,0.70) | (0.50,0.70,0.90) | (0.90,1.00,1.00) | (0.50,0.70,0.90) | (0.56,0.78,1.00) | (0.33,0.50,0.70) | (0.70,0.90,1.00) | (0.70,0.90,1.00) |
| 7 | (0.90,1.00,1.00) | (0.70,0.90,1.00) | (0.50,0.70,0.90) | (0.70,0.90,1.00) | (0.70,0.90,1.00) | (0.10,0.30,0.50) | (0.00,0.10,0.30) | (0.10,0.30,0.50) | (0.00,0.00,0.11) | (0.33,0.50,0.70) | (0.33,0.50,0.70) | (0.50,0.70,0.90) |
| 8 | (0.50,0.70,0.90) | (0.50,0.70,0.90) | (0.70,0.90,1.00) | (0.00,0.10,0.30) | (0.33,0.50,0.70) | (0.10,0.30,0.50) | (0.50,0.70,0.90) | (0.50,0.70,0.90) | (0.37,0.56,0.78) | (0.33,0.50,0.70) | (0.10,0.30,0.50) | (0.00,0.00,0.10) |
| 9 | (0.10,0.30,0.50) | (0.70,0.90,1.00) | (0.50,0.70,0.90) | (0.33,0.50,0.70) | (0.33,0.50,0.70) | (0.33,0.50,0.70) | (0.90,1.00,1.00) | (0.00,0.10,0.30) | (0.11,0.33,0.56) | (0.10,0.30,0.50) | (0.70,0.90,1.00) | (0.70,0.90,1.00) |
| 10 | (0.70,0.90,1.00) | (0.90,1.00,1.00) | (0.00,0.10,0.30) | (0.00,0.10,0.30) | (0.33,0.50,0.70) | (0.00,0.00,0.10) | (0.50,0.70,0.90) | (0.50,0.70,0.90) | (0.56,0.78,1.00) | (0.10,0.30,0.50) | (0.33,0.50,0.70) | (0.70,0.90,1.00) |

**Table 18** Weighted Normalized fuzzy decision matrix for lean tools

| Suppliers | C1 | C2 | C3 | C4 | C5 | C6 | C7 | C8 | C9 | C10 | C11 | C12 |
|---|---|---|---|---|---|---|---|---|---|---|---|---|
| 1 | (0.05,0.13,0.25) | (0.09,0.18,0.28) | (0.01,0.04,0.07) | (0.03,0.07,0.15) | (0.01,0.03,0.08) | (0.00,0.00,0.00) | (0.00,0.00,0.01) | (0.00,0.01,0.02) | (0.00,0.01,0.01) | (0.03,0.05,0.08) | (0.04,0.07,0.13) | (0.01,0.03,0.05) |
| 2 | (0.16,0.26,0.36) | (0.12,0.20,0.28) | (0.02,0.05,0.09) | (0.07,0.13,0.21) | (0.04,0.07,0.14) | (0.00,0.00,0.01) | (0.00,0.01,0.01) | (0.00,0.01,0.02) | (0.00,0.00,0.01) | (0.01,0.02,0.05) | (0.03,0.05,0.11) | (0.00,0.01,0.03) |
| 3 | (0.00,0.03,0.11) | (0.01,0.06,0.14) | (0.04,0.07,0.11) | (0.01,0.04,0.10) | (0.02,0.05,0.11) | (0.00,0.00,0.01) | (0.00,0.00,0.01) | (0.00,0.01,0.02) | (0.00,0.01,0.02) | (0.03,0.05,0.08) | (0.05,0.08,0.13) | (0.00,0.00,0.02) |
| 4 | (0.16,0.26,0.36) | (0.09,0.18,0.28) | (0.01,0.04,0.07) | (0.05,0.10,0.19) | (0.02,0.05,0.11) | (0.00,0.00,0.01) | (0.00,0.00,0.00) | (0.00,0.00,0.01) | (0.00,0.00,0.00) | (0.01,0.02,0.05) | (0.03,0.05,0.11) | (0.01,0.02,0.04) |
| 5 | (0.09,0.18,0.32) | (0.04,0.10,0.20) | (0.03,0.06,0.11) | (0.00,0.01,0.06) | (0.05,0.09,0.15) | (0.00,0.00,0.01) | (0.00,0.00,0.01) | (0.00,0.01,0.02) | (0.00,0.01,0.01) | (0.02,0.04,0.08) | (0.05,0.08,0.13) | (0.02,0.03,0.06) |
| 6 | (0.02,0.08,0.18) | (0.00,0.02,0.08) | (0.01,0.04,0.07) | (0.03,0.07,0.15) | (0.01,0.03,0.08) | (0.01,0.01,0.02) | (0.00,0.01,0.02) | (0.00,0.00,0.01) | (0.01,0.01,0.02) | (0.02,0.03,0.07) | (0.02,0.04,0.09) | (0.01,0.02,0.04) |
| 7 | (0.05,0.13,0.25) | (0.09,0.18,0.28) | (0.00,0.00,0.01) | (0.05,0.10,0.19) | (0.00,0.01,0.05) | (0.00,0.01,0.02) | (0.00,0.01,0.01) | (0.01,0.01,0.03) | (0.01,0.01,0.02) | (0.00,0.00,0.01) | (0.04,0.07,0.13) | (0.02,0.03,0.06) |
| 8 | (0.13,0.23,0.36) | (0.12,0.20,0.28) | (0.00,0.02,0.05) | (0.03,0.07,0.15) | (0.00,0.01,0.05) | (0.00,0.00,0.00) | (0.00,0.01,0.01) | (0.01,0.01,0.03) | (0.01,0.01,0.02) | (0.00,0.00,0.02) | (0.00,0.01,0.04) | (0.01,0.03,0.05) |
| 9 | (0.00,0.00,0.04) | (0.00,0.02,0.08) | (0.01,0.04,0.07) | (0.01,0.04,0.10) | (0.02,0.05,0.11) | (0.01,0.01,0.02) | (0.01,0.01,0.02) | (0.01,0.02,0.03) | (0.01,0.01,0.02) | (0.01,0.02,0.05) | (0.01,0.02,0.06) | (0.01,0.02,0.04) |
| 10 | (0.13,0.23,0.36) | (0.04,0.10,0.20) | (0.03,0.06,0.11) | (0.03,0.07,0.15) | (0.01,0.03,0.08) | (0.00,0.00,0.01) | (0.00,0.01,0.01) | (0.00,0.01,0.02) | (0.00,0.01,0.02) | (0.02,0.04,0.08) | (0.04,0.07,0.13) | (0.01,0.02,0.04) |

$A^* = \{(0.16,0.26,0.36), (0.12,0.20,0.28), (0.04,0.07,0.11), (0.07,0.13,0.21), (0.05,0.09,0.15), (0.01,0.01,0.02), (0.01,0.01,0.02), (0.01,0.02,0.03), (0.01,0.01,0.02), (0.03,0.05,0.08), (0.05,0.08,0.13), (0.02,0.03,0.06)\}$

$A^- = \{(0.00,0.00,0.04), (0.00,0.02,0.08), (0.00,0.00,0.01), (0.00,0.01,0.06), (0.00,0.01,0.05), (0.00,0.00,0.00), (0.00,0.00,0.00), (0.00,0.00,0.01), (0.00,0.00,0.00), (0.00,0.00,0.01), (0.00,0.01,0.04), (0.00,0.00,0.02)\}$

**Table 19** Weighted normalized fuzzy decision matrix for green tools

| Suppliers | C13 | C14 | C15 | C16 | C17 | C18 | C19 | C20 | C21 | C22 | C23 | C24 |
|---|---|---|---|---|---|---|---|---|---|---|---|---|
| 1 | (0.08,0.18,0.33) | (0.05,0.13,0.29) | (0.02,0.04,0.10) | (0.05,0.11,0.26) | (0.02,0.06,0.14) | (0.00,0.00,0.01) | (0.00,0.00,0.01) | (0.00,0.00,0.00) | (0.00,0.00,0.01) | (0.02,0.05,0.11) | (0.03,0.07,0.14) | (0.01,0.03,0.07) |
| 2 | (0.08,0.18,0.33) | (0.05,0.13,0.29) | (0.02,0.04,0.10) | (0.00,0.02,0.09) | (0.01,0.03,0.10) | (0.00,0.01,0.03) | (0.00,0.00,0.01) | (0.01,0.02,0.04) | (0.00,0.01,0.02) | (0.02,0.05,0.11) | (0.00,0.00,0.01) | (0.01,0.04,0.09) |
| 3 | (0.04,0.10,0.23) | (0.04,0.09,0.22) | (0.03,0.07,0.14) | (0.03,0.08,0.20) | (0.03,0.08,0.17) | (0.00,0.01,0.03) | (0.01,0.01,0.03) | (0.01,0.02,0.05) | (0.00,0.01,0.02) | (0.00,0.02,0.05) | (0.01,0.04,0.10) | (0.01,0.03,0.07) |
| 4 | (0.00,0.02,0.10) | (0.01,0.06,0.16) | (0.02,0.04,0.10) | (0.07,0.15,0.29) | (0.06,0.11,0.19) | (0.01,0.02,0.04) | (0.00,0.01,0.02) | (0.01,0.01,0.03) | (0.00,0.00,0.01) | (0.00,0.02,0.05) | (0.00,0.02,0.07) | (0.00,0.00,0.01) |
| 5 | (0.04,0.10,0.23) | (0.05,0.13,0.29) | (0.00,0.02,0.07) | (0.07,0.15,0.29) | (0.04,0.10,0.19) | (0.00,0.01,0.02) | (0.00,0.00,0.01) | (0.01,0.01,0.03) | (0.00,0.00,0.00) | (0.02,0.04,0.09) | (0.04,0.08,0.14) | (0.00,0.01,0.03) |
| 6 | (0.00,0.00,0.03) | (0.00,0.02,0.10) | (0.00,0.01,0.04) | (0.03,0.08,0.20) | (0.02,0.06,0.14) | (0.01,0.01,0.03) | (0.01,0.02,0.03) | (0.01,0.02,0.04) | (0.01,0.01,0.03) | (0.01,0.03,0.07) | (0.03,0.07,0.14) | (0.02,0.05,0.09) |
| 7 | (0.10,0.20,0.33) | (0.07,0.17,0.32) | (0.02,0.06,0.13) | (0.07,0.15,0.29) | (0.04,0.10,0.19) | (0.00,0.01,0.02) | (0.00,0.00,0.01) | (0.00,0.01,0.02) | (0.00,0.00,0.00) | (0.01,0.03,0.07) | (0.01,0.04,0.10) | (0.01,0.04,0.09) |
| 8 | (0.05,0.14,0.30) | (0.05,0.13,0.29) | (0.03,0.07,0.14) | (0.00,0.02,0.09) | (0.02,0.06,0.14) | (0.00,0.01,0.02) | (0.00,0.01,0.02) | (0.01,0.02,0.04) | (0.00,0.01,0.02) | (0.01,0.03,0.07) | (0.00,0.02,0.07) | (0.00,0.00,0.01) |
| 9 | (0.01,0.06,0.17) | (0.07,0.17,0.32) | (0.02,0.06,0.13) | (0.03,0.08,0.20) | (0.02,0.06,0.14) | (0.00,0.01,0.03) | (0.01,0.02,0.03) | (0.00,0.00,0.01) | (0.00,0.01,0.02) | (0.00,0.02,0.05) | (0.03,0.07,0.14) | (0.02,0.05,0.09) |
| 10 | (0.08,0.18,0.33) | (0.10,0.18,0.32) | (0.00,0.01,0.04) | (0.00,0.02,0.09) | (0.02,0.06,0.14) | (0.00,0.00,0.00) | (0.00,0.01,0.02) | (0.01,0.02,0.04) | (0.01,0.01,0.03) | (0.00,0.02,0.05) | (0.01,0.04,0.10) | (0.02,0.05,0.09) |

$A^* = \{(0.10,0.20,0.33), (0.10,0.18,0.32), (0.03,0.07,0.14), (0.07,0.15,0.29), (0.06,0.11,0.19), (0.01,0.02,0.04), (0.01,0.02,0.03), (0.01,0.02,0.05), (0.01,0.01,0.03), (0.02,0.05,0.11), (0.04,0.08,0.14), (0.02,0.05,0.09)\}$

$A^- = \{(0.00,0.00,0.03), (0.00,0.02,0.10), (0.00,0.01,0.04), (0.00,0.02,0.09), (0.01,0.03,0.10), (0.00,0.00,0.00), (0.00,0.00,0.01), (0.00,0.00,0.00), (0.00,0.00,0.00), (0.00,0.02,0.05), (0.00,0.00,0.01), (0.00,0.00,0.01)\}$

**Table 20** Distances of the ratings of each alternative from $A^+$ with respect to each criterion for lean tools

| Suppliers | C1 | C2 | C3 | C4 | C5 | C6 | C7 | C8 | C9 | C10 | C11 | C12 |
|---|---|---|---|---|---|---|---|---|---|---|---|---|
| $d(A_1, A^+)$ | 0.12 | 0.02 | 0.03 | 0.06 | 0.06 | 0.02 | 0.01 | 0.01 | 0.01 | 0.00 | 0.01 | 0.01 |
| $d(A_2, A^+)$ | 0.00 | 0.00 | 0.02 | 0.00 | 0.02 | 0.01 | 0.01 | 0.01 | 0.01 | 0.02 | 0.02 | 0.02 |
| $d(A_3, A^+)$ | 0.22 | 0.13 | 0.00 | 0.09 | 0.04 | 0.01 | 0.01 | 0.01 | 0.00 | 0.00 | 0.00 | 0.03 |
| $d(A_4, A^+)$ | 0.00 | 0.02 | 0.03 | 0.02 | 0.04 | 0.01 | 0.01 | 0.02 | 0.01 | 0.02 | 0.02 | 0.01 |
| $d(A_5, A^+)$ | 0.06 | 0.09 | 0.01 | 0.12 | 0.00 | 0.01 | 0.01 | 0.01 | 0.01 | 0.00 | 0.00 | 0.00 |
| $d(A_6, A^+)$ | 0.17 | 0.17 | 0.03 | 0.06 | 0.06 | 0.00 | 0.00 | 0.02 | 0.00 | 0.01 | 0.04 | 0.01 |
| $d(A_7, A^+)$ | 0.12 | 0.02 | 0.07 | 0.02 | 0.08 | 0.01 | 0.01 | 0.00 | 0.00 | 0.05 | 0.01 | 0.00 |
| $d(A_8, A^+)$ | 0.03 | 0.00 | 0.05 | 0.06 | 0.08 | 0.02 | 0.01 | 0.00 | 0.00 | 0.04 | 0.07 | 0.01 |
| $d(A_9, A^+)$ | 0.26 | 0.17 | 0.03 | 0.09 | 0.04 | 0.00 | 0.00 | 0.00 | 0.00 | 0.02 | 0.05 | 0.01 |
| $d(A_{10}, A^+)$ | 0.03 | 0.09 | 0.01 | 0.06 | 0.06 | 0.01 | 0.01 | 0.01 | 0.00 | 0.00 | 0.01 | 0.01 |

**Table 21** Distances of the ratings of each alternative from $A^-$ with respect to each criterion for lean tools

| Suppliers | C1 | C2 | C3 | C4 | C5 | C6 | C7 | C8 | C9 | C10 | C11 | C12 |
|---|---|---|---|---|---|---|---|---|---|---|---|---|
| $d(A_1, A^-)$ | 0.15 | 0.16 | 0.04 | 0.06 | 0.02 | 0.02 | 0.04 | 0.01 | 0.03 | 0.03 | 0.05 | 0.01 |
| $d(A_2, A^-)$ | 0.26 | 0.17 | 0.06 | 0.12 | 0.07 | 0.02 | 0.04 | 0.00 | 0.03 | 0.01 | 0.04 | 0.01 |
| $d(A_3, A^-)$ | 0.04 | 0.04 | 0.07 | 0.03 | 0.04 | 0.01 | 0.05 | 0.01 | 0.03 | 0.03 | 0.06 | 0.02 |
| $d(A_4, A^-)$ | 0.26 | 0.16 | 0.04 | 0.09 | 0.04 | 0.02 | 0.05 | 0.00 | 0.04 | 0.01 | 0.04 | 0.01 |
| $d(A_5, A^-)$ | 0.20 | 0.08 | 0.07 | 0.00 | 0.08 | 0.01 | 0.05 | 0.01 | 0.03 | 0.03 | 0.06 | 0.02 |
| $d(A_6, A^-)$ | 0.09 | 0.00 | 0.04 | 0.06 | 0.02 | 0.01 | 0.04 | 0.00 | 0.03 | 0.02 | 0.02 | 0.01 |
| $d(A_7, A^-)$ | 0.24 | 0.16 | 0.00 | 0.09 | 0.00 | 0.01 | 0.04 | 0.01 | 0.03 | 0.02 | 0.05 | 0.02 |
| $d(A_8, A^-)$ | 0.24 | 0.17 | 0.03 | 0.06 | 0.00 | 0.02 | 0.04 | 0.01 | 0.03 | 0.01 | 0.01 | 0.01 |
| $d(A_9, A^-)$ | 0.00 | 0.00 | 0.04 | 0.03 | 0.04 | 0.01 | 0.04 | 0.02 | 0.03 | 0.01 | 0.01 | 0.01 |
| $d(A_{10}, A^-)$ | 0.24 | 0.08 | 0.07 | 0.06 | 0.02 | 0.02 | 0.04 | 0.01 | 0.03 | 0.03 | 0.05 | 0.01 |

**Table 22** Distances of the ratings of each alternative from $A^+$ with respect to each criterion for green tools

| Suppliers | C13 | C14 | C15 | C16 | C17 | C18 | C19 | C20 | C21 | C22 | C23 | C24 |
|---|---|---|---|---|---|---|---|---|---|---|---|---|
| $d(A_1, A^+)$ | 0.02 | 0.04 | 0.03 | 0.03 | 0.05 | 0.02 | 0.01 | 0.03 | 0.01 | 0.00 | 0.01 | 0.02 |
| $d(A_2, A^+)$ | 0.02 | 0.04 | 0.03 | 0.14 | 0.08 | 0.01 | 0.01 | 0.00 | 0.00 | 0.00 | 0.09 | 0.01 |
| $d(A_3, A^+)$ | 0.09 | 0.08 | 0.00 | 0.07 | 0.03 | 0.01 | 0.00 | 0.00 | 0.01 | 0.04 | 0.04 | 0.02 |
| $d(A_4, A^+)$ | 0.18 | 0.13 | 0.03 | 0.00 | 0.00 | 0.00 | 0.01 | 0.01 | 0.01 | 0.04 | 0.05 | 0.06 |
| $d(A_5, A^+)$ | 0.09 | 0.04 | 0.05 | 0.00 | 0.01 | 0.01 | 0.01 | 0.01 | 0.02 | 0.01 | 0.00 | 0.05 |
| $d(A_6, A^+)$ | 0.21 | 0.17 | 0.07 | 0.07 | 0.05 | 0.00 | 0.00 | 0.00 | 0.00 | 0.02 | 0.01 | 0.00 |
| $d(A_7, A^+)$ | 0.00 | 0.02 | 0.01 | 0.00 | 0.01 | 0.01 | 0.01 | 0.02 | 0.02 | 0.02 | 0.04 | 0.01 |
| $d(A_8, A^+)$ | 0.05 | 0.04 | 0.00 | 0.14 | 0.05 | 0.01 | 0.00 | 0.00 | 0.00 | 0.02 | 0.05 | 0.06 |
| $d(A_9, A^+)$ | 0.13 | 0.02 | 0.01 | 0.07 | 0.05 | 0.01 | 0.00 | 0.02 | 0.01 | 0.04 | 0.01 | 0.00 |
| $d(A_{10}, A^+)$ | 0.02 | 0.00 | 0.07 | 0.14 | 0.05 | 0.02 | 0.00 | 0.00 | 0.00 | 0.04 | 0.04 | 0.00 |

**Table 23** Distances of the ratings of each alternative from $A^-$ with respect to each criterion for green tools

| Suppliers | C13 | C14 | C15 | C16 | C17 | C18 | C19 | C20 | C21 | C22 | C23 | C24 |
|---|---|---|---|---|---|---|---|---|---|---|---|---|
| $d(A_1, A^-)$ | 0.21 | 0.13 | 0.04 | 0.12 | 0.03 | 0.01 | 0.05 | 0.02 | 0.04 | 0.02 | 0.05 | 0.02 |
| $d(A_2, A^-)$ | 0.21 | 0.13 | 0.04 | 0.00 | 0.00 | 0.01 | 0.05 | 0.01 | 0.04 | 0.02 | 0.04 | 0.01 |
| $d(A_3, A^-)$ | 0.13 | 0.09 | 0.07 | 0.08 | 0.05 | 0.01 | 0.04 | 0.01 | 0.04 | 0.03 | 0.02 | 0.02 |
| $d(A_4, A^-)$ | 0.04 | 0.04 | 0.04 | 0.14 | 0.08 | 0.01 | 0.04 | 0.01 | 0.04 | 0.03 | 0.01 | 0.05 |
| $d(A_5, A^-)$ | 0.13 | 0.13 | 0.02 | 0.14 | 0.07 | 0.01 | 0.05 | 0.01 | 0.05 | 0.01 | 0.05 | 0.04 |
| $d(A_6, A^-)$ | 0.00 | 0.00 | 0.00 | 0.08 | 0.03 | 0.01 | 0.04 | 0.01 | 0.03 | 0.01 | 0.05 | 0.01 |
| $d(A_7, A^-)$ | 0.21 | 0.16 | 0.06 | 0.14 | 0.07 | 0.01 | 0.05 | 0.01 | 0.05 | 0.01 | 0.02 | 0.01 |
| $d(A_8, A^-)$ | 0.18 | 0.13 | 0.07 | 0.00 | 0.03 | 0.01 | 0.04 | 0.01 | 0.04 | 0.01 | 0.01 | 0.05 |
| $d(A_9, A^-)$ | 0.08 | 0.16 | 0.06 | 0.08 | 0.03 | 0.01 | 0.04 | 0.02 | 0.04 | 0.03 | 0.05 | 0.01 |
| $d(A_{10}, A^-)$ | 0.21 | 0.17 | 0.00 | 0.00 | 0.03 | 0.02 | 0.04 | 0.01 | 0.03 | 0.03 | 0.02 | 0.01 |

## References


Akman, G., & Alkan, A. (2006). Tedarik Zinciri Yönetiminde Bulanık AHP yöntemi kullanılarak tedarikçilerin performansının ölçülmesi: Otomotiv Yan Sanayiinde bir uygulama. *İstanbul Ticaret Üniversitesi Fen Bilimleri Dergisi, 5*(9), 23–46.

Amid, A., Ghodsypour, S. H., & O'Brien, C. (2006). Fuzzy multiobjective linear model for supplier selection in a supply chain. *International Journal of Production Economics, 104*(2), 394–407.

Amid, A., Ghodsypour, S. H., & O'Brien, C. (2011). A weighted max–min model for fuzzy multi-objective supplier selection in a supply chain. *International Journal of Production Economics, 131*(1), 139–145.

Anvari, A., Zulkifli, N., Sorooshian, S., & Boyerhassani, O. (2014). An integrated design methodology based on the use of group AHP-DEA approach for measuring lean tools efficiency with undesirable output. *The International Journal of Advanced Manufacturing Technology, 70*(9–12), 2169–2186.

Banaeian, N., Mobli, H., Nielsen, I. E., & Omid, M. (2015). Criteria definition and approaches in green supplier selection–a case study for raw material and packaging of food industry. *Production & Manufacturing Research, 3*(1), 149–168.

Behera, S. K., & Nayak, J. R. (2011). Solution of multi-objective mathematical programming problems in fuzzy approach. *International Journal on Computer Science and Engineering, 3*(12), 3790.

Bellman, R. E., & Zadeh, L. A. (1970). Decision-making in a fuzzy environment. *Journal of Management Science, 17*(4), 141–164.

Buckley, J. J. (1985). Fuzzy hierarchical analysis. *Fuzzy Sets Systems, 17*(1), 233–247.

Büyüközkan, G., & Çifçi, G. (2012). A novel hybrid MCDM approach based on fuzzy DEMATEL, fuzzy ANP and fuzzy TOPSIS to evaluate green suppliers. *Expert Systems with Applications, 39*(3), 3000–3011.

Carvalho, H., Govindan, K., Azevedo, S. G., & Cruz-Machado, V. (2017). Modelling green and lean supply chains: an eco-efficiency perspective. *Resources, Conservation and Recycling, 120*, 75–87.

Chen, C. T. (2000). Extensions of the TOPSIS for group decision-making under fuzzy environment. *Fuzzy Sets System, 114*, 1–9.

Chen, C.-T., Lin, C.-T., & Huang, S.-F. (2006). A fuzzy approach for supplier evaluation and selection in supply chain management. *International Journal of Production Economics, 102*(2), 289–301.

Dağdeviren, M., & Erarslan, E. (2008). PROMETHEE sıralama yöntemi ile tedarikçi seçimi. *Gazi Üniversitesi Mühendislik-Mimarlık Fakültesi Dergisi, 23*(1).
Dobos, I., & Vörösmarty, G. (2014). Green supplier selection and evaluation using DEA-type composite indicators. *International Journal of Production Economics, 157*, 273–278.
Erginel, N., & Gecer, A. (2016). Fuzzy multi-objective decision model for calibration supplier selection problem. *Computers & Industrial Engineering, 102*, 166–174.
Galankashi, M. R., Hisjam, M., & Helmi, S. A. (2016). Lean supplier selection: a data envelopment analysis (DEA) approach. In *Proceedings of the sixth international conference on industrial engineering and operations management (IEOM)*. Kuala Lumpur.
Govindan, K., Rajendran, S., Sarkis, J., & Murugesan, P. (2015). Multi criteria decision making approaches for green supplier evaluation and selection: a literature review. *Journal of Cleaner Production, 98*, 66–83.
Hamdan, S., & Cheaitou, A. (2017). Supplier selection and order allocation with green criteria: an MCDM and multi-objective optimization approach. *Computers & Operations Research, 81*, 282–304.
Hsieh, T. Y., Lu, S. T., & Tzeng, G. H. (2004). Fuzzy MCDM approach for planning and design tenders selection in public office buildings. *International Journal of Project Management, 22*(7), 573–584.
Jadidi, O., Hong, T. S., Firouzi, F., Yusuff, R. M., & Zulkifli, N. (2008). TOPSIS and fuzzy multi-objective model integration for supplier selection problem. *Journal of Achievements in Materials and Manufacturing Engineering, 31*(2), 762–769.
Junior, F. R. L., Osiro, L., & Carpinetti, L. C. R. (2014). A comparison between fuzzy AHP and fuzzy TOPSIS methods to supplier selection. *Applied Soft Computing, 21*, 194–209.
Kannan, D., Khodaverdi, R., Olfat, L., Jafarian, A., & Diabat, A. (2013). Integrated fuzzy multi criteria decision making method and multi-objective programming approach for supplier selection and order allocation in a green supply chain. *Journal of Cleaner Production, 47*, 355–367.
Kannan, D., de Sousa Jabbour, A. B. L., & Jabbour, C. J. C. (2014). Selecting green suppliers based on GSCM practices: using fuzzy TOPSIS applied to a Brazilian electronics company. *European Journal of Operational Research, 233*(2), 432–447.
Kannan, D., Govindan, K., & Rajendran, S. (2015). Fuzzy axiomatic design approach based green supplier selection: a case study from Singapore. *Journal of Cleaner Production, 96*, 194–208.
Kargi, V. S. A. (2016). Supplier selection for a textile company using the fuzzy TOPSIS method. *Manisa Celal Bayar University the Faculty of Economic and Administrative Sciences Journal, 789*.
Kumar, M., Vrat, P., & Shankar, R. (2006). A fuzzy programming approach for vendor selection problem in a supply chain. *International Journal of Production Economics, 101*(2), 273–285.
Kumar, D., Rahman, Z., & Chan, F. T. (2017). A fuzzy AHP and fuzzy multi-objective linear programming model for order allocation in a sustainable supply chain: a case study. *International Journal of Computer Integrated Manufacturing, 30*(6), 535–551.
Kuo, R. J., Wang, Y. C., & Tien, F. C. (2010). Integration of artificial neural network and MADA methods for green supplier selection. *Journal of Cleaner Production, 18*(12), 1161–1170.
Lee, A. H., Kang, H. Y., Hsu, C. F., & Hung, H. C. (2009). A green supplier selection model for high-tech industry. *Expert Systems with Applications, 36*(4), 7917–7927.
Liao, C. N., Fu, Y. K., & Wu, L. C. (2016). Integrated FAHP, ARAS-F and MSGP methods for green supplier evaluation and selection. *Technological and Economic Development of Economy, 22*(5), 651–669.
Lin, R. H. (2012). An integrated model for supplier selection under a fuzzy situation. *International Journal of Production Economics, 138*(1), 55–61.
Ng, W. L. (2008). An efficient and simple model for multiple criteria supplier selection problem. *European Journal of Operational Research, 186*(3), 1059–1067.
Pandya, M., Patel, V., & Nilesh, P. (2017). A review of lean tools & techniques for cycle time reduction. *REST Journal on Emerging Trends in Modelling and Manufacturing, 3*(1), 2455–4537.

Semra, B. B. (2003). A case study of supplier selection for lean supply by using a mathematical model. *Logistics Information Management, 16*(6), 451–459.

Şenocak, A. A., & Gören, H. G. (2016). An integrated approach for sustainable supplier selection in fuzzy environment. *International Journal of Intelligent Systems and Applications in Engineering*, 130–135.

Shaik, M., & Abdul-Kader, W. (2011). Green supplier selection generic framework: a multi-attribute utility theory approach. *International Journal of Sustainable Engineering, 4*(1), 37–56.

Shaw, K., Shankar, R., Yadav, S. S., & Thakur, L. S. (2012). Supplier selection using fuzzy AHP and fuzzy multi-objective linear programming for developing low carbon supply chain. *Expert Systems with Applications, 39*(9), 8182–8192.

Shen, L., Olfat, L., Govindan, K., Khodaverdi, R., & Diabat, A. (2013). A fuzzy multi criteria approach for evaluating green suppliers performance in green supply chain with linguistic preferences. *Resources, Conservation and Recycling, 74*, 170–179.

Talluri, S., & Narasimhan, R. (2003). Vendor evaluation with performance variability: a max–min approach. *European Journal of Operational Research, 146*(3), 543–552.

Tiwari, R. N., Dharmahr, S., & Rao, J. R. (1987). Fuzzy goal programming-an additive model. *Fuzzy Sets and Systems, 24*(1), 27–34.

Wang, T. Y., & Yang, Y. H. (2009). A fuzzy model for supplier selection in quantity discount environments. *Expert Systems with Applications, 36*(10), 12179–12187.

Wu, C., & Barnes, D. (2016). An integrated model for green partner selection and supply chain construction. *Journal of Cleaner Production, 112*, 2114–2132.

Yildiz, A. (2013). *Bulanık Çok Kriterli Karar Verme Yöntemleri ile Tedarikçi Seçimi ve Ekonomik Sipariş Miktarının Tespiti: Otomotiv Sektöründe Bir Uygulama.* Marmara Üniversitesi, Fen Bilimleri Enstitüsü: Doctora Tezi.

Zadeh, L. A. (1965). Fuzzy sets. *Information and Control, 8*(3), 338–353.

Zimmermann, H. J. (1978). Fuzzy programming and linear programming with several objective functions. *Fuzzy Sets and System, 1*, 44–55.

# The Impact of Routing on $CO_2$ Emissions at a Retail Grocery Store Chain: A GIS-Based Solution Approach

**Çağrı Koç, Mehmet Erbaş, and Eren Özceylan**

## 1 Introduction

In the last decade, the impact of environmental issues in the context of the road freight transportation has been extensively studied. Minimizing the fuel consumption and $CO_2$ emissions cost in the Vehicle Routing Problem (VRP) may result in different customer assignments and route selections than the traditional version. The classical VRP aims to determine an optimal routing plan for a fleet of homogeneous vehicles to serve a set of customers, such that each vehicle route starts and ends at the depot, each customer is visited once by one vehicle, and some side constraints are satisfied. Many variants and extensions of the VRP have been intensively studied in the literature (see Laporte, 2009; Toth and Vigo, 2014).

Bektaş and Laporte (2011) introduced the Pollution-Routing Problem (PRP) which is an extension of the classical VRP with time windows. It is assumed that load and speed may change from one arc to another, but all parameters will remain constant on a given arc. The objective approximates the total amount of energy consumed on a given road segment, which directly translates into fuel consumption and further into GHG emissions. An adaptive large neighbourhood search (ALNS) metaheuristic is proposed by Demir et al. (2012) for the PRP. This metaheuristic first applies the classical ALNS scheme to construct vehicle

Ç. Koç (✉)
Department of Business Administration, Social Sciences University of Ankara, Ankara, Turkey
e-mail: cagri.koc@asbu.edu.tr

M. Erbaş
General Directorate of Mapping, Ministry of National Defense, Ankara, Turkey

E. Özceylan
Department of Industrial Engineering, Gaziantep University, Gaziantep, Turkey



T. Paksoy et al. (eds.), *Lean and Green Supply Chain Management*, International Series in Operations Research & Management Science 273,
https://doi.org/10.1007/978-3-319-97511-5_5

routes, and then applies the speed optimization algorithm (SOA) to compute the speed on each arc. In a later study, the Bi-Objective PRP is introduced by same authors (Demir et al., 2014), which jointly minimizes fuel consumption and driving time. The authors developed an adapted version of their ALNS metaheuristic, and compared four a posteriori methods, namely the weighting method, the weighting method with normalization, the epsilon-constraint method and a new hybrid method, using a scalarization of the two objective functions. The Time-Dependent PRP is introduced by Franceschetti et al. (2013) under a two-stage planning horizon. The authors developed an explicit congestion model in addition to the PRP objectives and presented a mathematical formulation in which vehicle speeds are optimally selected from a set of discrete values. Koç et al. (2014) introduced the Fleet Size and Mix PRP where the objective is a linear combination of vehicle, fixed cost, fuel cost and $CO_2$ emissions, and driver cost. The authors formally defined the problem, presented a mathematical model and developed a hybrid evolutionary metaheuristic which several algorithmic features. Kramer et al. (2015) developed a metaheuristic for the PRP, as well as for the Fuel Consumption VRP and the Energy Minimizing VRP. The authors integrated the iterated local search with a set partitioning procedure and the SOA. Koç et al. (2016b) studied and analyzed the combined impact of depot location, fleet composition and routing on vehicle emissions in city logistics. The authors introduced a variant of the PRP without time windows and developed an ALNS metaheuristic to solve it. Koç and Karaoglan (2016) studied the green VRP and developed an exact algorithm based on branch-and-cut. Baykasoğlu and Subulan (2016) developed a novel mathematical programming model for a real-life intermodal freight transportation planning problem of a large-scaled international logistics company in Turkey. This study also handles various critical decisions such as transportation mode selection and fleet consolidation. Furthermore, environmental conscious load plans are generated by minimizing the amounts of total $CO_2$ emitted by the different transportation modes. Dabia et al. (2017) developed a branch-and-price algorithm for a variant of the PRP. The authors considered the master problem as a set-partitioning problem, and the pricing problem as a speed and start time elementary shortest path problem with resource constraints. Column generation and a tailored labeling algorithms are used to solve the master and pricing problems, respectively.

Fleet composition is a common problem in practice and its decisions predominantly involve choosing the number, types, and capacities of vehicles to be used. These are affected by several market variables such as transportation rates, transportation costs and expected demand. Heterogeneous fleet of vehicles are used to serve for customer demands (see Baldacci et al., 2008; Koç et al., 2016a). Another practical side of the VRP considers multi-trip where vehicles can perform several trips per day, because of their limited number and capacity (see Cattaruzza et al., 2016; Olivera and Viera, 2007). Prins (2002) studied the multi-trip heterogeneous fixed fleet VRP and developed sequential heuristics, a new merge heuristic, steepest descent local search and tabu search. The method is applied to the case of a furniture manufacturer located near Nantes on the Atlantic coast of France.

In recent years, several authors have used GIS-based solution methods to solve the VRP and its variants. Bozkaya et al. (2010) studied the competitive multi-facility location-routing problem. A hybrid heuristic method which combines genetic algorithm and tabu search is developed. Genetic algorithm is used for the location part, and tabu search of GIS-based solution method is used for the VRP part. The method is applied on a case study arising at a supermarket store chain in the city of Istanbul. Yanik et al. (2014) later studied the capacitated VRP with multiple pickup, single delivery and time windows. A hybrid metaheuristic approach is developed which integrates a genetic algorithm for vendor selection and allocation, and a GIS-based solution method which uses a modified savings algorithm for the routing part. Krichen et al. (2014) used a GIS-based method to solve the VRP with loading and distance constraints.

Fleet composition, multi-trip and vehicle routing decisions all bear on $CO_2$ emissions in road freight transportation. Some of their interactions are already well documented (see Koç et al., 2014; Koç, 2016; Kramer et al., 2015). To our knowledge, the interplay between multi-trip, fleet composition and vehicle routing decisions and their influence on environment has not yet been investigated. Our purpose is to analyze these interrelated components within a unified framework.

This paper makes three main scientific contributions. Its first contribution is to formally define and mathematically model this new problem and solve it by means of a GIS-based solution method using a tabu search heuristic which can be used to store, analyze and visualize all data as well as model solutions in geographic format. Its second contribution is to carry out extensive computational experiments and analyses on a real dataset of a company which is one of the major retail grocery store chain in Turkey in order to gain a deep understanding into the interactions between the components of the problem. Its third contribution is to provide several managerial and policy insights by exploring the trade-offs between various parameters such as number of vehicles and shop demand.

The remainder of this paper is structured as follows. Section 2 present the problem setting. Section 3 presents the solution method, a case study, and computational experiments. Finally, conclusions are presented in Sect. 4.

# 2 Problem Setting

We now briefly provide our fuel consumption and $CO_2$ emissions model in Sect. 2.1, followed by the problem definition and mathematical formulation in Sect. 2.2.

## *2.1 Fuel Consumption and $CO_2$ Emissions*

The comprehensive emissions model of Barth et al. (2005), Scora and Barth (2006), and Barth and Boriboonsomsin (2008) is used and applied to the PRP (Bektaş and

Laporte, 2011; Demir et al., 2012) and to some of its extensions (Franceschetti et al., 2013; Demir et al., 2014; Koç et al., 2014, 2016b). This model estimates fuel consumption and emissions at a given time instant. We now briefly recall the modified heterogeneous fleet version of the model where the index set of vehicle types is denoted by $\mathcal{H}$. The fuel consumption rate $FR^h$ (L/s) of a vehicle of type $h \in \mathcal{H}$ is given by

$$FR^h = \xi(kNV + P^h/\eta)/\kappa, \tag{1}$$

where the variable $P^h$ is the second-by-second engine power output (in kW) of vehicle type $h$. Variable $P^h$ can be calculated as

$$P^h = P^h_{tract}/n_{tf} + P_{acc}, \tag{2}$$

where the engine power demand $P_{acc}$ is associated with the running losses of the engine and the operation of vehicle accessories such as air conditioning and electrical loads. We assume that $P_{acc} = 0$. The total tractive power requirement $P^h_{tract}$ (in kW) for a vehicle of type $h$ is

$$P^h_{tract} = (M^h\tau + M^h g \sin\theta + 0.5C_d\rho A v^2 + M^h g C_r \cos\theta)v/1000, \tag{3}$$

where $M^h$ is the total vehicle weight (in kg) and $v$ is the vehicle speed (m/s). The fuel consumption $F^h$ (L) of vehicle type $h$ over a distance $d$ traveling at constant speed $v$ is calculated as

$$F^h = kNV\lambda d/v + P^h\lambda\gamma d/v, \tag{4}$$

where $\lambda = \xi/\kappa\psi$ and $\gamma = 1/1000n_{tf}\eta$ are constants. Let $\beta = 0.5C_d\rho A$ be a vehicle-specific constant and $\alpha = \tau + g\sin\theta + gC_r\cos\theta$ be a vehicle-arc specific constant. Therefore, $F^h$ can be rewritten as

$$F^h = \lambda(kNVd/v + M^h\gamma\alpha d + \beta\gamma d v^2) \tag{5}$$

where the first term $kNVd/v$ is the engine module, the second term $M^h\gamma\alpha d$ is the weight module, and the third term $\beta\gamma d v^2$ is the speed module.

## 2.2 Problem Definition and Mathematical Formulation

The problem is defined on a complete directed graph $\mathcal{G} = (\mathcal{N}, \mathcal{A})$, where $\mathcal{A} = \{(i, j) : i \in \mathcal{N}, j \in \mathcal{N}, i \neq j\}$ is the set of arcs. $\mathcal{N} = 0 \cup \mathcal{N}_c$ is a set of nodes in which "0" is the depot node and $\mathcal{N}_c$ is a set of shop nodes. Each arc $(i, j) \in \mathcal{A}$ has a nonnegative distance $d_{ij}$. The distance matrix is asymmetric $d_{ij} \neq d_{ji}$ between $i$

and $j$. Each arc $(i, j) \in \mathcal{A}$ has a nonnegative travel time $c_{ij}$. Each shop $i \in \mathcal{N}_c$ has a demand $q_i$ and a service time $p_i$. The index set of vehicle types is denoted by $\mathcal{H}$ and for each vehicle type a fixed number of limited identical vehicles $m^h$ are available. The index set of routes is denoted by $\mathcal{R} = \{1, \ldots, r, \ldots\}$. The capacity of a vehicle of type $h \in \mathcal{H}$ is denoted by $Q^h$. The curb weight of a vehicle of type $h \in \mathcal{H}$ is denoted by $w^h$. The maximum allowed working duration is $T_{max}$ for each vehicle. Fuel and $CO_2$ emissions cost is denoted by $f_c$. Driver wage is denoted by $f_d$.

The objective of the problem is to minimize a total cost function encompassing fuel and $CO_2$ emissions costs and driver costs. A feasible solution contains a set of routes for a heterogeneous fleet of vehicles that meet the demands of all shops. Each shop is visited once by a single vehicle, each vehicle must depart from and return to the depot, to serve a quantity of demand that does not exceed its capacity. Furthermore, vehicles can perform several trips per day.

To formulate the problem, we define the following decision variables. Let $x^h_{ijr}$ be equal to 1 if a vehicle of type $h \in \mathcal{H}$ travels directly from node $i$ to node $j$ on route $r \in \mathcal{R}$. Let $f^h_{ijr}$ be the amount of commodity flowing on arc $(i, j) \in \mathcal{A}$ by a vehicle of type $h \in \mathcal{H}$ on route $r \in \mathcal{R}$. Therefore, the total load of vehicle of type $h$ on arc $(i, j)$ is $w^h + f^h_{ij}$. We now present the mathematical formulation for the problem:

$$\text{Minimize} \sum_{(i,j)\in\mathcal{A}} f_c \lambda k N V d_{ij}/v \tag{6}$$

$$+ \sum_{h\in\mathcal{H}} \sum_{(i,j)\in\mathcal{A}} \sum_{r\in\mathcal{R}} f_c \lambda \gamma \alpha_{ij} d_{ij} (w^h x^h_{ijr} + f^h_{ijr}) \tag{7}$$

$$+ \sum_{(i,j)\in\mathcal{A}} f_c \lambda \beta \gamma d_{ij} v^2 \tag{8}$$

$$+ \sum_{h\in\mathcal{H}} \sum_{(i,j)\in\mathcal{A}} \sum_{r\in\mathcal{R}} f_d c_{ij} x^h_{ijr} \tag{9}$$

subject to

$$\sum_{j\in\mathcal{N}} x^h_{0j1} \le m^h \qquad h \in \mathcal{H} \tag{10}$$

$$\sum_{h\in\mathcal{H}} \sum_{j\in\mathcal{N}} \sum_{r\in\mathcal{R}} x^h_{ijr} = 1 \qquad i \in \mathcal{N}_c \tag{11}$$

$$\sum_{h\in\mathcal{H}} \sum_{i\in\mathcal{N}} \sum_{r\in\mathcal{R}} x^h_{ijr} = 1 \qquad j \in \mathcal{N}_c \tag{12}$$

$$\sum_{h\in\mathcal{H}} \sum_{j\in\mathcal{N}_c} x^h_{0jr} \ge \sum_{h\in\mathcal{H}} \sum_{j\in\mathcal{N}_c} x^h_{0j,r+1} \qquad r \in \mathcal{R} : r < |\mathcal{R}| \tag{13}$$

$$\sum_{h\in\mathcal{H}} \sum_{j\in\mathcal{N}} \sum_{r\in\mathcal{R}} f^h_{jir} - \sum_{h\in\mathcal{H}} \sum_{j\in\mathcal{N}} \sum_{r\in\mathcal{R}} f^h_{ijr} = q_i \qquad i \in \mathcal{N}_c \tag{14}$$

$$q_j x^h_{ijr} \leq f^h_{ijr} \leq (Q^h - q_i) x^h_{ijr} \qquad (i,j) \in \mathcal{A}, h \in \mathcal{H}, r \in \mathcal{R} \quad (15)$$

$$x^h_{ijr} + \sum_{q \in \mathcal{H}, q \neq h} \sum_{u \in \mathcal{N}, j \neq u} x^q_{jur} \leq 1 \qquad h \in \mathcal{H}, i \in \mathcal{N}, j \in \mathcal{N}_c, i \neq j, r \in \mathcal{R} \quad (16)$$

$$\sum_{(i,j) \in \mathcal{A}} \sum_{r \in \mathcal{R}} c_{ij} x^h_{ijr} \leq T_{max} \qquad h \in \mathcal{H} \quad (17)$$

$$x^h_{ijr} \in \{0,1\} \qquad (i,j) \in \mathcal{A}, h \in \mathcal{H}, r \in \mathcal{R} \quad (18)$$

$$f^h_{ijr} \geq 0 \qquad (i,j) \in \mathcal{A}, h \in \mathcal{H}, r \in \mathcal{R}. \quad (19)$$

In the objective function, the first three terms represent the cost of fuel consumption and of $CO_2$ emissions. In particular, term (6) computes the cost induced by the engine module, term (7) reflects the cost induced by the weight module, and term (8) measures the cost induced by the speed module. Finally, term (9) computes the total driver wage.

Constraints (10) bounds the number of vehicles for each type. Constraints (11) and (12) ensure that each customer is visited exactly once. Constraints (13) impose that a vehicle cannot start route $r + 1$ before finish route $r$. Constraints (14) and (15) define the flows. Constraints (16) state that each customer is assigned to only one vehicle type. Constraints (17) ensure that the total travel time cannot exceed the maximum allowed working duration. Finally, constraints (18) and (19) enforce the integrality and nonnegativity restrictions on the variables.

# 3 A Case Study

This section presents the ArcGIS solution method in Sect. 3.1, case study and its input data in Sect. 3.2, and the associated results obtained with the ArcGIS in Sect. 3.3.

## *3.1 ArcGIS Solution Approach*

Many commercial software packages are available to solve the VRP. In our paper, we use the ArcGIS 10.2 commercial GIS package to solve our problem and build our GIS-based decision support framework. This package is used in many broad areas where spatially-enabled data need to be stored, retrieved, analyzed, visualized and even served online.

ArcGIS stores all problem data in geographic format and visualize them as well as the solutions obtained through the metaheuristic approach. It uses the Network Analyst Extension (NAE) repeatedly to solve the problem. The NAE defines the problem and specify its parameters such as network restrictions, costs, demands, vehicle capacities, and type of output. The ArcGIS uses a tabu search metaheuristic which follows the classical tabu search principles such as non-improving solutions are accepted along the way, but cycling of solutions are avoided using tabu lists and tabu tenure parameters (see Glover and Laguna, 1998).

The ArcGIS first creates an origin-destination matrix of shortest travel costs between all nodes that must be visited by a route. In the second step, the ArcGIS generates a feasible initial solution by inserting each location one at a time into the most suitable route. In the third phase, the ArcGIS aims to obtain better solution by applying the following three procedures.

1. Changing the sequence nodes on a single route.
2. Moving a single node from its current route to a better route.
3. Swapping two nodes between their respective routes.

For further details about the ArcGIS, the NAE and its application on the VRP, we refer the reader to the ArcGIS (2017).

## 3.2 *Input Data*

We consider the input data of BİM Inc. which is one of the major retail grocery store chain in Turkey. The company sells a wide range of food items and consumer goods at competitive prices. It limits its product portfolio to approximately 600 items and aims at having diverse private label products in such a way that they satisfy 80% of the basic daily requirements of a household. Products are displayed on pallets and not on shelves in such a way as to provide easy access for the customer. Each store has simple systems and minimalist interior design. Store sizes, lighting systems, and positioning are designed to reflect a cozy and comfortable shopping atmosphere which provides a non-stressful and no hassle environment for customers. The company is the pioneer of discount store model in Turkey where it began in 1995 with 21 stores and by the end of 2016 had reached 5530 stores (BİM Inc., 2017).

A well-designed logistic network between depots and stores is the one of the main of the company to speed up decision-making and implementation processes. The company has a depot and 90 stores in the city of Gaziantep that is an important commercial and industrial center for Turkey and considered stores are located in two districts cover 85% of total population of Gaziantep. Figure 1 presents an illustration of the depot and shops locations.

Four types of vehicles, A, B, C and D, each with a fixed number, are operating in the city. A list of and values for the common parameters for all vehicle types and specific parameters for each vehicle type are given in Tables 1 and 2, respectively. The demand of each store is expressed in kg and each vehicle are designed to satisfy these specific store demands. In total, there are 10 benchmark instances which includes from 36 to 75 shops. Each shop has a service time which includes the unloading time (6 min per 500 kg), and parking time and handling paperwork for

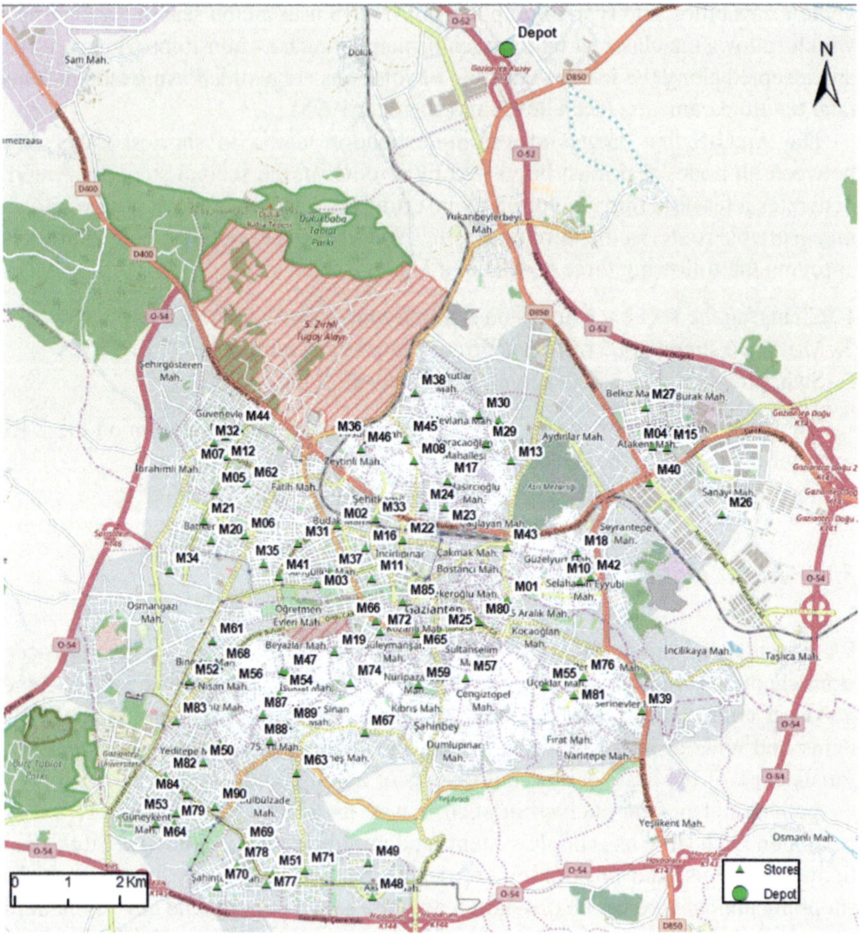


**Fig. 1** Illustration of the depot and shops locations

shipment (30 min per shop). As a company policy, the maximum allowed working duration for each vehicle is 8 h. The vehicle speed is fixed at 50 km/h. In the original PRP, the road angle for each arc is considered as zero. However, we consider a specific real road angle for each arc. OpenStreetMap (2018) is built by a community of mappers that contribute and maintain data about roads, trails, cafs, railway stations, and much more, all over the world. In this study, Gaziantep road network is downloaded from OpenStreetMap (2018) data. Figure 2 presents an illustration of elevation with roads.

**Table 1** Common parameters for all vehicle types

| | | |
|---|---|---|
| $\xi$ | Fuel-to-air mass ratio | 1 |
| $g$ | Gravitational constant (m/s$^2$) | 9.81 |
| $\rho$ | Air density (kg/m$^3$) | 1.2041 |
| $C_r$ | Coefficient of rolling resistance | 0.01 |
| $\eta$ | Efficiency parameter for diesel engines | 0.45 |
| $f_c$ | Fuel and $CO_2$ emissions cost (£/l) | 1.4 |
| $f_d$ | Driver wage (£/s) | 0.0022 |
| $\kappa$ | Heating value of a typical diesel fuel (kj/g) | 44 |
| $\psi$ | Conversion factor (g/s to L/s) | 737 |
| $n_{tf}$ | Vehicle drive train efficiency | 0.45 |
| $\tau$ | Acceleration (m/s$^2$) | 0 |
| $k$ | Engine friction factor (kj/rev/l) | 0.2 |
| $N$ | Engine speed (rev/s) | 36.67 |
| $V$ | Engine displacement (l) | 6.9 |
| $C_d$ | Coefficient of aerodynamics drag | 0.7 |
| $A$ | Frontal surface area (m$^2$) | 8 |

**Table 2** Vehicle specific parameters

| Type | $w^h$—curb weight (kg) | $Q^h$—max payload (kg) | # of vehicles |
|---|---|---|---|
| A | 2830 | 5100 | 2 |
| B | 3235 | 9800 | 2 |
| C | 3615 | 12,500 | 8 |
| D | 4513 | 15,700 | 6 |

## 3.3 *Computational Experiments and Analyses*

This section presents the results of computational experiments. We conduct all experiments on a server with an Intel Core i7 CPU 3.07 Ghz processor. We consider fixed parameters during the analyses of the ArcGIS for the case study. However, we will explore the effects of possible changes in the values of the main parameters on the total cost of the system. In particular, we will investigate whether the solution method yields cost-effective solutions when there is increase or decrease in the values of these parameters.

The aim of the computational experiments is three; (1) to solve the problem described in Sect. 2.2, and in particular to empirically calculate the effect of any changes in: (2) the shop demand, and (3) number of vehicles.

### 3.3.1 Results Obtained on the Instances

Table 3 presents the results obtained on 10 benchmark instances of the company. The first column shows the instance name, and the other columns display the number of shops ($|N_c|$), fleet composition, the total distance (km), $CO_2$ (kg), fuel and $CO_2$ emissions cost (£), driver cost (£), total cost (£), and finally total CPU time (s).

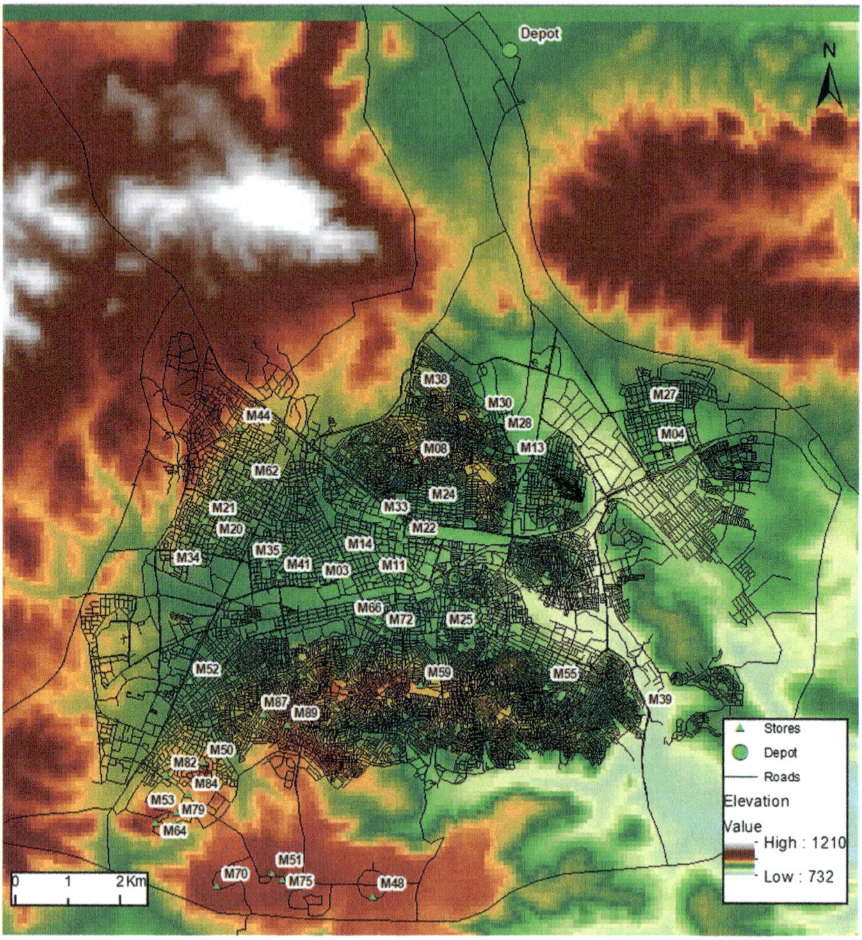


**Fig. 2** Illustration of roads with elevation

The column "Fleet composition" shows the actual number of vehicles used in the solution. The letters A–D correspond to the vehicle types and the number denote the number of each type of vehicle used. For example, "A1 B2" indicates that one vehicles of type A and two vehicles of type B are used in the solution. To evaluate the environmental impact of the solutions, we also report the average amount of $CO_2$ emissions (kg) based on the assumption that 1 l of gasoline contains 2.32 kg of $CO_2$ (Coe, 2005).

Table 3 shows that $CO_2$ emissions ranges between 225.11 and 293.73 kg for instances Gaziantep-BIM-1,2,3,4,5,6,7 and 8. However, in instance Gaziantep-BIM-9 vehicles emit 467.17 kg of $CO_2$, and in instance Gaziantep-BIM-10 vehicles emit 705.26 kg of $CO_2$, which are quite higher than the first eight instances. For instances Gaziantep-BIM-1,2,3,4,5,6,7 and 8, C and D vehicle types frequently used which

**Table 3** Detailed results on ten instances

| Instance | $\|N_c\|$ | Fleet composition | Total distance (km) | $CO_2$ (kg) | Fuel and $CO_2$ emissions cost (£) | Driver cost (£) | Total cost (£) | CPU time (s) |
|---|---|---|---|---|---|---|---|---|
| Gaziantep-BIM-1 | 41 | B1 C2 D6 | 293.73 | 395.71 | 238.79 | 392.62 | 631.41 | 0.08 |
| Gaziantep-BIM-2 | 37 | C1 D6 | 231.54 | 302.79 | 182.72 | 343.17 | 525.89 | 0.07 |
| Gaziantep-BIM-3 | 37 | C1 D6 | 231.88 | 318.68 | 192.31 | 344.02 | 536.33 | 0.07 |
| Gaziantep-BIM-4 | 37 | C1 D6 | 225.11 | 301.40 | 181.88 | 345.32 | 527.20 | 0.07 |
| Gaziantep-BIM-5 | 36 | A1 B1 D6 | 251.76 | 336.97 | 203.34 | 344.00 | 547.34 | 0.07 |
| Gaziantep-BIM-6 | 40 | C2 D6 | 264.13 | 360.58 | 217.59 | 378.43 | 596.02 | 0.08 |
| Gaziantep-BIM-7 | 42 | C1 D6 | 237.59 | 326.91 | 197.27 | 359.18 | 556.45 | 0.07 |
| Gaziantep-BIM-8 | 50 | B1 C3 D6 | 279.16 | 385.41 | 232.58 | 452.88 | 685.46 | 0.08 |
| Gaziantep-BIM-9 | 70 | A1 B2 C8 D5 | 467.17 | 608.94 | 367.47 | 651.36 | 1018.82 | 0.12 |
| Gaziantep-BIM-10 | 75 | A2 B2 C8 D6 | 705.26 | 893.30 | 539.06 | 860.93 | 1399.99 | 0.14 |

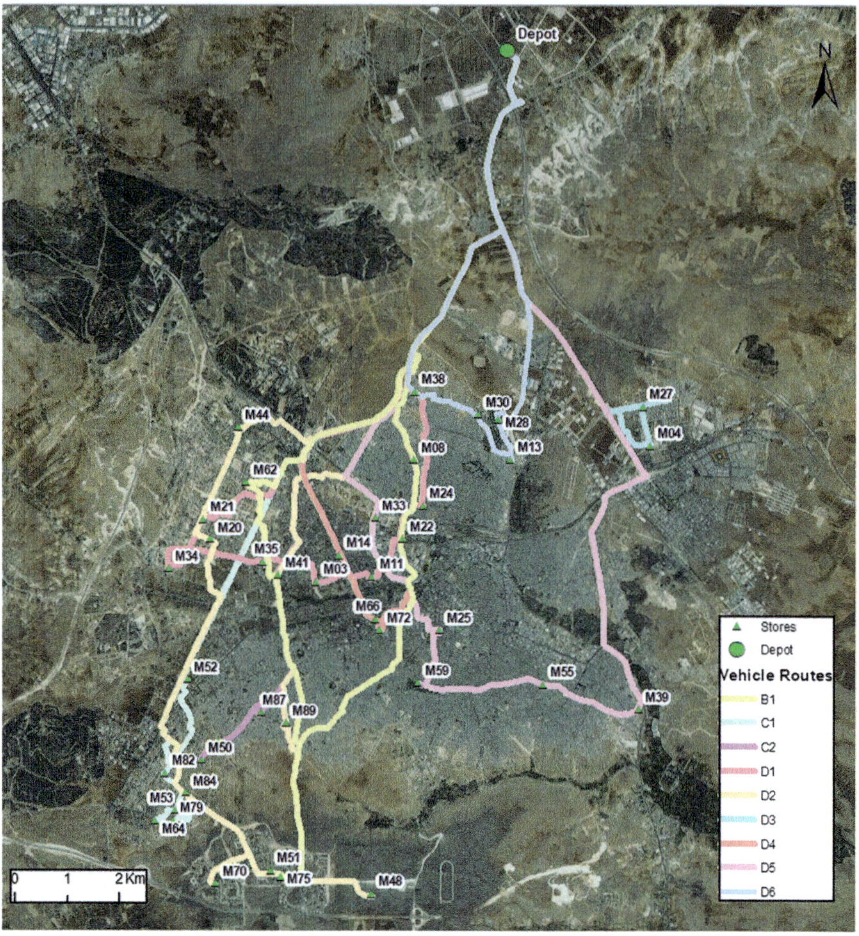


**Fig. 3** Illustration of the routes of the instance Gaziantep-BIM-1

indicates that types of vehicles are more suitable for medium-size instances. For large-size instances, Gaziantep-BIM-9 and Gaziantep-BIM-10, all vehicle types are used to satisfy the shop demands. One can see that all computation times are below 0.14 s, which is fast for our problem. Figure 3 presents an illustration of the routes of the instance Gaziantep-BIM-1.

### 3.3.2 The Effect of the Shop Demand

This section investigates the effect of the shop demand on solution quality. To do so, we decrease and increase the demand of each shop by 10 and 20%. Table 4 presents the average results for these experiments. The detailed results of these experiments can be found in Appendix. Columns $Dev_{TD}$, $Dev_{CO_2}$, $Dev_{FCE}$, $Dev_{DC}$, and

**Table 4** The effect of the shop demand

| Change in shop demand (%) | $Dev_{TD}$ (%) | $Dev_{CO_2}$ (%) | $Dev_{FCE}$ (%) | $Dev_{DC}$ (%) | $Dev_{TC}$ |
|---|---|---|---|---|---|
| −20 | −16.61 | −15.70 | −15.70 | −11.09 | −12.74 |
| −10 | −11.28 | −10.77 | −10.77 | −6.04 | −7.74 |
| 0 | 0.00 | 0.00 | 0.00 | 0.00 | 0.00 |
| +10 | 9.87 | 9.99 | 9.99 | 5.42 | 7.06 |
| +20 | 18.02 | 16.34 | 16.34 | 11.24 | 13.07 |

$Dev_{TC}$ shows the percentage deviations in total distance, $CO_2$ emissions, fuel and $CO_2$ emissions cost, driver cost, and total cost of each scenario over the base case (0%), respectively.

Table 4 shows that $CO_2$ emissions are decreased by −15.70% and −10.77%, when we decrease the shop demand by −20% and −10%, respectively. When we increase the shop demand by 10% and 20%, $CO_2$ emissions increased by 9.99% and 16.34%, respectively. Driver costs are decreased by −11.09% and −6.04%, when we decrease the shop demand by −20% and −10%, respectively. On the other hand, driver costs are increased by 5.42% and 11.24%, when we increase the shop demand by 10% and 20%, respectively. Total costs are decreased by −12.74% and −7.74%, when we decrease the shop demand by −20% and −10%, respectively. On the other hand, total costs are increased by 7.06% and 13.07%, when we increase the shop demand by 10% and 20%, respectively. It is important to note that 10 and 20% increases in shop demand result in infeasible solutions for a large-size (i.e., Gaziantep-BIM-10) instance. In other words, the vehicle fleet becomes insufficient when the total shop demand is higher than the base case for benchmark instances which has 75 shops or more. As can be clearly seen in Table 4, any changes in shop demand directly effects $CO_2$ emissions, fuel and $CO_2$ emissions costs, driver costs and total costs as well as traveled distance in similar manner.

### 3.3.3 The Effect of the Number of Vehicles

We now analyze the effect of the number of vehicles on solution quality. To do so, we decrease and increase the fixed numbers of each vehicle type by "1" and "2". Table 5 presents the average results for these experiments. The detailed results of each experiments can be found in Appendix.

Table 5 shows that $CO_2$ emissions are increased by 3.91% and 4.04%, when we decrease the number of vehicles by −2 and −1, respectively. When we increase the number of vehicles by +1 and +2, $CO_2$ emissions are decreased by −2.96% and −1.04%, respectively. Driver costs are increased by 0.54% and 0.31%, when we decrease the number of vehicles by −2 and −1, respectively. On the other hand, driver costs are decreased by −0.14% and −0.85%, when we increase the number of vehicles by +1 and +2, respectively. Total costs are increased by 1.74% and 1.65%, when we decrease the number of vehicles by −2 and −1, respectively. On the other

**Table 5** The effect of the number of vehicles

| Change in # of vehicles | $Dev_{TD}$ (%) | $Dev_{CO_2}$ (%) | $Dev_{FCE}$ (%) | $Dev_{DC}$ (%) | $Dev_{TC}$ (%) |
|---|---|---|---|---|---|
| −2 | 4.98 | 3.91 | 3.91 | 0.54 | 1.74 |
| −1 | 2.90 | 4.04 | 4.04 | 0.31 | 1.65 |
| 0 | 0.00 | 0.00 | 0.00 | 0.00 | 0.00 |
| +1 | −1.26 | −2.96 | −2.96 | −0.14 | −1.14 |
| +2 | −3.13 | −1.04 | −1.04 | −0.85 | −0.92 |

hand, total costs are decreased by −1.14% and −0.92%, when we increase the number of vehicles by +1 and +2, respectively. The detailed results show that −2 and −1 decreases in the number of each vehicle type result in infeasible solutions for instances with have 70 or more shops. It is clear that the vehicle fleet becomes insufficient to satisfy the shop demands when we decrease the number of vehicles. These results demonstrates it is required to increase the fixed number of vehicles for solving the instances which has more than 70 shops. Furthermore, our results show that any increase in fixed number of vehicles reduces the total cost. On the other hand, any decrease in fixed number of vehicles increases the total cost.

## 4 Conclusions

We have formally and mathematically modeled a new problem variant of the Pollution-Routing Problem. We have solved it by means of a GIS-based solution method using a tabu search heuristic which can be used to store, analyze and visualize all data as well as model solutions in geographic format. We have carried out extensive computational experiments and analyses on a real dataset of a company which is one of the major retail grocery store chain in Turkey in order to gain a deep understanding into the interactions between the components of the problem. We have finally provided several managerial and policy insights by exploring the trade-offs between number of vehicles and shop demand. Beyond the computational comparisons we have just made, we showed the flexibility of ArcGIS which is a commercial decision support tool and capable of analyzing the trade-offs that can be established between various parameters within quite short computation time.

**Acknowledgements** The authors are indebted to two anonymous referees whose comments have greatly helped to improve the paper.

## Appendix

Table 6 presents the detailed results of the effect of changes in the shop demand. Table 7 presents the detailed results of the effect of changes in number of vehicles.

**Table 6** Detailed results of the effect of changes in the shop demand

| Instance | Change in shop demand (%) | $\|N_c\|$ | # of unserved shops | Total distance (km) | $CO_2$ (kg) | Fuel and $CO_2$ emissions cost (£) | Driver cost (£) | Total cost (£) |
|---|---|---|---|---|---|---|---|---|
| Gaziantep-BIM-1 | −20 | 41 | – | 229.54 | 331.36 | 199.96 | 345.63 | 545.59 |
| Gaziantep-BIM-2 | −20 | 37 | – | 196.95 | 261.67 | 157.90 | 305.75 | 463.66 |
| Gaziantep-BIM-3 | −20 | 37 | – | 201.82 | 261.79 | 157.98 | 307.18 | 465.16 |
| Gaziantep-BIM-4 | −20 | 37 | – | 195.58 | 266.12 | 160.59 | 308.57 | 469.16 |
| Gaziantep-BIM-5 | −20 | 36 | – | 200.09 | 282.74 | 170.62 | 303.48 | 474.10 |
| Gaziantep-BIM-6 | −20 | 40 | – | 213.23 | 307.12 | 185.33 | 334.86 | 520.20 |
| Gaziantep-BIM-7 | −20 | 42 | – | 211.19 | 275.60 | 166.31 | 323.19 | 489.50 |
| Gaziantep-BIM-8 | −20 | 50 | – | 239.13 | 324.11 | 195.58 | 404.30 | 599.89 |
| Gaziantep-BIM-9 | −20 | 70 | – | 382.33 | 502.94 | 303.50 | 577.60 | 881.09 |
| Gaziantep-BIM-10 | −20 | 75 | – | 497.65 | 636.01 | 383.80 | 749.11 | 1132.91 |
| Gaziantep-BIM-1 | −10 | 41 | – | 262.30 | 353.49 | 213.31 | 368.51 | 581.82 |
| Gaziantep-BIM-2 | −10 | 37 | – | 210.12 | 285.21 | 172.11 | 323.81 | 495.92 |
| Gaziantep-BIM-3 | −10 | 37 | – | 200.32 | 267.38 | 161.349 | 321.99 | 483.34 |
| Gaziantep-BIM-4 | −10 | 37 | – | 218.97 | 304.84 | 183.95 | 327.46 | 511.41 |
| Gaziantep-BIM-5 | −10 | 36 | – | 213.96 | 293.29 | 176.99 | 320.86 | 497.84 |
| Gaziantep-BIM-6 | −10 | 40 | – | 237.92 | 325.07 | 196.16 | 355.27 | 551.43 |
| Gaziantep-BIM-7 | −10 | 42 | – | 210.19 | 292.39 | 176.44 | 338.74 | 515.18 |
| Gaziantep-BIM-8 | −10 | 50 | – | 252.39 | 333.86 | 201.46 | 426.86 | 628.33 |
| Gaziantep-BIM-9 | −10 | 70 | – | 395.98 | 522.32 | 315.20 | 609.33 | 924.52 |
| Gaziantep-BIM-10 | −10 | 75 | – | 586.29 | 767.25 | 463.00 | 801.83 | 1264.82 |
| Gaziantep-BIM-1 | +10 | 41 | – | 308.87 | 427.08 | 257.72 | 414.16 | 671.88 |
| Gaziantep-BIM-2 | +10 | 37 | – | 254.11 | 346.53 | 209.11 | 362.72 | 571.84 |
| Gaziantep-BIM-3 | +10 | 37 | – | 255.61 | 327.24 | 197.474 | 364.81 | 562.28 |
| Gaziantep-BIM-4 | +10 | 37 | – | 225.11 | 301.40 | 181.88 | 345.32 | 527.20 |
| Gaziantep-BIM-5 | +10 | 36 | – | 269.74 | 386.71 | 233.36 | 364.01 | 597.37 |
| Gaziantep-BIM-6 | +10 | 40 | – | 308.67 | 406.01 | 245.00 | 404.49 | 649.50 |
| Gaziantep-BIM-7 | +10 | 42 | – | 253.83 | 347.82 | 209.89 | 377.86 | 587.75 |
| Gaziantep-BIM-8 | +10 | 50 | – | 321.37 | 434.35 | 262.11 | 481.35 | 743.46 |
| Gaziantep-BIM-9 | +10 | 70 | – | 529.59 | 693.79 | 418.67 | 692.00 | 1110.67 |
| Gaziantep-BIM-10 | +10 | 75 | 3 | – | – | – | – | – |
| Gaziantep-BIM-1 | +20 | 41 | – | 327.90 | 425.68 | 256.88 | 434.86 | 691.74 |
| Gaziantep-BIM-2 | +20 | 37 | – | 278.48 | 361.95 | 218.42 | 382.55 | 600.97 |
| Gaziantep-BIM-3 | +20 | 37 | – | 267.42 | 359.73 | 217.076 | 381.73 | 598.80 |
| Gaziantep-BIM-4 | +20 | 37 | – | 280.45 | 378.67 | 228.51 | 386.17 | 614.67 |
| Gaziantep-BIM-5 | +20 | 36 | – | 278.81 | 373.21 | 225.22 | 380.63 | 605.84 |
| Gaziantep-BIM-6 | +20 | 40 | – | 314.13 | 419.67 | 253.25 | 421.86 | 675.11 |
| Gaziantep-BIM-7 | +20 | 42 | – | 290.14 | 382.98 | 231.11 | 399.32 | 630.42 |
| Gaziantep-BIM-8 | +20 | 50 | – | 335.13 | 450.28 | 271.72 | 503.99 | 775.71 |
| Gaziantep-BIM-9 | +20 | 70 | – | 556.89 | 730.49 | 440.81 | 725.89 | 1166.70 |
| Gaziantep-BIM-10 | +20 | 75 | 9 | – | – | – | – | – |

**Table 7** Detailed results of the effect of changes in number of vehicles

| Instance | Change in # of vehicles | $\lvert N_c \rvert$ | # of unserved shops | Total distance (km) | $CO_2$ (kg) | Fuel and $CO_2$ emissions cost (£) | Driver cost (£) | Total cost (£) |
|---|---|---|---|---|---|---|---|---|
| Gaziantep-BIM-1 | −2 | 41 | – | 282.34 | 382.07 | 230.56 | 390.82 | 621.38 |
| Gaziantep-BIM-2 | −2 | 37 | – | 253.63 | 333.63 | 201.33 | 346.67 | 548.00 |
| Gaziantep-BIM-3 | −2 | 37 | – | 249.25 | 334.77 | 202.02 | 346.77 | 548.79 |
| Gaziantep-BIM-4 | −2 | 37 | – | 249.64 | 330.66 | 199.53 | 349.21 | 548.74 |
| Gaziantep-BIM-5 | −2 | 36 | – | 255.48 | 332.85 | 200.86 | 344.59 | 545.45 |
| Gaziantep-BIM-6 | −2 | 40 | – | 291.79 | 387.36 | 233.75 | 382.81 | 616.56 |
| Gaziantep-BIM-7 | −2 | 42 | – | 242.53 | 332.41 | 200.59 | 359.96 | 560.55 |
| Gaziantep-BIM-8 | −2 | 50 | – | 290.64 | 401.35 | 242.19 | 454.70 | 696.90 |
| Gaziantep-BIM-9 | −2 | 70 | 11 | – | – | – | – | – |
| Gaziantep-BIM-10 | −2 | 75 | 30 | – | – | – | – | – |
| Gaziantep-BIM-1 | −1 | 41 | – | 287.32 | 390.51 | 235.65 | 391.61 | 627.26 |
| Gaziantep-BIM-2 | −1 | 37 | – | 250.32 | 328.83 | 198.43 | 346.15 | 544.58 |
| Gaziantep-BIM-3 | −1 | 37 | – | 236.19 | 325.19 | 196.23 | 344.70 | 540.94 |
| Gaziantep-BIM-4 | −1 | 37 | – | 247.40 | 334.52 | 201.87 | 348.85 | 550.72 |
| Gaziantep-BIM-5 | −1 | 36 | – | 246.70 | 331.91 | 200.29 | 343.20 | 543.49 |
| Gaziantep-BIM-6 | −1 | 40 | – | 276.76 | 379.40 | 228.95 | 380.43 | 609.38 |
| Gaziantep-BIM-7 | −1 | 42 | – | 240.28 | 348.18 | 210.11 | 359.61 | 569.71 |
| Gaziantep-BIM-8 | −1 | 50 | – | 288.43 | 400.16 | 241.47 | 454.35 | 695.83 |
| Gaziantep-BIM-9 | −1 | 70 | – | 472.05 | 623.36 | 376.16 | 652.13 | 1028.29 |
| Gaziantep-BIM-10 | −1 | 75 | 10 | – | – | – | – | – |
| Gaziantep-BIM-1 | 1 | 41 | – | 264.23 | 387.38 | 233.76 | 387.95 | 621.71 |
| Gaziantep-BIM-2 | 1 | 37 | – | 231.54 | 302.79 | 182.72 | 343.17 | 525.89 |
| Gaziantep-BIM-3 | 1 | 37 | – | 231.88 | 227.27 | 137.15 | 344.02 | 481.17 |
| Gaziantep-BIM-4 | 1 | 37 | – | 225.11 | 314.77 | 189.95 | 345.32 | 535.27 |
| Gaziantep-BIM-5 | 1 | 36 | – | 251.76 | 336.97 | 203.34 | 344.00 | 547.34 |
| Gaziantep-BIM-6 | 1 | 40 | – | 259.88 | 356.82 | 215.32 | 377.76 | 593.08 |
| Gaziantep-BIM-7 | 1 | 42 | – | 237.59 | 326.91 | 197.27 | 359.18 | 556.45 |
| Gaziantep-BIM-8 | 1 | 50 | – | 287.45 | 394.89 | 238.30 | 454.20 | 692.50 |
| Gaziantep-BIM-9 | 1 | 70 | – | 446.03 | 604.67 | 364.89 | 648.01 | 1012.90 |
| Gaziantep-BIM-10 | 1 | 75 | – | 645.65 | 823.94 | 497.21 | 851.49 | 1348.70 |
| Gaziantep-BIM-1 | 2 | 41 | – | 264.23 | 387.38 | 233.76 | 387.95 | 621.71 |
| Gaziantep-BIM-2 | 2 | 37 | – | 231.54 | 302.79 | 182.72 | 343.17 | 525.89 |
| Gaziantep-BIM-3 | 2 | 37 | – | 231.88 | 318.68 | 192.31 | 344.02 | 536.33 |
| Gaziantep-BIM-4 | 2 | 37 | – | 225.11 | 314.77 | 189.95 | 345.32 | 535.27 |
| Gaziantep-BIM-5 | 2 | 36 | – | 251.76 | 336.97 | 203.34 | 344.00 | 547.34 |
| Gaziantep-BIM-6 | 2 | 40 | – | 237.59 | 326.91 | 197.27 | 359.18 | 556.45 |
| Gaziantep-BIM-7 | 2 | 42 | – | 237.59 | 326.91 | 197.27 | 359.18 | 556.45 |
| Gaziantep-BIM-8 | 2 | 50 | – | 272.04 | 385.55 | 232.66 | 451.76 | 684.41 |
| Gaziantep-BIM-9 | 2 | 70 | – | 424.26 | 595.09 | 359.11 | 644.56 | 1003.67 |
| Gaziantep-BIM-10 | 2 | 75 | – | 594.21 | 799.06 | 482.19 | 843.34 | 1325.53 |

# References


ArcGIS (2017). *ArcGIS network analyst tutorial.* <http://desktop.arcgis.com/en/arcmap/latest/extensions/network-analyst/about-the-network-analyst-tutorial-exercises.htm>.

Baldacci, R., Battarra, M., & Vigo, D. (2008). Routing a heterogeneous fleet of vehicles. In B. L. Golden, S. Raghavan, & E. A. Wasil (Eds.), *The vehicle routing problem: Latest advances and new challenges* (pp. 1–25). New York: Springer.

Barth, M., & Boriboonsomsin, K. (2008). *Real-world $CO_2$ impacts of traffic congestion.* Technical report. Paper for the 87th Annual Meeting of Transportation Research Board. <http://www.uctc.net/papers/846.pdf>

Barth, M., Younglove, T., & Scora, G. (2005). *Development of a heavy-duty diesel modal emissions and fuel consumption model.* Technical report UCB-ITS-PRR-2005-1. California PATH Program, Institute of transportation Studies, University of California at Berkeley.

Baykasoğlu, A., & Subulan, K. (2016). A multi-objective sustainable load planning model for intermodal transportation networks with a real-life application. *Transportation Research Part E, 95*, 207–247.

Bektaş, T., & Laporte, G. (2011). The pollution-routing problem. *Transportation Research Part B, 45*, 1232–1250.

BİM Inc. (2017). *BİM Inc.'s hard-discount concept.* <http://english.bim.com.tr/Category/627/bim-philosophy.aspx>

Bozkaya, B., Yanik, S., & Balcisoy, S. (2010). A GIS-based optimization framework for competitive multi-facility location-routing problem. *Networks and Spatial Economics, 10*, 297–320.

Cattaruzza, D., Absi, N., & Feillet, D. (2016). Vehicle routing problems with multiple trips. *4OR, 14*, 1–37.

Coe, E. (2005). *Average carbon dioxide emissions resulting from gasoline and diesel fuel, United States Environmental Protection Agency.* Technical report. <http://www.epa.gov/otaq/climate/420f05001.pdf>

Dabia, S., Demir, E. & van Woensel, T. (2017). An exact approach for a variant of the pollution routing problem. *Transportation Science, 51*, 607–628.

Demir, E., Bektaş, T., & Laporte, G. (2012). An adaptive large neighborhood search heuristic for the pollution-routing problem. *European Journal of Operational Research, 223*, 346–359.

Demir, E., Bektaş, T., & Laporte, G. (2014). The bi-objective pollution-routing problem. *European Journal of Operational Research, 232*, 464–478.

Franceschetti, A., Honhon, D., Van Woensel, T., Bektaş, T., & Laporte, G. (2013). The time-dependent pollution-routing problem. *Transportation Research Part B, 56*, 265–293.

Glover, F.W., & Laguna, M. (1998). *Tabu search.* Boston: Kluwer Academic.

Koç, Ç. (2016). A unified-adaptive large neighborhood search metaheuristic for periodic location-routing problems. *Transportation Research Part C, 68*, 265–284.

Koç, Ç., Bektaş, T., Jabali, O., & Laporte, G. (2014). The fleet size and mix pollution-routing problem. *Transportation Research Part B, 70*, 239–254.

Koç, Ç., Bektaş, T., Jabali, O., & Laporte, G. (2016a). Thirty years of heterogeneous vehicle routing. *European Journal of Operational Research, 249*, 1–21.

Koç, Ç., Bektaş, T., Jabali, O., & Laporte, G. (2016b). The impact of depot location, fleet composition and routing on emissions in city logistics. *Transportation Research Part B, 84*, 81–102.

Koç, Ç., & Karaoglan, I. (2016). The green vehicle routing problem: A heuristic based exact solution approach. *Applied Soft Computing, 39*, 154–164.

Kramer, R., Subramanian, A., Vidal, T., & Lucídio dos Anjos, F. C. (2015). A matheuristic approach for the pollution-routing problem. *European Journal of Operational Research, 243*, 523–539.

Krichen, S., Faiz, S., Tlili, T., & Tej, K. (2014). Tabu-based GIS for solving the vehicle routing problem. *Expert Systems with Applications, 41*, 6483–6493.

Laporte, G. (2009). Fifty years of vehicle routing. *Transportation Science, 43*, 408–416.

Olivera, A., & Viera, O. (2007). Adaptive memory programming for the vehicle routing problem with multiple trips. *Computers & Operations Research, 34*, 28–47.

OpenStreetMap. (2018). <https://www.openstreetmap.org>

Prins, C. (2002). Efficient heuristics for the heterogeneous fleet multitrip VRP with application to a large-scale real case. *Journal of Mathematical Modelling and Algorithms, 1*, 135–150.

Scora, M., & Barth, G. (2006). Comprehensive Modal Emission Model (CMEM), Version 3.01, Users Guide. Technical report. <http://www.cert.ucr.edu/cmem/docs/CMEM_User_Guide_v3.01d.pdf>

Toth, P., & Vigo, D. (Eds.) (2014). *Vehicle routing: Problems, methods, and applications. MOS-SIAM series on optimization*. Philadelphia: SIAM.

Yanik, S., Bozkaya, B., & deKervenoael, R. (2014). A new VRPPD model and a hybrid heuristic solution approach for e-tailing. *European Journal of Operational Research, 236*, 879–890.

# A Simulated Annealing Algorithm Based Solution Method for a Green Vehicle Routing Problem with Fuel Consumption

**Kenan Karagul, Yusuf Sahin, Erdal Aydemir, and Aykut Oral**

## 1 Introduction

Logistics is a term describing essential components of a production process such as transportation, storage, and handling of products from raw material to the final consumption points (McKinnon 2010). Among these components, transport activity is the most important. It is related to the movement and coordination of goods. During the transport of goods, different vehicles can be used depending on the type and specification of the product, the underlying technology, the relevant infrastructure and the nature of the respective operations (Bektas 2017). Transportation allows the carriage of loads from door-to-door, that is, between the starting and arriving points, which is a significant effect as a complementary element in other modes of transport.

Environmental pollution, which is widely spreading on a large area all over the world, is one of the most critical threats to the survival of living beings. Because of the processes such as electric power generation, industrial processes, heating, and transportation, millions of kilograms of carbon dioxide ($CO_2$) are spreading to the atmosphere everyday. Electricity and industrial production areas are suitable

K. Karagul
Honaz Vocational High School, Pamukkale University, Denizli, Turkey

Y. Sahin
Business Administration Department, Mehmet Akif Ersoy University, Burdur, Turkey

E. Aydemir (✉)
Industrial Engineering Department, Süleyman Demirel University, Isparta, Turkey
e-mail: erdalaydemir@sdu.edu.tr

A. Oral
Department of Economics, Pamukkale University, Denizli, Turkey

 
T. Paksoy et al. (eds.), *Lean and Green Supply Chain Management*, International Series in Operations Research & Management Science 273,
https://doi.org/10.1007/978-3-319-97511-5_6

for filtration systems due to high $CO_2$ emission values. Numerous small energy consumption points and vehicles used in transportation define other sectors that are not currently available for filtration. In the coming years, the technological changes in production technologies and the characteristics of transportation fuels may enable the reduction of carbon dioxide emissions. Green logistics practices are also eco-friendly business processes. Environmental management of the supply chain and logistics activities (green logistics) is now becoming increasingly popular as an enterprise practice.

The Vehicle Routing Problem (VRP) is a favorite and NP-Hard class combinatorial optimization problem. Depending on a set of constraints, it is a generalized version of the Travelling Salesman Problem and defined as the problem of determining the lowest-cost routes for deliveries to geographically separate customers from a warehouse. VRP is the primary element of many operational research problems and has many variations. The exact solution methods and many common heuristics have been proposed to solve VRPs.

In recent years, significant emphasis has been placed on sustainable logistics practices in order to overcome environmental concerns. The version that focuses on sustainable business practices is called Green Vehicle Routing Problem (G-VRP) in general. Due to the fact that it is an important area within the logistics activities, studies related to G-VRP has shown a significant increase in recent years. In this type of VRP, unlike the original VRP, the effect of the routing on the environment is being tried to be minimized. It is known that there is a direct effect of energy (fuel) consumption on the $CO_2$ emissions depending on the movement and the load of vehicles.

Most of the vehicles in road transport work with diesel or gasoline fuel in Turkey. Today, although vehicles that use alternative energy sources (LPG, electricity, biodiesel etc.) have been developed, gasoline and diesel vehicles continue to be used predominantly in this sector. Emissions from freight transport largely depend on the type of fuel used (Piecyk 2010). Higher use of roads in freight and passenger transport has led to increased traffic intensity and air pollution as well as transportation costs (Aydemir and Cubuk 2016). According to the statistics of the Turkish Statistical Institute (TSI) in 2014, total Green House Gas (GHG) emissions as $CO_2$ equivalent increased by 125% in 2014 compared to the emissions in 1990 and the share of the transportation sector in total greenhouse gas emission values is about 19%. Figure 1 shows the change in emission values between 1990 and 2014. Despite declining in some years, there is a tendency to increase in greenhouse gas emissions per capita in general. The $CO_2$ equivalent emission per capita was 6.08 tonnes in 2014, while it was 3.77 tonnes for the year 1990 (TSI 2014).

Investments and promotions that aim to direct drivers to public transport and freight transport from road to rail have been the result of efforts to reduce the use of fossil fuels. Although the effort being undertaken in this direction is obviously beneficial, achieving the significant decline will only be possible with a multi-perspective approach (Erdoğan and Miller-Hooks 2012). Governments should continue to increase their promotional and financial support for the use of greener energy resources. On the other hand, both public institutions and private sector

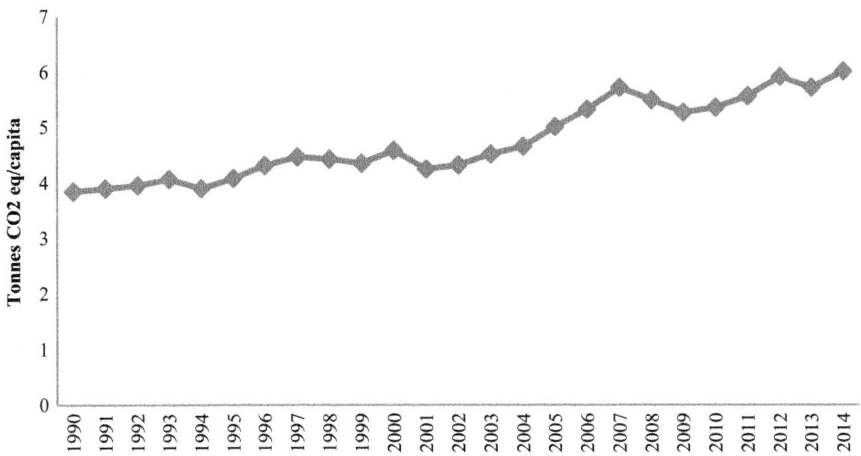


**Fig. 1** GHG emissions per capita in Turkey, 1990–2014 (TSI 2014)

companies should prefer eco-friendly energy sources in vehicles they use for transportation.

Along with the increased emphasis on supply chain management in the environmental sense, the need to develop models to reduce environmental pollution has emerged. At this point, sustainable transport, which aims to efficiently transfer goods, services and sustainable transport and delivery systems, has become an important issue for companies. The development of theoretical models is the first step in these studies. Factors such as travel distance, vehicle weight and vehicle speed which affect the emission value produced by the vehicle should be controlled. Another critical factor that determines the emission value is the total vehicle weight, which is the sum of the cargo and the vehicle's empty weight (Ayadi et al. 2014). Improvements to the above factors will help to achieve sustainable transport goals.

In this section, three analytical models were used to identify the G-VRP. A Simulated Annealing-based solution algorithm is preferred for the solution of large-scale test instances. The results obtained from this solution algorithm are compared with the results of the classical G-VRP model. Although there are many studies in the literature for G-VRP, the current contribution of the study is to propose new mathematical models alongside the classical models used for G-VRP and solve them quickly and efficiently with the Simulated Annealing method.

## 2 Background

Vehicle Routing Problem (VRP), first described by Dantzig and Ramser (1959), is a generalized version of the Travelling Salesman Problem and defined as the problem of determining the lowest-cost routes for deliveries to geographically separated customers from geographical locations, depending on a number of factors (Cetin and Gencer 2010). The main components of VRP are the road network, warehouses, and vehicles. VRP is generally describes by vehicle capacity or route distance constraints. If only capacity-related constraints are identified, the problem is named as Capacitated Vehicle Routing Problem (CVRP).

CVRP has been a problem that has attracted many researchers since its first definition. Many exact solution methods have been developed taking into account the capacity constraints. Some of them are applied with the necessary modifications to distance limited problems. However, the heuristics methods that are developed often add accountability to both constraints (Barhant and Laporte 2006). One of the first methods proposed for this problem is Clarke and Wright (1964) (C–W) Savings Algorithm. They introduced the savings concept which is based on the computation of savings for combining two customers into the same route (Pichpibul and Kawtummachai 2013). Since then, many different types have been introduced by adding new constraints to VRP, and many models and algorithms have been developed for these types of problems. In order to obtain different types of VRP, each component may be supplemented with different constraints and states, or each can achieve specific goals.

The literature on VRP is extensive and to date, many different VRP variations have been developed. The main types of VRP are Capacitated VRP (e.g., Bräysy and Gendreau 2005), VRP with Time Windows (e.g., Alvarenga et al., 2007), VRP with Backhauls (e.g., Gribkovskaia et al. 2008), VRP with Pick-up and Delivery (e.g., Ganesh and Narendran 2007), VRP with Heterogeneous Vehicle Fleet (e.g., Salhi et al. 2014), Open VRP (e.g., Özyurt et al. 2006), Periodic VRP (e.g., Cacchiani et al. 2014) and Stochastic VRP (e.g., Yan et al. 2006). The version that focuses on sustainable business practices is named as Green Vehicle Routing in general. Research on VRP, in which environmental issues such as fuel consumption and greenhouse gases (GHG) were assessed, has begun quite recently. Since this study is related to G-VRP, the following section only refers to the literature related to this type of problems.

In the above-mentioned types of vehicle routing problems, variations of the G-VRP arise when the objective function is formed by considering the fuel consumption or emission values. During the last ten years period, a series of literature reviews have been conducted on the G-VRP. Lin et al. (2014) presented an extensive literature review of G-VRP. They provide a classification of G-VRP that categorizes G-VRP into Green-VRP as Pollution-Routing Problem, VRP in Reverse Logistics, and suggested research gaps between its state and more luxurious models describing the complexity in real-world cases. Eglese and Bektaş (2014), described the current fuel consumption and emission models in the literature and the ways in which these models can be integrated into existing formulations or approaches for VRP. As an extension of the classical VRP, the Pollution-Routing Problem (PRP) accounts the amount of greenhouse emissions, fuel, travel times and costs (Bektaş and Laporte 2011; Koç et al. 2014; Kramer et al. 2015). Demir et al. (2014) provide a review of recent research on green road freight transportation. In this research, they focus on the scientific literature related to fuel reduction in road freight transportation by means of operations research techniques. Park and Chae (2014) focus on solution approaches for papers related to G-VRP which were published after 2000. Zhang et al. (2014) reviewed 115 studies published over the last 2 decades and observed the practice of Swarm Intelligence (SI) in Green Logistics (GL). The integration of GL and SI has been systematically classified into GL and SI categories by analyzing the context of the problem and the methodology used. Toro et al. (2016) conducted a literature survey including variations, solution models and methodologies of the VRP, in which the G-VRP is included. Recent studies related to G-VRP are summarised in the following sections.

## 2.1 Exact Solution Methods for G-VRP

It is possible to obtain an optimal solution in small-scale problems with exact solution methods. However, it is difficult and time-consuming to find optimal solutions in large-scale problems as the energy consumption or emission values assigning a specific customer to a tour depend on the other customers that are assigned to the tour. Studies using exact solution methods such as branch and bound dynamic programming for solving the problem are included in the literature. Kara et al. (2007) proposed a new cost function based on distance and load of the vehicle for the Energy Minimizing Vehicle Routing Problem (EMVRP). They developed an Integer Programming (IP) formulation with $O(n^2)$ binary variables and $O(n^2)$ constraints. Bektaş and Laporte (2011) offered a new integer programming formulation for the Pollution-Routing Problem which minimizes a total cost function composed of labor, fuel and emission costs expressed as a function of load, speed, and other parameters. The proposed model is solved by the branch-cut method, which is an advantageous method for integer programming problems. Huang et al. (2012) studied a G-VRP with simultaneous pickup and delivery problem (G-VRPSPD). They suggested a linear integer programming model including $CO_2$ emission and fuel consumption costs for G-VRPSPD, modified from the commodity flow based VRPSPD formulation.

Treitl et al. (2014) proposed an integer programming model for Inventory Routing Problem to minimize total transport costs as well as costs for $CO_2$ emissions from transport activities and warehousing activities over the planning horizon. Ramos et al. (2012) developed a modular and innovative solution approach for the multi-depot VRP and applied to a real case-study in order to restructure the current operation and achieve a more environmental-friendly solution. The primary goal of the research is to define service areas and vehicle routes that minimize the $CO_2$ emissions of a logistics system with multiple products and depots. They solved the decomposition solution method using the branch-and-bound algorithm.

Pan et al. (2013) adopted the emissions functions in a Mixed Integer Linear Programming (MILP) to minimize the $CO_2$ emissions related to freight transport in two extensive supply chains. Franceschetti et al. (2013) described an integer linear programming formulation of The Time-Dependent Pollution-Routing Problem

(TDPRP) consists of routing a fleet of vehicles in order to serve a set of customers and determining the speeds on each leg of the routes. Taha et al. (2014) presented an integer programming based exact solution model for the small size G-VRP. Alkawaleet et al. (2014) investigated the effect of $CO_2$ emissions on the inventory and routing decisions determined over a given time horizon. They formulate a mixed integer programming for the inventory routing problem of a product distributed to a number of customers from a single distribution center. Andelmin and Bartolini (2017) developed an exact algorithm for the G-VRP based on a set partitioning formulation using a multigraph. They weakened their formulations by adding weak subset sequence inequalities, subset sequence inequalities, and k-path cuts.

## 2.2 *Heuristic and Meta-Heuristic Algorithms for G-VRP*

Real-life optimization problems are very complex and require analysis of large data sets. Although an exact solution model has been developed for G-VRP, it is impossible to obtain a solution in an acceptable time. It is often enough to find an approximate solution to real-life problems. Therefore, there are many heuristic and meta-heuristic methods developed for G-VRP solutions. When the literature is examined, it is seen that the proposed heuristic methods are based on the route construction and neighborhood search. The Clark and Wright's Saving Algorithm (Clark and Wright 1964) is the most commonly used method to construct routes in G-VRP. Faulin et al. (2011) and Ubeda et al. (2011) constructed some algorithms with environmental criteria based on the Saving Algorithm to address the need for solutions to real problems in delivery companies or logistic carriers. Aranda et al. (2012) also developed an environmental performance method based on Life Cycle Assessment (LCA) and complemented with the Saving Algorithm to qualify the environmental performance of the end of life tyres (ELTs) management system, in terms of $CO_2$ emissions. Peiying et al. (2013) suggested a heuristic method called Bi-directional Optimization Heuristic Algorithm (BOHA) to reduce the most cost based on low carbon emissions. Zhou and Lee (2017) proposed a nonlinear mixed integer programming model with vehicle speeds, vehicle weights, road grades, and vehicle routes as decision variables. They used sweep algorithm for the initial solution and 2-opt local search algorithm for the improvements to find vehicle routes of G-VRP.

In recent years, many meta-heuristic methods are used to solve the G-VRP. The most preferred methods in the G-VRP solution are Genetic Algorithms (GA), Tabu Search (TS) and Simulated Annealing (SA). In addition to these methods, Scatter Search (SS), Near-Exact (NEA), Local Search (LS), Iterative Route Construction and Improvement (IRCI), Artificial Bee Colony (ABC) algorithms are also used in the literature. Maden et al. (2010) described a TS based heuristic algorithm for vehicle routing and scheduling problems to minimize the total travel time, where the time required for a vehicle to travel along any road in the network varies according to the time of travel. Kuo and Lin (2010) proposed a model to calculate the total fuel consumption when given a routing plan. They consider three factors that affect fuel consumption and used a simple Tabu Search to optimize the routing plan. Jabali et al. (2012) presented a model that considers travel time, fuel consumption, and $CO_2$ emissions costs accounting for time-dependent travel times between customers. They solved the model using a tabu search procedure. Li (2012) presented a mathematical model for the VRP with time windows (VRPTW) with a new objective function of minimizing the total fuel consumption and solved the problem using a novel TS algorithm with a random variable neighborhood descent procedure (RVND). This algorithm uses an adaptive parallel route construction heuristic, introduces six neighborhood search methods and employs a random neighborhood ordering and shaking mechanisms. Kwon et al. (2013) adopted a mixed integer-programming model for the objective of minimizing the sum of variable operation costs, including a cost-benefit assessment of acquiring carbon rights under a cap-and-trade regime. They deployed TS algorithms were deployed together with three neighborhood generation methods. Úbeda et al. (2014) proposed a TS algorithm based on Gendreau et al.'s (1994) approach to solving the green distance CVRP. They applied this approach on a set of instances obtained from the company Eroski producing the cleanest solutions in all cases. Ene et al. (2016) presented a SA and TS-based hybrid metaheuristic algorithm to analyze the effect of a heterogeneous fleet on reducing fuel consumption. They used a time-oriented nearest neighbour heuristic to generate the initial solution for the algorithm and preferred local search method to generate neighbours.

Kuo (2010) suggested a model for calculating total fuel consumption for the time-dependent vehicle routing problem (TDVRP). Then a SA algorithm is used to find the vehicle routing with the lowest total fuel consumption. Suzuki (2011) developed an approach to the time-constrained, multiple-stop, truck-routing problem that minimizes the fuel consumption and pollutants emission. They used the enumeration technique to find the optimal solutions for instances with $n = 5$ or $n = 10$ and the compressed annealing (Ohlmann and Thomas 2007) for experiments in which $n = 15$. Xiao et al. (2012) presented a mathematical model for Fuel Consumption Rate (FCR) considered CVRP (FCVRP) in which the fuel consumption rate is added to the CVRP. They developed SA algorithm for the proposed model. In this algorithm, Swap, Relocation, 2-opt, and Hybrid exchange rules are used. Yasin and Vincent (2013) adopted a model using the mathematical model of Erdoğan and Miller-Hooks (2012) and developed a SA based solution method for the G-VRP. Küçükoğlu and Öztürk (2015) formulated G-VRP with time windows (G-VRPTW) using a mixed integer linear programming model. They adopted a memory structure SA (MSA-SA) meta-heuristic algorithm due to the high complexity of the proposed problem. To calculate fuel consumption and $CO_2$ emissions, they integrated proposed an algorithm with a calculation procedure. Koç and Karaoglan (2016) proposed a SA heuristic based exact solution approach to solve the G-VRP by considering a limited driving range of vehicles in conjunction with limited refueling infrastructure. They used branch-and-cut algorithm based exact algorithm to improve lower bounds and a heuristic algorithm based on SA to obtain upper bounds. Vincent et al. (2017) generated a mathematical model to minimize the total cost of travel by driving Plug-in Hybrid Electric Vehicle (PHEV) for the hybrid vehicle routing problem (HVRP), which is an extension of the G-VRP.

Urquhart et al. (2010) used an Evolutionary Multi-Objective Algorithm to investigate the trade-off between $CO_2$ savings, distance and number of vehicles used in a typical vehicle routing problem with Time Windows (VRPTW). Omidvar and Tavakkoli-Moghaddam (2012) used SA and GA methods along with a partial heuristic method and an exact algorithm for solving small-scale problems. They aimed to minimize the total cost of vehicles, traveled distance, travel time and emissions solving time-dependent VRP. Jemai et al. (2012) implemented the NSGA-II evolutionary algorithm to the bi-objective G-VRP to minimize the total traveled distance and the total $CO_2$ emissions with respect to classical routing constraints. Ayadi et al. (2014) proposed a mathematical model for the G-VRP with multiple trips developed a solution method by combining a GA with a local search procedure to solve it. Oliveira et al. (2017) used GA that incorporates elements of local and population search to minimize $CO_2$ emission per route for G-VRP. Bouzekri et al. (2014) defined the bi-objective G-VRP (bi-GVRP) in the context of sustainable transportation and applied the genetic algorithm to solve bi-GVRP benchmarks. Hsueh (2016) proposed a mathematical model considering heterogeneous fleet which is affected several factors such as vehicle types and conditions, travel speeds, roadway gradients, and payloads. They developed a customized GA for solving the model. Cooray and Rupasinghe (2017) implemented a GA to solve the Energy-Minimizing VRP and used machine learning techniques to determine the parameters of the developed GA. Tunga et al. (2017) developed a mathematical model to minimize the total energy consumed and balancing the routes, and proposed GA for finding out a solution to the G-VRP with different constraints.

# 3 Mathematical Models

## *3.1 A Classical Green VRP Model*

VRPs have many different forms; however, most of them minimize the distance cost while visiting each customer once and with respect to vehicles capacity constraints. Actually, the consumed fuel amount is more important than the traveled distance for fuel cost savings (Xiao et al. 2012). In G-VRP, a set of delivery routes are determined to satisfy the demand with a minimum distance costs and the minimum volume of emitted $CO_2$. The G-VRP is also an *NP*-hard problem due to the fact that it is an extension of the standard VRP considering with green supply chain preferences.

Consider a G-VRP defined over a directed graph G = (V, A ) where V = { 0, 1, 2, ..., $n$} the node set where is V=0 is the depot and A = { $(i,j)$ : $i,j \in$ V, $i \neq j$} is the set of arcs the components of which are given as:

$d_{ij}$ Distance between nodes $i$and $j$
$q_i$ Non-negative weight (demand and supply) of node $i$
$c_{ij}$ Traveling cost between nodes $i$ and $j$
$Q$ Capacity of a vehicle (truck)
$k$ Number of identical vehicles
$K$ Number of vehicles
$n$ Number of customers

An unlimited number of the homogeneous vehicle fleet is available at the depot to serve customers with fuel tank capacity $Q$ (liters) and fuel consumption rate $r$ (liters per km). The problem is to determine the corresponding vehicle routes so as to minimize the total cost subject to the following assumptions (Kara et al. 2007; Koç and Karaoglan 2016):

- Each vehicle is used for at most one route,
- Each route starts and ends at the depot,
- Each node is served exactly by one vehicle,
- Fuel level at the vehicle's tank must be greater than or equal to the fuel consumption between any two nodes,
- The amount of fuel in a vehicle's tank is sufficient to be able to visit any pair of nodes,
- The load of a vehicle does not exceed its capacity $Q$.

The decision variables are:

- $q_{ij}$: the amount filled to the vehicle $k$ between nodes $i$ and $j$
- $x_{ij}^k = \begin{cases} 1 & \text{if vehicle } k \text{ drives from customer } i \text{ to customer j} \\ 0 & \text{otherwise} \end{cases}$
- $y_i^k = \begin{cases} 1 & \text{if vehicle } k \text{ visits customer } i \\ 0 & \text{otherwise} \end{cases}$

In the solution of a VRP, a matrix representation is used for distance, time and cost parameters between nodes $i$ and $j$. In G-VRP, a matrix representation also requires showing $CO_2$ emissions based on the estimation of $CO_2$ emitted between nodes $i$ and $j$ (Palmer 2007). About the linear formulation of emission volume, considering the delivery with a distance for Heavy Duty Vehicle (HDV) which has the average speed 80 km/h and fully loaded 25 tons (Hassel and Samaras 1999) Eq. (1) is given as follows (Elbouzekri et al. 2013):

$$E_{ij}\,(q,d) = d_{ij} \times \left[ \left( \frac{e_f - e_e}{Q} \right) q_{ij} + e_e \right] \tag{1}$$

Where:

$E_{ij}(q,d)$: the $CO_2$ emissions from a vehicle in kg/km with the variable of load q in ton and d in km

$e_f$: the $CO_2$ emissions of a fully loaded vehicle (1.096 kg/km for a HDV truck)

$e_e$: the $CO_2$ emissions of an empty vehicle (0.772 kg/km for a HDV truck)

The objectives are to find a set of m vehicle routes of minimum total cost (distance) and minimum total emitted $CO_2$ emission level. The mathematical model is given as follows (modified from Xiao et al. 2012; Bouzekri and Alaoui 2014):

$$\min TotalCost\ (f) = \sum_{i=0}^{n}\sum_{j=0}^{n}\sum_{k=1}^{K} c_{ij}.x_{ij}^{k} \tag{2}$$

$$\min CO_2Emission\ (g) = \sum_{i=0}^{n}\sum_{j=0}^{n}\sum_{k=1}^{K} d_{ij}.\left[\left(\frac{e_{f-}e_e}{Q}\right).q_{ij}^{k} + e_{e1}.x_{ij}^{k}\right] \tag{3}$$

Subject to:

$$\sum_{i=0}^{n} x_{0i}^{k} \le 1 \qquad \forall k = 1,\ldots,K \tag{4}$$

$$\sum_{i=0}^{n} x_{i0}^{k} \le 1 \qquad \forall k = 1,\ldots,K \tag{5}$$

$$\sum_{i=1}^{n} y_i^k \le M.\sum_{j=1}^{n} x_{0j}^{k} \qquad \forall k = 1,\ldots,K \tag{6}$$

$$\sum_{i=1}^{n} y_i^k \le M.\sum_{j=1}^{n} x_{j0}^{k} \qquad \forall k = 1,\ldots,K \tag{7}$$

$$\sum_{k=1}^{n} y_i^k \le 1 \qquad \forall k = 1,\ldots,K \tag{8}$$

$$\sum_{j=1}^{n} x_{ji}^{k} = \sum_{j=1}^{n} x_{ij}^{k} \qquad \begin{matrix}\forall k = 1,\ldots,K \\ \forall i = 1,\ldots,K\end{matrix} \tag{9}$$

$$\sum_{i=0}^{n} \sum_{\substack{j=1 \\ i \ne j}}^{n} q_{ij}^{k} \le Q.x_{ij}^{k} \qquad \forall k = 1,\ldots,K \tag{10}$$

$$\sum_{\substack{j=0 \\ i \neq j}}^{n} q_{ij}^{k} - \sum_{\substack{j=0 \\ i \neq j}}^{n} q_{ji}^{k} = D_i \qquad \forall i = 1, \ldots, n \tag{11}$$

The $f$ function is related to minimize total traveling cost from $i$ and $j$can be expressed as $c_{ij} = c_0 \times d_{ij}$, where $c_0$ is the unit fuel cost from $i$ to $j$. About the $g$ function, it is to minimize the sum of the total vehicle emitted $CO_2$ emission level considering as a green VRP. The mathematical model has two types of constraint set which are routing and capacity. Constraints (4–9) are related to routing, Constraints (4–7) of them have ensured that each vehicle tour begins and ends at the depot. Constraint (8) also guarantees that each node (except the depot) is visited by a single vehicle and by Constraint (9) each node is linked only with a pair of nodes also except the depot which respects the Kirchhoff Law. Constraints (10 and 11) are the capacity constraints which ensure that no vehicle can be over-loaded and limits the maximal load when $x_{ij}^{k} = 0$ respectively. As another feature of Constraint (11), it indicates the reduced cargo of the vehicle and also doesn't permit any illegal sub-tours.

## 3.2 *Proposed G-VRP Model*

The proposed G-VRP model contains some differences compared to the classical model described earlier. First, a new emission calculation equation for the $G$ function is proposed. Accordingly, the weight of the vehicle's fuel deposit will decrease as the vehicle travels, thus changing the overall weight of the vehicle. As a result, it is predicted that the amount of emitted emissions will also vary. The new emission equation is called $G$ and it is given Eq. (12). In Eq. (12), an assumption of fuel consumption for a truck per km is averagely obtained as 0.3 L from the actual manufacturer's technical datasets. Generally, this situation is also an assumption of the mathematical models. However, using this approach can be founded in some studies with different ways (Xiao et al. 2012; Koç and Karaoglan 2016).

$$\min CO_2Emission(G) = \sum_{i=0}^{n}\sum_{j=0}^{n}\sum_{k=1}^{K} d_{ij}.\left[\left(\frac{e_{f-e_e}}{Q}\right).q_{ij}^{k} - 0.3\left(\frac{e_{f-e_e}}{Q}\right) + e_e\right] \tag{12}$$

Then, the second important difference of the proposed models in terms of objective functions is the minimization the distance instead of a cost function which is formulated as:

$$\min ObjFunc\,(F) = \sum_{i=0}^{n}\sum_{j=0}^{n}\sum_{k=1}^{K} d_{ij}^{k} \tag{13}$$

### 3.3 Convex Composition Model

In fact, the proposed G-VRP mathematical model also has a bi-objective form. At this phase, a convex composition approach is applied in order to obtain an aggregation objective function as a single objective function that is given as follows:

$$\min ObjFunc\,(H) = \frac{1}{3}F + \frac{2}{3}G \tag{14}$$

According to $H$ function, it composes with two objective functions that the extended objective function ($G$) of emitted $CO_2$ emission level affects twice as much than the extended objective function ($F$) of total distance cost.

## 4 Solution Algorithm: Simulated Annealing

The determination of the optimal solution for the VRP using analytical methods is a difficult task. Depending on this feature, metaheuristic methods are generally referred for solving these problems. The purpose of these methods is to investigate the solution space efficiently and to provide useful solutions close to the optimal solution expeditiously. One of the heuristic methods that can be used to solve vehicle routing problems is also SA, first applied with success on the Ising spin glass problem by Kirkpatrick et al. (1983). It is the algorithmic counterpart to this physical annealing process, using the well-known Metropolis algorithm as its inner loop (Johnson et al. 1989). This method has an extensive use in solving advanced optimization problems similar to other metaheuristic algorithms.

The basic of the algorithm is derived from the solid annealing principle. The first step of the method consists of "melting" the system at a high and efficient temperature. When it is heated, an internal particle of solid rise into disordered shape with the high temperature. At the second step, the temperature of the system is reduced until there is no change. At each temperature, particles reach an equilibrium state. At the last, the temperature reaches the ground state in the room and then internal energy is reduced to the minimum (Lin and Fei 2012). The method has four parameters: the initial temperature, the number of solutions to be produced at each temperature, the temperature reduction function, and the stop criterion. The SA based solution algorithm is given as follows in Table 1 which includes Pseudo-Code and the values of the run parameters in detail.

**Table 1** The SA based solution algorithm with proposed $CO_2$ emission model

| |
|---|
| **// Problem Definition** |
| **// Inputs** |
| [ Read Problem ] |
| 1- n customers with position coordinates and demand |
| 2- m vehicles with capacity |
| 3- one depot with position coordinates |
| **// Select Cost Function (Given in Mathematical Definitions)** |
| Cost Function-1 // Bozuekri and Alaoui (2014) Model |
| Cost Function-2 // Proposed Model-1 |
| Cost Function-3 // Proposed Model-2 |
| **// Simulated Annealing Parameters** |
| MaxIt = 1200 // Maximum Number of Iterations |
| MaxIt2 = 80 // Maximum Number of Inner Iterations |
| T0 = 100 // Initial Temperature |
| Alpha = 0.98 // Cooling ratio |
| **// Initialization** |
| **// Create Initial Solution** |
| solutionRandom = CreateRandomSolution(Inputs) |
| Cost, Solution = CostFunction(solutionRandom) |
| **// Update Best Solution Ever Found** |
| BestSol = Cost, Solution |
| **// Array to Hold Best Cost Values** |
| BestCostHistory = CreateZeroVector(MaxIt) |
| **// Set Initial Temperature** |
| T = T0 |
| **// Simulated Annealing Main Loop** |
| for it = 1:MaxIt |
| for it2 = 1:MaxIt2 |
| **// Create Neighborhoods** |
| newSolution = Shake(solutionRandom) **// Create Neighbors** |
| newCost, newSolution = CostFunction(newSolution) **// Calculate Solutions Costs** |
| if newCost<=Cost |
| **// new is better, so it is accepted** |
| solution = newSolution |
| Else |
| **// new is not better, so it is accepted conditionally** |
| delta = newCost-Cost |
| p = exp(-delta/T) |
| if rand<=p |
| solution = newSolution |
| End |
| End |

(continued)

**Table 1** (continued)

| |
|---|
| **// Update Best Solution** |
| if Cost<=BestSolutionCost |
| BestSolution = newSolution |
| End |
| End |
| **// Store Best Cost** |
| BestCost[ it ] = BestSolutionCost |
| **// Display Iteration Information** |
| iterationNumber, BestCost |
| **// Reduce Temperature** |
| T = alpha*T |
| End |
| **// Results** |
| [ Outputs ] |
| 1- Minimum Distance |
| 2- Minimum CO2 (Xiao et al. 2012; Bouzekri and Alaoui 2014) |
| 3- Minimum CO2 (Proposed Model) |
| 4- Routes |
| 5- Solution Graph |
| 6- Performance Graph |

# 5 Experimental Study

This section describes computational experiments obtained to investigate the performance of the proposed model by using SA algorithm. The algorithm was coded in MATLAB® and run on a computer with i7 2.9 GHz CPU and 8 GB RAM. In this study, three models are used for the instances which are C1–C14 from Christofides et al. (1979) and Set A–B–P from Augerat (1995) as CVRPs.

The investigated models are:

- **Model 1:** GVRP Model (Xiao et al. 2012; Bouzekri and Alaoui 2014)
- **Model 2:** Proposed GVRP Model
- **Model 3:** Convex Composition Model

First of all, in order to show the efficiency of the proposed G-VRP models, an illustrative small example is given in below by Jaramillo (2011). The small instance considers ten customers that will be served from a single depot and the usage of two vehicles, each with a capacity of 12 tons, and a curb weight of 8 tons. Table 2 includes depot and customer locations coordinates, customer demands in tons, and distances between locations in miles.

The results obtained for the sample data set detailed above are summarized in Table 3. In addition, the graphics of solution performance and routes are combined as in Fig. 2 for visualizing the small G-VRP's best solutions from Model 3 in Table 3.

**Table 2** Small VRP instance dataset (Jaramillo 2011)

| | Coordinates | | | $d_{ij}$ | | | | | | | | | | |
|---|---|---|---|---|---|---|---|---|---|---|---|---|---|---|
| L | x | y | $q_{ij}$ | 0 | 1 | 2 | 3 | 4 | 5 | 6 | 7 | 8 | 9 | 10 |
| 0 | 15 | 6 | 0 | 0 | 84 | 75 | 90 | 23 | 59 | 57 | 42 | 33 | 100 | 74 |
| 1 | 85 | 54 | 2 | | 0 | 50 | 43 | 61 | 61 | 39 | 43 | 61 | 23 | 81 |
| 2 | 90 | 4 | 2 | | | 0 | 89 | 54 | 89 | 66 | 42 | 72 | 73 | 111 |
| 3 | 56 | 87 | 3 | | | | 0 | 74 | 38 | 34 | 63 | 57 | 31 | 48 |
| 4 | 37 | 15 | 1 | | | | | 0 | 51 | 40 | 18 | 25 | 79 | 71 |
| 5 | 24 | 65 | 2 | | | | | | 0 | 24 | 50 | 27 | 63 | 21 |
| 6 | 46 | 54 | 3 | | | | | | | 0 | 30 | 25 | 46 | 46 |
| 7 | 53 | 24 | 3 | | | | | | | | 0 | 30 | 62 | 71 |
| 8 | 26 | 38 | 2 | | | | | | | | | 0 | 71 | 45 |
| 9 | 86 | 77 | 4 | | | | | | | | | | 0 | 78 |
| 10 | 8 | 80 | 1 | | | | | | | | | | | 0 |

According to results presented in Table 3, two $CO_2$ emission calculation methods are used for the comparison of the Models 1–3. One of them is developed by Bouzekri and Alaoui (2014) and the other one is proposed by this study in Eq. (12). In fact, Eq. (12) is also used as an objective function for the Model 2 and 3. However, it is to be an emission calculation method that is used in Bouzekri and Alaoui (2014) model (Model 1) within SA algorithm.

Computational results of the problems were compared using One-way ANOVA and Tukey tests. The averages of the statistically significant results were tested with the Tukey Test and the differences between the different models were investigated. At first, the box-plot graphic is given for the small G-VRP example in Fig. 3 and Differences between models were tested by one-way ANOVA that is given in Table 4. The hypotheses established for the analysis are presented below. $H_0$ hypothesis is that the averages of the models are equal to each other and, the $H_1$ hypothesis is that the averages of the models are not equal to each other and they are given Eqs. (14 and 15).

$$H_0 : \mu_1 = \mu_2 = \mu_3 \tag{14}$$

$$H_1 : \mu_1 \neq \mu_2 \neq \mu_3 \tag{15}$$

According to the Table 4, the difference between the models for all parameters is statistically significant for the small G-VRP problem. Before testing whether the difference between the models is statistically significant, the Bartlett test was used to test whether the variances of the models were homogeneous. The Bartlett's test results for the Small G-VRP problem are given in Table 5 .

$$H_0 : \sigma_1 = \sigma_2 = \sigma_3 \tag{16}$$

$$H_1 : \sigma_1 \neq \sigma_2 \neq \sigma_3 \tag{17}$$

**Table 3** SA solution results for small G-VRP

| | Model 1 | | | Model 2 | | | Model 3 | | |
|---|---|---|---|---|---|---|---|---|---|
| | Dist (miles) | $CO_2$ (kg) | Time (s) | Dist (miles) | $CO_2$ (kg) | Time (s) | Dist (miles) | $CO_2$ (kg) | Time (s) |
| Best | 395.66 | 356.61 | 5.58 | 392.55 | 356.25 | 5.48 | 392.55 | 352.36 | 5.80 |
| Worst | 500.72 | 450.05 | 5.52 | 459.76 | 408.97 | 5.55 | 430.45 | 386.20 | 5.55 |
| Average | 437.60 | 394.25 | 5.57 | 426.02 | 383.83 | 5.54 | 406.27 | 365.38 | 5.68 |
| B & A (2014)[a] | | 397.08 | | | 386.59 | | | 368.03 | |
| Emission savings (kg) | | 2.83 | | | 2.77 | | | 2.65 | |

[a]Bozuekri and Alaoui (2014)

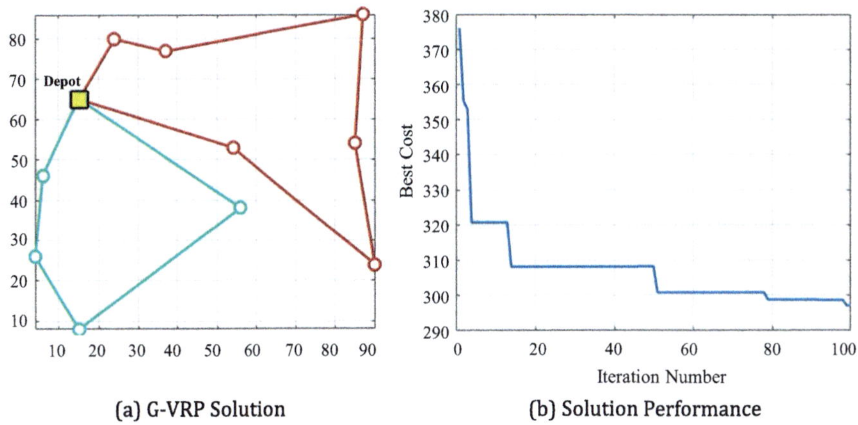


**Fig. 2** The best solution for small G-VRP instance

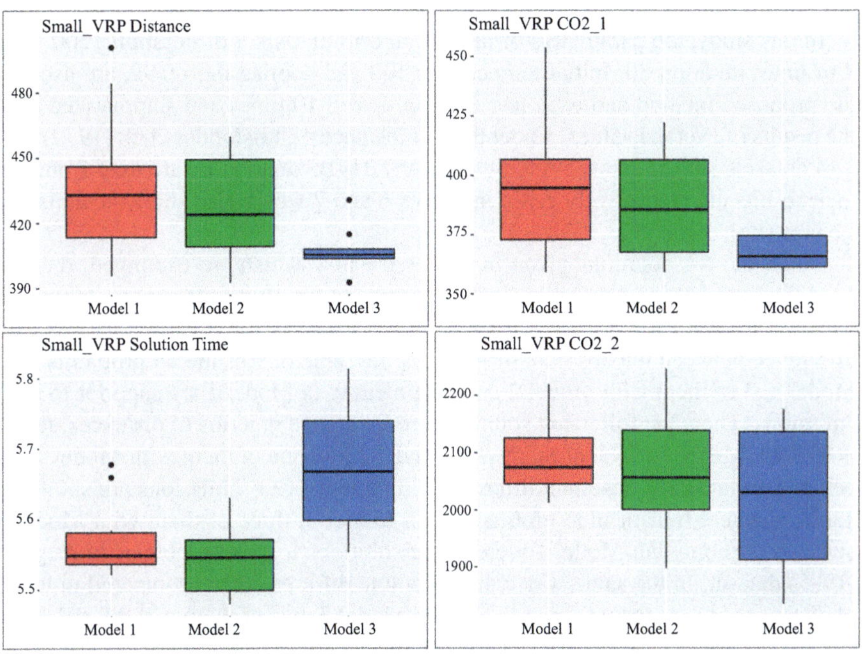


**Fig. 3** Box-plot for the small G-VRP instance

As can be seen from Table 5, the hypothesis that the variances of the model for distance, $CO_2$_1, and $CO_2$_2 are homogeneous is rejected. The hypothesis that the variances are homogeneous for the solution time (Sol_Time) is accepted.

**Table 4** Results of one-way ANOVA

| Problem name | Parameter | $df_1$ | $df_2$ | F value | P(>F) | alfa | F-Crt | $H_0$ |
|---|---|---|---|---|---|---|---|---|
| Small_G-VRP | Distance | 2 | 27 | 4.047 | 0.029 | 0.05 | 4.242 | Reject[a] |
| Small_G-VRP | $CO_2$_1 | 2 | 27 | 4.161 | 0.027 | 0.05 | 4.242 | Reject[a] |
| Small_G-VRP | $CO_2$_2 | 2 | 27 | 4.193 | 0.026 | 0.05 | 4.242 | Reject[a] |
| Small_G-VRP | Sol_Time | 2 | 27 | 11.195 | 0.000 | 0.05 | 4.242 | Reject[a] |

[a]Hypothesis was rejected significantly in two-tailed test at the 0.05 level

**Table 5** Variance homogeneity test results (Bartlett test)

| Problem | Parameter | df | Chisq value | P value | alfa | Chisq-crt | H0 |
|---|---|---|---|---|---|---|---|
| Small_G-VRP | Distance | 2 | 9.108 | 0.011 | 0.05 | 5.991 | Reject[a] |
| Small_G-VRP | $CO_2$_1 | 2 | 8.623 | 0.013 | 0.05 | 5.991 | Reject[a] |
| Small_G-VRP | $CO_2$_2 | 2 | 8.625 | 0.013 | 0.05 | 5.991 | Reject[a] |
| Small_G-VRP | Sol_Time | 2 | 4.243 | 0.12 | 0.05 | 5.991 | Not reject |

[a]Hypothesis was rejected significantly in two-tailed test at the 0.05 level

In this study, the parameters of the SA algorithm, max # of iteration 1200, max # of inner iteration 80, Initial temperature 100 and cooling ratio 0.99, are used on the proposed method and each test instance is run 10 times and summarized with the results of average values. About the test instances, Christofides et al. (1979) C1–C14 datasets and Augerat (1995) Set A1–A3, B1–B3 and P1–P3 are used from and their results are respectively given in Tables 6 and 7 which also show the emission savings (kg).

When the SA solutions given in Tables 6 and 7 which are examined, there is no difference between solution times in terms of absolute differences. However, when examined from the point of view of $CO_2$ emissions and distances, there is a difference between the absolute differences. In Table 6, 9 of the 14 problems with Model 3, 3 of them with Model 2 and 2 of them with Model 1 are superior to $CO_2$ emissions. Then, the following solutions are compared in terms of distances: ten of them with Model 3, three of them with Model 2 and one of them with Model 1 are better. Similarly, the absolute differences in terms of $CO_2$ emissions are shown in Table 7, where four of nine problems with Model 3, three of them with Model 2 and two of them with Model 1 were better solutions for nine problems in Augerat (1995) dataset. In the same way, when absolute differences are compared in terms of distances, it is seen that six of them with Model 3, two of them with Model 1 and one of them with Model 1 have better solutions, respectively.

The "t-test" is used to check whether the difference between the models' averages is statistically significant. The hypothesizes are given below and the results presented in Table 8.

$$H_0 : \mu_1 = \mu_2 \tag{18}$$

$$H_1 : \mu_1 \neq \mu_2 \tag{19}$$

**Table 6** Computational results for Christofides et al. (1979) C1–C14 instances

| Data sets | | Model 1 | | | Model 2 | | | Model 3 | | |
|---|---|---|---|---|---|---|---|---|---|---|
| | | Dist (miles) | $CO_2$ (kg) | Time (s) | Dist (miles) | $CO_2$ (kg) | Time (s) | Dist (miles) | $CO_2$ (kg) | Time (s) |
| C1 | Average | 595.20 | 538.99 | 60.47 | 585.84 | 529.20 | 60.38 | 577.19 | 524.29 | 60.60 |
| | B & A (2014)[a] | | 539.30 | | | 529.50 | | | 524.59 | |
| | Emission savings (kg) | | 0.31 | | | 0.30 | | | 0.30 | |
| C2 | Average | 961.32 | 864.69 | 75.30 | 973.10 | 875.37 | 75.36 | 968.99 | 872.36 | 75.42 |
| | B & A (2014) | | 865.22 | | | 875.90 | | | 872.90 | |
| | Emission savings (kg) | | 0.53 | | | 0.53 | | | 0.54 | |
| C3 | Average | 921.31 | 825.79 | 72.44 | 936.47 | 836.38 | 70.57 | 908.03 | 815.51 | 70.79 |
| | B & A (2014) | | 826.18 | | | 836.77 | | | 815.89 | |
| | Emission savings (kg) | | 0.39 | | | 0.39 | | | 0.38 | |
| C4 | Average | 1333.26 | 1197.51 | 82.21 | 1312.29 | 1177.59 | 82.42 | 1243.68 | 1124.86 | 82.59 |
| | B & A (2014) | | 1198.05 | | | 1178.13 | | | 1125.38 | |
| | Emission savings (kg) | | 0.54 | | | 0.54 | | | 0.54 | |
| C5 | Average | 2294.22 | 2084.83 | 106.30 | 2218.40 | 2019.73 | 106.40 | 2255.39 | 2058.43 | 107.00 |
| | B & A (2014) | | 2085.79 | | | 2020.67 | | | 2059.39 | |
| | Emission savings (kg) | | 0.96 | | | 0.94 | | | 0.96 | |
| C6 | Average | 584.88 | 529.13 | 61.02 | 580.92 | 525.20 | 60.80 | 602.37 | 545.54 | 61.05 |
| | B & A (2014) | | 529.43 | | | 525.47 | | | 545.85 | |
| | Emission savings (kg) | | 0.30 | | | 0.27 | | | 0.31 | |
| C7 | Average | 990.88 | 888.61 | 75.09 | 982.81 | 884.25 | 74.92 | 940.85 | 848.44 | 75.09 |
| | B & A (2014) | | 889.15 | | | 899.23 | | | 848.97 | |
| | Emission savings (kg) | | 0.54 | | | 0.94 | | | 0.53 | |
| C8 | Average | 921.31 | 825.79 | 72.03 | 936.47 | 836.38 | 70.07 | 908.03 | 815.51 | 69.88 |
| | Bouzekri and Alaoui (2014) | | 826.18 | | | 836.77 | | | 815.89 | |
| | Emission savings (kg) | | 0.39 | | | 0.39 | | | 0.38 | |

(continued)

**Table 6** (continued)

| Data sets | | Model 1 | | | Model 2 | | | Model 3 | | |
|---|---|---|---|---|---|---|---|---|---|---|
| | | Dist (miles) | $CO_2$ (kg) | Time (s) | Dist (miles) | $CO_2$ (kg) | Time (s) | Dist (miles) | $CO_2$ (kg) | Time (s) |
| C9 | Average | 1336.72 | 1203.06 | 81.93 | 1342.86 | 1205.94 | 81.87 | 1260.15 | 1138.32 | 81.62 |
| | B & A (2014) | | 1203.61 | | | 1206.48 | | | 1138.84 | |
| | Emission savings (kg) | | 0.55 | | | 0.54 | | | 0.52 | |
| C10 | Average | 2288.61 | 2089.01 | 93.21 | 2265.09 | 2063.65 | 92.82 | 2212.79 | 2020.00 | 93.41 |
| | B & A (2014) | | 2089.98 | | | 2064.61 | | | 2020.93 | |
| | Emission savings (kg) | | 0.97 | | | 0.96 | | | 0.93 | |
| C11 | Average | 1520.22 | 1363.53 | 69.08 | 1459.61 | 1313.56 | 69.74 | 1418.97 | 1285.53 | 71.11 |
| | B & A (2014) | | 1364.14 | | | 1314.15 | | | 1286.12 | |
| | Emission savings (kg) | | 0.61 | | | 0.59 | | | 0.59 | |
| C12 | Average | 923.56 | 829.77 | 75.57 | 963.45 | 861.71 | 76.12 | 916.34 | 823.85 | 75.21 |
| | B & A (2014) | | 830.11 | | | 862.06 | | | 824.17 | |
| | Emission savings (kg) | | 0.34 | | | 0.35 | | | 0.32 | |
| C13 | Average | 1461.97 | 1320.24 | 69.22 | 1432.85 | 1291.66 | 69.49 | 1360.28 | 1233.39 | 69.27 |
| | B & A (2014) | | 1320.85 | | | 1292.25 | | | 1233.96 | |
| | Emission savings (kg) | | 0.61 | | | 0.59 | | | 0.57 | |
| C14 | Average | 924.96 | 832.06 | 75.72 | 923.27 | 829.35 | 75.94 | 960.46 | 863.62 | 76.00 |
| | B & A (2014) | | 832.38 | | | 829.68 | | | 863.98 | |
| | Emission savings (kg) | | 0.28 | | | 0.33 | | | 0.36 | |

[a] B & A (2014): Bouzekri and Alaoui (2014)

**Table 7** Computational results for Augerat (1995) A1–A3, B1–B3 and P1–P3 instances

| Data sets | | Model 1 | | | Model 2 | | | Model 3 | | |
|---|---|---|---|---|---|---|---|---|---|---|
| | | Dist (miles) | $CO_2$ (kg) | Time (s) | Dist (miles) | $CO_2$ (kg) | Time (s) | Dist (miles) | $CO_2$ (kg) | Time (s) |
| A1An37k5 | Average | 710.81 | 623.95 | 60.00 | 705.89 | 617.40 | 60.02 | 701.04 | 620.27 | 59.95 |
| | B & A (2014)[a] | | 624.50 | | | 617.94 | | | 620.78 | |
| | Emission Savings (kg) | | 0.55 | | | 0.54 | | | 0.61 | |
| A2An54k7 | Average | 1277.16 | 1145.31 | 65.46 | 1263.10 | 1129.68 | 66.91 | 1246.26 | 1118.18 | 68.72 |
| | B & A (2014) | | 1146.26 | | | 1130.60 | | | 1119.10 | |
| | Emission Savings (kg) | | 0.95 | | | 0.92 | | | 0.92 | |
| A3An80k10 | Average | 1966.76 | 1752.30 | 76.24 | 1950.21 | 1733.67 | 77.47 | 1957.22 | 1749.76 | 77.21 |
| | B & A (2014) | | 1753.60 | | | 1734.96 | | | 1751.09 | |
| | Emission Savings (kg) | | 1.30 | | | 1.29 | | | 1.33 | |
| B1Bn35k5 | Average | 986.23 | 865.08 | 62.57 | 987.32 | 868.30 | 62.36 | 988.45 | 871.77 | 62.22 |
| | B & A (2014) | | 865.68 | | | 868.90 | | | 872.37 | |
| | Emission Savings (kg) | | 0.60 | | | 0.60 | | | 0.60 | |
| B2Bn52k7 | Average | 697.75 | 623.61 | 67.84 | 700.88 | 625.56 | 67.44 | 705.85 | 632.11 | 67.54 |
| | B & A (2014) | | 624.03 | | | 625.98 | | | 632.54 | |
| | Emission Savings (kg) | | 0.42 | | | 0.42 | | | 0.43 | |
| B3Bn78k10 | Average | 1336.14 | 1186.99 | 76.48 | 1349.33 | 1197.03 | 76.35 | 1330.05 | 1181.73 | 76.49 |
| | B & A (2014) | | 1187.85 | | | 1197.91 | | | 1182.59 | |
| | Emission Savings (kg) | | 0.86 | | | 0.88 | | | 0.86 | |
| P1Pn16k8 | Average | 452.90 | 373.23 | 67.78 | 454.10 | 372.66 | 68.05 | 452.84 | 373.84 | 67.63 |
| | B & A (2014) | | 373.93 | | | 373.34 | | | 374.55 | |
| | Emission Savings (kg) | | 0.70 | | | 0.68 | | | 0.71 | |
| P2Pn51k10 | Average | 807.97 | 718.96 | 78.44 | 809.37 | 718.79 | 77.13 | 797.85 | 711.12 | 77.20 |
| | B & A (2014) | | 719.69 | | | 719.50 | | | 711.84 | |
| | Emission Savings (kg) | | 0.73 | | | 0.71 | | | 0.72 | |
| P3Pn101k4 | Average | 773.53 | 695.97 | 63.82 | 774.32 | 695.47 | 63.91 | 754.97 | 682.68 | 65.14 |
| | B & A (2014) | | 696.15 | | | 695.65 | | | 682.85 | |
| | Emission Savings (kg) | | 0.18 | | | 0.18 | | | 0.17 | |

[a] B & A (2014): Bouzekri and Alaoui (2014)

**Table 8** t-Test results

| Parameters | Means | Estimate | t-stat | df | p Value | Result | CI %95 |
|---|---|---|---|---|---|---|---|
| Distance | Model 1–Model 2 | 11.581 | 0.877 | 17 | 0.393 | Not reject | (−16.303; 39.466) |
| | Model 1–Model 3 | 31.329 | 2.827 | 11 | 0.017 | Reject[a] | (6.908; 55.75) |
| | Model 2–Model 3 | 19.748 | 2.281 | 12 | 0.041 | Not reject | (0.923; 38.573) |
| $CO_2$_1 | Model 1–Model 2 | 10.484 | 0.869 | 17 | 0.397 | Not reject | (−15.002; 35.97) |
| | Model 1–Model 3 | 29.049 | 2.845 | 11 | 0.016 | Reject[a] | (6.582; 51.516) |
| | Model 2–Model 3 | 18.565 | 2.35 | 13 | 0.036 | Not reject | (1.432; 35.698) |
| $CO_2$_2 | Model 1–Model 2 | 10.423 | 0.873 | 17 | 0.395 | Not reject | (−14.808; 35.653) |
| | Model 1–Model 3 | 28.873 | 2.859 | 11 | 0.016 | Reject[a] | (6.65; 51.095) |
| | Model 2–Model 3 | 18.565 | 2.35 | 13 | 0.036 | Not reject | (1.432; 35.698) |
| Solution time | Model 1–Model 2 | 0.03 | 1.306 | 18 | 0.208 | Not reject | (−0.018; 0.079) |
| | Model 1–Model 3 | −0.107 | −3.145 | 15 | 0.007 | Reject[a] | (−0.179; −0.034) |
| | Model 2–Model 3 | −0.137 | −4.155 | 18 | 0.001 | Reject[a] | (−0.206; −0.068) |

[a]Hypothesis was rejected significantly in two-tailed test at the 0.05 level

According to the test results, the hypothesis that the mean values of objective functions are equal for distance, $CO_2$_1, $CO_2$_2 for Model 1 and Model 2, and Model 2 and Model 3 is accepted. On the other hand, the difference between Model 1 and Model 3 for these three objective functions was found to be statistically significant at the 5% significance level for the two-tailed test. The difference between Model 1and Model 3, and Model 2 and Model 3 for solution time is found to be statistically significant at the 5% significance level and two-tailed test. The $H_0$ hypothesis cannot be rejected for the difference between Model 1 and Model 2. Consequently, the proposed G-VRP model is statistically significant and has more efficient solutions. In addition, it can be predicted that it provides some opportunities for challenges on the green supply chain and encourages the researchers and industrial practitioners.

## 6 Conclusions and Further Research

The results of the research conducted were analyzed in terms of absolute differences and statistical analysis. Two alternative models are proposed in the literature. The Model 1 based on the literature is obtained by adding the effect of fuel consumption as an extension of Model 2 from the proposed models. The Model 3 is obtained by convex composition of two objective functions based on distance minimization with Model 2.

The C1–C14 instances from Christofides et al. (1979) and set A, B and P instances from Augerat (1995) were solved with SA algorithm. Each problem was run 10 times and the best, worst and average solutions were summarized. The results are compared with other solutions in the literature especially Bouzekri and Alaoui (2014) solutions.

The solutions of Model 1, Model 2 and Model 3 with SA were statistically analyzed in detail with absolute values. In absolute difference analysis, a total of 23 problems were compared with respect to $CO_2$ emissions. Thirteen of them with Model 3, 6 of them with Model 2 and 4 of them with Model 1 were better solutions. About the total distances, 16 of them with Model 3, 4 of them with Model 2 and 3 of them with Model 1 were the best solutions.

In this study, the analysis was made using theoretical VRP test problems. However, the proposed model for G-VRP is tried to be solved depending on vehicle load and total distance relation. It is our expectation that this objective function is considered to provide a more efficient $CO_2$ emission minimization on real industrial problems and the proposed models are suggested to address green thinking to the researchers. As a further research, the load effect of these models in real life problems, such as slope, altitude, load intensity, etc can be applied as different extensions of G-VRP.

## References

Alkawaleet, N., Hsieh, Y. F., & Wang, Y. (2014). Inventory routing problem with CO2 emissions consideration. In *Logistics operations, supply chain management and sustainability* (pp. 611–619). Cham: Springer.

Alvarenga, G. B., Mateus, G. R., & De Tomi, G. (2007). A genetic and set partitioning two-phase approach for the vehicle routing problem with time windows. *Computers and Operations Research, 34*(6), 1561–1584.

Andelmin, J., & Bartolini, E. (2017). An exact algorithm for the green vehicle routing problem. *Transportation Science, 51*(4), 1288–1303.

Aranda, U. A., Ferreira, G., Bribián, Z. I., & Vásquez, Z. D. (2012). Study of the environmental performance of end-of-life tyre recycling through a simplified mathematical approach. *Thermal Science, 16*(3), 889–899.

Augerat, P. (1995). VRP problem instances set A-B-P. Retrieved July 27, 2017, from http://vrp.atd-lab.inf.puc-rio.br/index.php/en/

Ayadi, R., ElIdrissi, A. E., Benadada, Y., & Alaoui, A. E. H. (2014). Evolutionary algorithm for a green vehicle routing problem with multiple trips. In *2014 International conference on logistics and operations management (GOL)* (pp. 148–154). IEEE.

Aydemir, H., & Cubuk, M. K. (2016). The investigation of the general situation of highways in turkey with recent changes and recommendations on future strategies. *Gazi Journal of Engineering Sciences, 2*(3), 129–146.

Barnhart, C., & Laporte, G. (Eds.). (2006). *Handbooks in operations research and management science: Transportation* (Vol. 14). Amsterdam: Elsevier.

Bektas, T. (2017). *Freight transport and distribution: Concepts and optimisation models*. Boca Raton, FL: CRC Press.

Bektaş, T., & Laporte, G. (2011). The pollution-routing problem. *Transportation Research Part B: Methodological, 45*(8), 1232–1250.

Bouzekri, E. E. A., Aahmed, E. A., & Youssef, B. (2014). The green capacitated vehicle routing problem: Optimizing of emissions of greenhouse gas. In *2014 International conference on logistics and operations management (GOL)* (pp. 161–167). IEEE.

Bouzekri, E. A., Elhassania, M. E. S. S. O. U. D., & Alaoui, A. E. H. (2013). A hybrid ant colony system for green capacitated vehicle routing problem in sustainable transport. *Journal of Theoretical and Applied Information Technology, 53*(2), 198–208.

Bouzekri, E. A., & Alaoui, A. E. H. (2014). A hybrid ant colony system for green capacitated vehicle routing problem in sustainable transport. *International Journal of Scientific and Engineering Research, 5*(9), 70–77.

Bräysy, O., & Gendreau, M. (2005). Vehicle routing problem with time windows, part I: Route construction and local search algorithms. *Transportation Science, 39*(1), 104–118.

Cacchiani, V., Hemmelmayr, V. C., & Tricoire, F. (2014). A set-covering based heuristic algorithm for the periodic vehicle routing problem. *Discrete Applied Mathematics, 163*, 53–64.

Cetin, S., & Gencer, C. (2010). Vehicle routing problems with hard time windows and simultaneous pick up and delivery: A mathematical model. *Journal of the Faculty of Engineering and Architecture of Gazi University, 25*(3), 579–585.

Christofides, N., Mingozzi, A., & Toth, P. (1979). The vehicle routing problem. In N. Christofides, A. Mingozzi, P. Toth, & C. Sandi (Eds.), *Combinatorial optimization* (pp. 315–338). Chichester: Wiley.

Clarke, G., & Wright, J. W. (1964). Scheduling of vehicles from a central depot to a number of delivery points. *Operations Research, 12*(4), 568–581.

Cooray, P. L. N. U., & Rupasinghe, T. D. (2017). Machine learning-based parameter tuned genetic algorithm for energy minimizing vehicle routing problem. *Journal of Industrial Engineering, 2017*, 13.

Dantzig, G. B., & Ramser, J. H. (1959). The truck dispatching problem. *Management Science, 6*(1), 80–91.

Demir, E., Bektaş, T., & Laporte, G. (2014). A review of recent research on green road freight transportation. *European Journal of Operational Research, 237*(3), 775–793.

Eglese, R., & Bektaş, T. (2014). Chapter 15: Green vehicle routing. In P. Toth & D. Vigo (Eds.), *Vehicle routing: Problems, methods, and applications* (2nd ed., pp. 437–458). Philadelphia, PA: Society for Industrial and Applied Mathematics.

Ene, S., Küçükoğlu, I., Aksoy, A., & Öztürk, N. (2016). A hybrid metaheuristic algorithm for the green vehicle routing problem with a heterogeneous fleet. *International Journal of Vehicle Design, 71*(1–4), 75–102.

Erdoğan, S., & Miller-Hooks, E. (2012). A green vehicle routing problem. *Transportation Research Part E: Logistics and Transportation Review, 48*(1), 100–114.

Faulin, J., Juan, A., Lera, F., & Grasman, S. (2011). Solving the capacitated vehicle routing problem with environmental criteria based on real estimations in road transportation: A case study. *Procedia-Social and Behavioral Sciences, 20*, 323–334.

Franceschetti, A., Honhon, D., Van Woensel, T., Bektaş, T., & Laporte, G. (2013). The time-dependent pollution-routing problem. *Transportation Research Part B: Methodological, 56*, 265–293.

Ganesh, K., & Narendran, T. T. (2007). CLASH: A heuristic to solve vehicle routing problems with delivery, pick-up and time windows. *International Journal of Services and Operations Management, 3*(4), 460–477.

Gendreau, M., Hertz, A., & Laporte, G. (1994). A tabu search heuristic for the vehicle routing problem. *Management Science, 40*(10), 1276–1290.

Gribkovskaia, I., Laporte, G., & Shyshou, A. (2008). The single vehicle routing problem with deliveries and selective pickups. *Computers and Operations Research, 35*(9), 2908–2924.

Hassel, H., & Samaras, J. (1999). *Methodology for calculating transport emissions and energy consumption (report for the projet MEET)*. Edinburgh: Transport Research Laboratory.

Hsueh, C. F. (2016). The green vehicle routing problem with stochastic travel speeds. In *CICTP 2016* (pp. 1–12).

Huang, Y., Shi, C., Zhao, L., & Van Woensel, T. (2012). A study on carbon reduction in the vehicle routing problem with simultaneous pickups and deliveries. In *2012 IEEE international conference on service operations and logistics, and informatics (SOLI)* (pp. 302–307). IEEE.

Jabali, O., Woensel, T., & De Kok, A. G. (2012). Analysis of travel times and CO2 emissions in time-dependent vehicle routing. *Production and Operations Management, 21*(6), 1060–1074.

Jaramillo, J. R. (2011). The green vehicle routing. In *Proceedings, informs annual meeting*, October 5–7, Myrtle Beach, South Carolina, USA.

Jemai, J., Zekri, M., & Mellouli, K. (2012). An NSGA-II algorithm for the green vehicle routing problem. In *Evolutionary computation in combinatorial optimization* (pp. 37–48). Berlin: Springer.

Johnson, D. S., Aragon, C. R., McGeoch, L. A., & Schevon, C. (1989). Optimization by simulated annealing: An experimental evaluation; part I, graph partitioning. *Operations Research, 37*(6), 865–892.

Kara, I., Kara, B., & Yetis, M. K. (2007). Energy minimizing vehicle routing problem. In *Combinatorial optimization and applications* (pp. 62–71). Berlin: Springer.

Kirkpatrick, S., Gelatt, C. D., & Vecchi, M. P. (1983). Optimization by simulated annealing. *Science, 220*(4598), 671–680.

Koç, Ç., & Karaoglan, I. (2016). The green vehicle routing problem: A heuristic based exact solution approach. *Applied Soft Computing, 39*, 154–164.

Koç, Ç., Bektaş, T., Jabali, O., & Laporte, G. (2014). The fleet size and mix pollution-routing problem. *Transportation Research Part B: Methodological, 70*, 239–254.

Kramer, R., Subramanian, A., Vidal, T., & Lucídio dos Anjos, F. C. (2015). A matheuristic approach for the pollution-routing problem. *European Journal of Operational Research, 243*(2), 523–539.

Kuo, R. J., & Lin, L. M. (2010). Application of a hybrid of genetic algorithm and particle swarm optimization algorithm for order clustering. *Decision Support Systems, 49*(4), 451–462.

Kuo, Y. (2010). Using simulated annealing to minimize fuel consumption for the time-dependent vehicle routing problem. *Computers and Industrial Engineering, 59*(1), 157–165.
Küçükoğlu, İ., & Öztürk, N. (2015). An advanced hybrid meta-heuristic algorithm for the vehicle routing problem with backhauls and time windows. *Computers and Industrial Engineering, 86*, 60–68.
Kwon, Y. J., Choi, Y. J., & Lee, D. H. (2013). Heterogeneous fixed fleet vehicle routing considering carbon emission. *Transportation Research Part D: Transport and Environment, 23*, 81–89.
Li, J. (2012). Vehicle routing problem with time windows for reducing fuel consumption. *Journal of Computers, 7*(12), 3020–3027.
Lin, C., Choy, K. L., Ho, G. T., Chung, S. H., & Lam, H. Y. (2014). Survey of green vehicle routing problem: Past and future trends. *Expert Systems with Applications, 41*(4), 1118–1138.
Lin, L., & Fei, C. (2012). The simulated annealing algorithm implemented by the MATLAB. *International Journal of Computer Science Issues (IJCSI), 9*(6), 357–360.
Maden, W., Eglese, R., & Black, D. (2010). Vehicle routing and scheduling with time-varying data: A case study. *Journal of the Operational Research Society, 61*(3), 515–522.
McKinnon, A. (2010). Environmental sustainability. In *Green logistics: Improving the environmental sustainability of logistics*. London.
Ohlmann, J. W., & Thomas, B. W. (2007). A compressed-annealing heuristic for the traveling salesman problem with time windows. *INFORMS Journal on Computing, 19*(1), 80–90.
Oliveira, P. R. D. C., Mauceri, S., Carroll, P., & Pallonetto, F. (2017). A genetic algorithm for a green vehicle routing problem. In *International network optimization conference 2017 (INOC 2017)*, Lisboa, Portugal, 26–28 February 2017.
Omidvar, A., & Tavakkoli-Moghaddam, R. (2012). Sustainable vehicle routing: Strategies for congestion management and refueling scheduling. In *2012 IEEE international energy conference and exhibition (ENERGYCON)* (pp. 1089–1094). IEEE.
Özyurt, Z., Aksen, D., & Aras, N. (2006). Open vehicle routing problem with time deadlines: Solution methods and an application. In *Operations research proceedings 2005* (pp. 73–78). Berlin: Springer.
Palmer, A. (2007). *The development of an integrated routing and carbon dioxide emissions model for goods vehicles*. PhD thesis, School of Management, Cranfield University, Cranfield.
Pan, S., Ballot, E., & Fontane, F. (2013). The reduction of greenhouse gas emissions from freight transport by pooling supply chains. *International Journal of Production Economics, 143*(1), 86–94.
Park, Y., & Chae, J. (2014). A review of the solution approaches used in recent G-VRP. *International Journal of Advanced Logistics, 3*(1–2), 27–37.
Peiying, Y., Jiafu, T., & Yang, Y. U. (2013). Based on low carbon emissions cost model and algorithm for vehicle routing and scheduling in picking up and delivering customers to airport service. In *2013 25th Chinese control and decision conference (CCDC)* (pp. 1693–1697). IEEE.
Pichpibul, T., & Kawtummachai, R. (2013). A heuristic approach based on clarke-wright algorithm for open vehicle routing problem. *The Scientific World Journal, 2013*, 874349.
Piecyk, M. (2010). Carbon auditing of companies, supply chains and products. In *2010 Green logistics: Improving the environmental sustainability of logistics* (pp. 49–67). Kogan Page.
Ramos, T. R. P., Gomes, M. I., & Barbosa-Póvoa, A. P. (2012). Minimizing CO2 emissions in a recyclable waste collection system with multiple depots. In *EUROMA/POMS joint conference* (pp. 1–5).
Salhi, S., Imran, A., & Wassan, N. A. (2014). The multi-depot vehicle routing problem with heterogeneous vehicle fleet: Formulation and a variable neighborhood search implementation. *Computers and Operations Research, 52*, 315–325.
Suzuki, Y. (2011). A new truck-routing approach for reducing fuel consumption and pollutants emission. *Transportation Research Part D: Transport and Environment, 16*(1), 73–77.
Taha, M., Fors, M. N., & Shoukry, A. A. (2014). An exact solution for a class of green vehicle routing problem. In *International conference on industrial engineering and operations management* (pp. 7–9).

Toro, O., Eliana, M., Escobar, Z., Antonio, H., & Granada, E. (2016). Literature review on the vehicle routing problem in the green transportation context. *Luna Azul, 42*, 362–387.

Treitl, S., Nolz, P. C., & Jammernegg, W. (2014). Incorporating environmental aspects in an inventory routing problem. A case study from the petrochemical industry. *Flexible Services and Manufacturing Journal, 26*(1–2), 143–169.

Tunga, H., Bhaumik, A. K., & Kar, S. (2017). A method for solving bi-objective green vehicle routing problem (g-vrp) through genetic algorithm. *Journal of the Association of Engineers, India, 87*(1–2), 33–48.

Turkish Statistical Institute. (2014). Greenhouse gas emissions inventory, annual statistics, Ankara, Turkey. Retrieved July 21, 2017, from http://www.turkstat.gov.tr/PreHaberBultenleri.do?id=21582

Ubeda, S., Arcelus, F. J., & Faulin, J. (2011). Green logistics at Eroski: A case study. *International Journal of Production Economics, 131*(1), 44–51.

Úbeda, S., Faulin, J., Serrano, A., & Arcelus, F. J. (2014). Solving the green capacitated vehicle routing problem using a tabu search algorithm. *Lecture Notes in Management Science, 6*, 141–149.

Urquhart, N., Scott, C., & Hart, E. (2010). Using an evolutionary algorithm to discover low CO2 tours within a travelling salesman problem. In *European conference on the applications of evolutionary computation* (pp. 421–430). Berlin: Springer.

Vincent, F. Y., Redi, A. P., Hidayat, Y. A., & Wibowo, O. J. (2017). A simulated annealing heuristic for the hybrid vehicle routing problem. *Applied Soft Computing, 53*, 119–132.

Xiao, Y., Zhao, Q., Kaku, I., & Xu, Y. (2012). Development of a fuel consumption optimization model for the capacitated vehicle routing problem. *Computers and Operations Research, 39*(7), 1419–1431.

Yan, S., Chi, C. J., & Tang, C. H. (2006). Inter-city bus routing and timetable setting under stochastic demands. *Transportation Research Part A: Policy and Practice, 40*(7), 572–586.

Yasin, M., & Vincent, F. Y. (2013). A simulated annealing heuristic for the green vehicle routing problem. In *Proceedings of the institute of industrial engineers Asian conference 2013*. Singapore: Springer.

Zhang, Z., Long, K., Wang, J., & Dressler, F. (2014). On swarm intelligence inspired self-organized networking: Its bionic mechanisms, designing principles and optimization approaches. *IEEE Communications Surveys and Tutorials, 16*(1), 513–537.

Zhou, Y., & Lee, G. M. (2017). A lagrangian relaxation-based solution method for a green vehicle routing problem to minimize greenhouse gas emissions. *Sustainability, 9*(5), 776.

# Development of a Web-Based Decision Support System for Strategic and Tactical Sustainable Fleet Management Problems in Intermodal Transportation Networks

**Adil Baykasoğlu, Kemal Subulan, A. Serdar Taşan, Nurhan Dudaklı, Murat Turan, Erdin Çelik, and Özgür Ülker**

## 1 Introduction

In recent years, modeling and solving fleet management problems in intermodal transportation networks have received an increasing attention by both researchers and practitioners in the global logistics sector. On the other hand, there are still a few papers on intermodal fleet management problems when compared to unimodal transportation systems. The fleet management problems need to be solved by handling various complex decisions concurrently such as transportation mode selection, load allocation, outsourcing, type and frequency of maritime and railway transportation services, empty vehicle repositioning, fleet allocation/relocation, fleet expansion/reduction decisions etc. which make the fleet management systems much more complex in intermodal transportation environments. In fact, fleet planning should be addressed in an integrated way with transportation mode selection, load planning, vehicle routing, periodic vehicle inventory control, Ro-Ro and train scheduling and empty vehicle repositioning issues. For this reason, fleet management problems are much more complicated in intermodal transportation systems than unimodal road freight transport which considers only trucking part of the overall transportation system. However, existing studies in the current literature handled these aforementioned problems in a separated way (Baykasoğlu and Subulan 2016; Baykasoğlu et al. 2015). Furthermore, empty vehicle repositioning problem should

A. Baykasoğlu (✉) · K. Subulan · A. Serdar Taşan · N. Dudaklı
Department of Industrial Engineering, Dokuz Eylül University, Izmir, Turkey
e-mail: adil.baykasoglu@deu.edu.tr

M. Turan · E. Çelik · Ö. Ülker
Ekol Logistics Inc., Istanbul, Turkey



T. Paksoy et al. (eds.), *Lean and Green Supply Chain Management*, International Series in Operations Research & Management Science 273,
https://doi.org/10.1007/978-3-319-97511-5_7

also be covered by the fleet management tools because of its positive effects on increases in customer service level by decreasing the awaiting time. Unfortunately, there is a lack of studies on dynamic positioning and repositioning problems of multiple assets or resources such as trucks, trailers, containers etc. Moreover, simultaneous planning of these multiple resources and integration of backward flows (empty truck repositions) into forward flows (loaded truck movements) will provide significant time saving and cost reduction. Additionally, multi-objective intermodal fleet management which considers trade-offs among the cost, time and environmental aspects may also find more research merit (SteadieSeifi et al. 2014). Real-life intermodal fleet management systems may be full of uncertainties such as random freight demands, fuzzy transport capacities, fuzzy transit times etc. Therefore, coping with these uncertainties such as stochasticity, dynamism and fuzziness related to the elements of an intermodal transportation system can be seen as the other major research challenges. According to a comprehensive literature survey on decision support models in intermodal freight transportation (Caris et al. 2013), there is still a lack of studies in the literature which utilized advanced operations research and artificial intelligence techniques into a DSS for intermodal fleet management problems. It was also emphasized by Schorpp (2011) that application of a computer-based dynamic planning of a DSS can provide potential savings in empty travelled distance and delay for international freight transportation. Moreover, it should be noted here that most of the available studies in the literature concentrated on developing DSS for only single mode or unimodal fleet management. For the last decade, because of increasing attention to intermodal transportation, DSS for intermodal fleet management problems have become an attractive research topic and may provide future opportunities for the researchers. Besides, many available DSS have just cover one of the strategic, tactical or operational fleet management decisions of single decision maker. However, it is needed to develop a DSS which covers various time horizons and integrates several fleet management problems under responsibility of multiple decision makers.

Based on these motivations, this paper presents a web-based DSS including a fuzzy-stochastic mathematical programming model for intermodal fleet management system of a large-scaled international logistics company in Turkey. In addition to determine fleet sizing and composition issues, transportation mode selection, load planning, outsourcing, fleet allocation, fleet expansion/reduction, empty truck repositioning, vehicle inventory control etc. are also incorporated into different modules of the proposed DSS. Apart from the literature, both of the import and export load flows are taken into account while satisfying the stochastic freight demands of the customers. Transportation planning of multiple assets such as trucks, trailers, Ro-Ro vessels and block trains is also included for performing importation and exportation transport operations. Additionally, fuzziness embedded in marine and railway transport capacities, transit times, vehicle availabilities, cost values etc. are also incorporated into the proposed DSS for handling different types of uncertainties simultaneously. Finally, it is able to generate sustainable fleet and freight plans by making use of the proposed DSS since inclusion of economic, time and environmental oriented objectives. In other words, compromise solutions can be produced with respect to transport cost, transit time and $CO_2$ emission objectives. To the best of our knowledge, there is no similar DSS which integrates several strategic and tactical level intermodal fleet management problems of different decision makers.

The remainder part of this paper is organized as follows. In Sect. 2, a comprehensive literature review on developing optimization-based DSS for fleet planning issues in transportation systems is given. Section 3 presents the details of intermodal fleet management with its relevant sub-problems. Framework and main components of the proposed DSS are described in Sect. 4. In this section, development phases of a fuzzy-stochastic mathematical programming model and database integration are also included. In Sect. 5, the proposed DSS is implemented on a real world case study of a large-scaled international logistics company. Moreover, web-based user interface for visualization of different fleet management modules is also presented. Finally, conclusions and future researches are discussed in Sect. 6.

## 2 Previous Related Work

This section presents a brief summary of the former researches on development of optimization-based DSS for fleet planning issues in transportation environments.

First attempt at developing a DSS was performed by Avramovich et al. (1982) based on a large scaled linear programming model for fleet management and configuration of a truck transportation company in North America. In addition to improve timeliness of the planning process and manpower savings, their proposed DSS increased the profitability of the company and quality of fleet decisions. After this research, Wayne (1988) developed a DSS framework for military vehicle fleet management in order to provide cost effectiveness in transportation operations. Saidane (2007) developed a fleet DSS to improve the management and control of a military tracked and wheeled vehicles. Moreover, object-oriented analysis can also be carried out within the DSS for vehicle assignment, maintenance, retirement and inquiry issues. Couillard (1993) utilized a multi-criteria decision making approach and a stochastic programming model for developing a DSS to solve vehicle fleet planning problem including fleet size and mix decisions.

Basnet et al. (1996) developed a DSS which is named as FleetManager for milk tanker routing in New Zealand dairy industry. In order to generate tanker routes, sweep algorithm was used to form supplier clusters and allocation of these clusters to the tankers according their capacity. Then, sequence of the suppliers which minimizes the travelled distance was achieved by applying farthest insertion algorithm. Another optimization-based DSS was proposed by Fagerholt (2004) to solve vessel fleet scheduling problem of industrial shipping companies. The proposed DSS incorporates a user-friendly graphical interface and an optimization routine composed of insertion heuristic and hybrid local search algorithms. An interactive DSS based on a two-stage exact solution methodology was proposed by Ruiz et al. (2004) to a real-life vehicle routing problem with heterogeneous fleet of trucks. In the first stage, all of the feasible routes were generated via an implicit enumeration algorithm whereas the best set of routes were selected in the second stage by using an integer programming model. Another interactive optimization-based DSS was developed by Fagerholt et al. (2007) for routing and scheduling of Ro-Ro vessels and determining fleet size/mix decisions. They also took into account negotiation between the DSS and user in the bidding process.

A DSS based on a maximal expected coverage location model was presented by Repede (1994) in order to determine the size of ambulance fleet and allocation and relocation of ambulances within their service area. Andersson (2005) developed decision support tools for dynamic fleet management problems with real-world applications in airline planning and ambulance logistics. In the first application, the proposed DSS was based on a multi-commodity network flow model with side constraints. As the solution methodology, column generation algorithm and path relinking meta-heuristic were applied for generating aircraft fleet scheduling. In the latter DSS, a simulation model was developed for dispatching and relocating the ambulances dynamically (Andersson and Värbrand 2007). A stochastic dynamic programming model based DSS was developed by El-Din et al. (2015) for automatically planning of aircraft fleet capacity. Indeed, it was intended to determine optimal fleet scheduling decisions related to purchasing, leasing and disposing of the aircrafts over a planning horizon.

Matsatsinis (2004) developed a DSS which uses heuristic algorithms for dynamic routing and scheduling of a heterogeneous vehicle fleet in a Greek concrete distribution company. Zografos and Androutsopoulos (2008) presented a GIS-based DSS for an integrated problem which combines distribution of hazardous materials by a fleet of trucks, truck routing and appropriate deployment of emergency service units for quick response. In the solution phase, two integer programming models were incorporated into DSS for truck routing and emergency response unit location problems and solved by a Lagrange relaxation heuristic algorithm.

A heuristic-based DSS which includes depot and fleet management issues was presented by Jang et al. (2008) to locate depots, design sectors, vehicle routes and fleet configuration of a transportation network in Missouri. They formulated a mathematical program for fleet configuration and scheduling which aims at serving routes with minimum number of vehicles. A three-phase optimization-trend-simulation decision support system was presented by Kek et al. (2009) for vehicle fleet in a car sharing system which aims at allocating the staff resource by minimizing the vehicle relocation costs. Due to the unbalanced import/export freight demands, Bandeira et al. (2009) proposed a network optimization-based DSS for distribution of empty and loaded containers in an intermodal logistics network. After decomposing the integrated network model into two sub-models, they first formulated an integer program as a static model in order to handle transshipment and allocation of empty/loaded containers. Then, a dynamic model based on heuristic methods was also formulated for prioritizing the loaded container demands and dispatching empty containers. A combined Monte Carlo simulation and mathematical programming based DSS was developed by Fagerholt et al. (2010) for strategic planning decisions, i.e., fleet size and mix, contract analysis etc. in maritime transportation system of a Norwegian shipping company.

Santos et al. (2011) offered a web-based spatial decision support system for multiple vehicle routing problems by using Google Maps. Since the NP-hard nature of the problem, they applied hybridization of a heuristic and ant-colony metaheuristic algorithm by considering the fleet capacity and shift time limitations. Because of the dynamic traffic conditions and variability in travel times, Grzybowska and Barceló (2012) focused on real-time fleet management by making use of a DSS depending on pickup/delivery vehicle routing with time windows and dynamic traffic simulation models. Polo et al. (2014) presented a DSS for vehicle fleet management based on Terrestrial Trunked Radio (TETRA) terminals geolocations. By using a popular metaheuristic algorithm, i.e., Particle Swarm Optimization (PSO), it was aimed to achieve the best distribution of vehicle fleet on the territory in case of multiple requests received by a control center. Inghels et al. (2016) presented a dynamic tactical planning model which also includes optimal number and mix of ships in a fleet for service network design problem of a multimodal municipal solid waste transport. Hill and Böse (2017) presented a DSS which is able to reduce truck waiting time, traffic congestion and air pollution of an integrated resource planning and truck routing problem. A web-based DSS was developed by Müller and Tierney (2017) for a liner shipping fleet repositioning problem. First, a mixed-integer programming formulation of the problem was proposed with time window constraints, then a simulated annealing algorithm was employed to achieve the near-optimal solutions. A mathematical programming model was also presented by Martínez-López et al. (2018) for sustainable optimization of container vessel fleets in intermodal transportation networks. In fact, they aimed to achieve the best container vessel fleets options.

Based on the previous papers, following research contributions are made in this paper. A web-based DSS was presented for a real-life intermodal fleet planning problem of a large-scaled international logistics company. Apart from the available studies in the literature, the proposed DSS based on a fuzzy-stochastic mathematical program which provides the simultaneous optimization of several interactive fleet planning problems such as freight or load planning, fleet sizing/composition, fleet deployment, vehicle inventory management, fleet expansion/reduction, empty vehicle repositioning etc. Moreover, sustainability objectives, multiple types of resources (trucks, trailers, Ro-Ro vessels and trains) and different types of uncertainties were also taken into account.

Finally, reviewed articles dealing with the intermodal freight transportation are categorized with respect to problem/modelling features in Table 1. In fact, fleet sizing/composition, empty repositioning, routing and scheduling issues in intermodal transportation are also considered by this table.

**Table 1** Summary of the articles focused on planning problems in intermodal transportation networks

| Article | Problem definition | Problem characteristics | Freight planning | Fleet sizing | Empty repositions | Routing/ scheduling | Outsource | Environment | Uncertainty conditions | Multiple objective | Solution methodology |
|---|---|---|---|---|---|---|---|---|---|---|---|
| Guelat et al. (1990) | Network assignment and freight plan | Static/ Deterministic | ✓ | – | – | – | – | – | – | – | Heuristic algorithm |
| Arnold et al. (2004) | Terminal location problem | Static/ Deterministic | ✓ | – | – | – | – | – | – | – | Heuristic procedure |
| Chang (2008) | Freight routing | Static/ Deterministic | ✓ | – | – | ✓ | – | – | – | ✓ | Lagrangean relaxation and decomposition |
| Puetmann and Stadtler (2010) | Decentralized planning | Dynamic/ Stochastic | ✓ | – | ✓ | – | ✓ | – | ✓ | – | Collaborative planning |
| Bauer et al. (2010) | Scheduled service network design | Static/ Deterministic | ✓ | ✓ | ✓ | ✓ | – | ✓ | – | ✓ | Integer linear programming |
| Verma and Verter (2010) | Planning rail-truck transportation | Static/ Deterministic | ✓ | – | – | ✓ | – | – | – | ✓ | Iterative decomposition |
| Yang et al. (2011) | Intermodal freight routing | Static/ Deterministic | – | – | – | ✓ | – | – | – | ✓ | Goal programming model |
| Macharis et al. (2011) | Transport policy making | Dynamic/ Stochastic | ✓ | – | – | ✓ | – | ✓ | ✓ | ✓ | Mathematical and simulation modelling |
| Bruns and Knust (2012) | Load planning of trains at terminals | Static/ Deterministic | ✓ | ✓ | – | – | – | – | – | ✓ | Integer programming |
| Cho et al. (2012) | Multimodal routing with time windows | Static/ Deterministic | – | – | – | ✓ | – | – | – | ✓ | Dynamic programming |
| Verma et al. (2012) | Rail-truck intermodal routing | Static/ Deterministic | ✓ | – | ✓ | ✓ | – | – | – | ✓ | Tabu search |
| Bierwirth et al. (2012) | Service type & terminal selection | Static/ Deterministic | ✓ | – | – | – | – | – | – | – | Mixed-integer programming |
| Kalinina et al. (2013) | Matching of goods | Static/ Stochastic | ✓ | – | ✓ | ✓ | – | ✓ | ✓ | ✓ | Chance constrained programming |
| Meisel et al. (2013) | Production and transportation planning | Static/ Deterministic | ✓ | – | – | – | – | ✓ | – | ✓ | Heuristic algorithm |

| | | | | | | | | | | | |
|---|---|---|---|---|---|---|---|---|---|---|---|
| Garcia et al. (2013) | Assignment of containers & trucks | Dynamic/ Deterministic | ✓ | – | ✓ | ✓ | – | – | – | – | Integrated linear program and automated planning |
| Bhattacharya et al. (2014) | Strategic transport planning | Dynamic/ Stochastic | ✓ | – | – | ✓ | – | – | ✓ | – | Mixed-integer programming |
| Assadipour et al. (2015) | Managing hazmat freights | Static/ Deterministic | ✓ | – | – | ✓ | – | – | – | ✓ | Genetic algorithm |
| Resat and Turkay (2015) | Transportation network design | Static/ Deterministic | ✓ | – | – | ✓ | – | – | – | ✓ | MILP and augmented $\varepsilon$-constraint method |
| Li et al. (2015) | Container flow control | Dynamic/ Stochastic | ✓ | – | – | ✓ | – | – | ✓ | ✓ | A receding horizon control |
| Demir et al. (2016) | Service network design problem | Static/ Stochastic | ✓ | – | – | ✓ | – | ✓ | ✓ | – | Chance-constrained stochastic program |
| Ezabadi and Vergara (2016) | Logistics network design with routing | Static/ Deterministic | ✓ | – | – | ✓ | – | – | ✓ | – | Decomposition-based search algorithm |
| Febbraro et al. (2016) | Short/medium term transport planning | Dynamic/ Stochastic | ✓ | – | – | ✓ | – | – | ✓ | – | Agent-based modeling |
| Baykasoğlu and Subulan (2016) | Multi-mode load transport planning | Static/ Deterministic | ✓ | – | – | – | ✓ | ✓ | ✓ | ✓ | MILP and fuzzy multi-objective optimization |
| Lin and Lin (2016) | Terminal location | Static/ Deterministic | ✓ | – | – | ✓ | – | – | – | – | Two-stage approach based on a matheuristic |
| Wang and Meng (2017) | Discrete intermodal network design | Static/ Deterministic | ✓ | – | – | ✓ | – | – | – | – | Mixed-integer nonlinear programming |
| Proposed study | Intermodal fleet planning | Static/ Fuzzy-stochastic | ✓ | ✓ | ✓ | – | ✓ | ✓ | ✓ | ✓ | Hybrid fuzzy-stochastic mixed-integer program |

## 3 Overview of Intermodal Fleet Management

International logistics companies generally offer numerous transportation solutions and services to its customers involving direct road freight transport, combined marine-road transport and intermodal transportation due to some criteria such as transit time, tonnage limitations and product characteristics etc. In direct road freight transportation option, loads are transferred from origin to the consolidation centers of these companies via highways and delivered again from these centers to the final destinations. Moreover, import and export freight transportation operations are generally carried out simultaneously in most of the real-life settings. For this reason, importation of the loads can begin with the transfer of the loads to the consolidation centers via highways for further transportation and then delivered to the relevant origins. Furthermore, the combined marine and road freight transport integrates two different transportation modes and serves as an alternative option for particularly high tonnage products since the tonnage limitations imposed in many countries. Firstly, the loads are transferred from the origins to the seaports by Ro-Ro vessels and then delivered from these ports to the relevant destinations via highways. Finally, intermodal is a transportation model wherein three distinct means of transport are utilized, i.e., maritime, railway and road freight transports. Intermodal transportation aims at transferring loads from one location to another in a most effective, punctual, affordable and environment-friendly manner. In intermodal transportation, trailers or containers arriving by Ro-Ro vessels are loaded to block trains at the seaports and then transferred to the rail stations and finally delivered to the destinations by road freight transport. A featured intermodal transportation service provides lots of advantages to the company such as timely delivery, cost effectiveness and environmental awareness. The configuration of intermodal logistics network for the examined multi-objective, multiple resource and multi-period intermodal fleet management problem can be given as depicted in Fig. 1. At most of the seaports or railway stations, block trains and Ro-Ro vessels arrive and depart in accordance with a fixed time schedule. The intermediate transshipments are also possible for both marine and rail transportation services based on the consolidation concept. In detail, some loads can be unloaded from the Ro-Ro vessel at a transshipment seaport and then shipped to the destination via road freight transport. Similarly, some extra loads can be loaded to the Ro-Ro vessel at a transshipment seaport and then transferred to origins by Ro-Ro vessels. In other words, intermediate stops may exist in marine transportation. Additionally, some loads can also be exchanged between rail stations which correspond to rail-rail exchange. In fact, some trailers can be unloaded from a block train at a transshipment rail station and loaded to a public train which operates on a different rail line.

The remaining trailers are delivered to the destinations via road freight transport. On the other hand, exchange operations may increase transit time and transport costs of the intermodal transportation system. For this reason, they should be performed rapidly and efficiently (Baykasoğlu and Subulan 2016). In the light

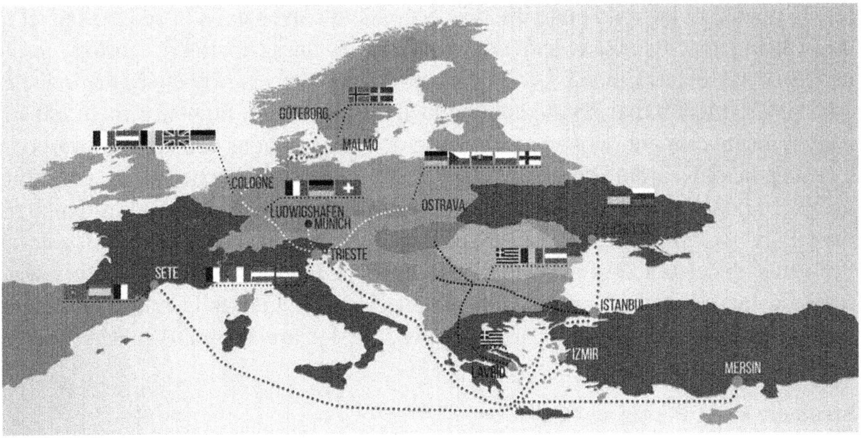


**Fig. 1** Intermodal logistics network configuration of a logistics company

of these information, determining the appropriate fleet size and composition of different kinds of resources is an essential problem encountered by the logistics companies while realizing international transportation services for both importation and exportation. In addition to fleet sizing, heterogeneous fleets are more common for these resources because of the different license plates of trucks and various types of trailers. When the fleet capacity of the companies is insufficient for satisfying the customers' freight demands, outsourcing option may be possible in road, marine and railway transportation services. For instance, companies may outsource the Ro-Ro vessels of the other logistics service providers and use the public train services instead of fixed-schedule block trains. Moreover, additional trucks and trailers can be hired in road freight transportation. Thus, outsourcing decisions may also affect the supplier selection and contracting processes. However, allocation of these resources (fleet allocation) to the hubs, routes and demand points is the other important issue. Because, effective and efficient fleet allocation can also decrease the empty or free movements. Actually, all of these fleet planning activities can be performed after determining the load plans. For this reason, multi-mode load planning which involves the transportation mode and service type (Standard, express and super-express) selection, load allocation to different modes of transport, frequency of the Ro-Ro vessel cruises and block train trips, intermodal route selection, outsourcing etc. should be first accomplished. Furthermore, empty vehicle repositioning issues may also arise due to the imbalance between the import and export transport operations within the same time period. Besides, difference between the fleet size and vehicle inventory at the time-space logistics network may also cause extra empty vehicle repositions. Actually, there is a strong interaction among fleet sizing, fleet allocation, vehicle inventory and empty vehicle repositioning issues. In detail, fleet allocation and vehicle inventories on the logistics network may affect the empty flows and this is also significantly connected with fleet size optimization. It should be noted here that fleet sizing and composition are highly related to the load plans and loaded vehicle movements. In addition to all of these, fleet capacity expansion/reduction decisions may come out according to the increases in freight demands of the customers and business volume of the logistics companies. Fleet expansion/reduction decisions may also influence the company budget for investment on new resources and these decisions can also be overcome by using the outsourcing options. Finally, all of the aforementioned issues constitute the intermodal fleet management problems and inappropriate fleet and freight plans may cause significant financial losses and decreases in customer service level. For better understanding, intermodal fleet management incorporates the following strategic, tactical and operational planning decisions (Dejax and Crainic 1987; Imai and Rivera 2001; Macharis and Bontekoning 2004; Caris et al. 2013; Herrera et al. 2017):

**Strategic Fleet Planning Decisions**

- Fleet sizing and composition of resources (Trucks, trailers, Ro-Ro vessels, trains etc.),
- Fleet expansion (purchasing new trucks/trailers) and fleet reduction decisions (sales of second-hand trucks/trailers).

**Tactical Fleet Planning Decisions**

- Intermodal freight transportation planning (multi-mode load planning),
- Loaded movements, resource allocations to different transportation modes and routes,
- Transportation mode and service type selection,
- Outsourcing alternatives for maritime and railway transportation services,
- Periodic resource inventories on the intermodal logistics network.

**Operational Fleet Planning Decisions**

- Empty truck/trailer repositioning problems,
- Load consolidation at transshipment seaports or rail stations,
- Truck routing and scheduling of block trains and Ro-Ro vessels.

In intermodal transportation, handling all the above fleet planning decisions simultaneously makes the fleet management much more complex than unimodal transportation systems. This is because all of these decisions should be undertaken in an integrated manner because of the high interaction among them. Moreover, intermodal fleet planning is composed of various sub-problems at different levels and also incorporates the decisions under the responsibility of more than one decision maker. These are the other vital reasons which cause complexity in intermodal fleet management. Based on these motivations, development of optimization-based DSS is needed to support the decision making processes in intermodal fleet planning. By making use of such a DSS, logistics companies may achieve optimal fleet size and allocation and minimal empty flows. In addition to possible cost and transit time reduction, logistics companies may also have environmental friendly freight and fleet plans.

# 4 Decision Support System Methodology

## *4.1 Framework and Main Components of the Proposed DSS*

In this section, a web-based DSS based on a fuzzy-stochastic mathematical programming model is introduced for a multi-objective, multi-resource and multi-period intermodal fleet planning problem. As highly recommended by the existing studies in the literature (Macharis and Bontekoning 2004; Caris et al. 2008, 2013; SteadieSeifi et al. 2014; Baykasoğlu and Subulan 2016), development of a DSS is quite needed to support the strategic, tactical and operational level fleet planning activities of different actors/players in intermodal transportation. For this reason, an architecture for the proposed DSS which incorporates a holistic optimization model is presented in Fig. 2. Similar to the other decision support systems, the proposed DSS consists of four main components.

1. *Model component*: This modeling and optimization component includes a mixed-integer mathematical programming (MIP) model of the examined intermodal fleet planning problem and the relevant fuzzy and stochastic model versions and

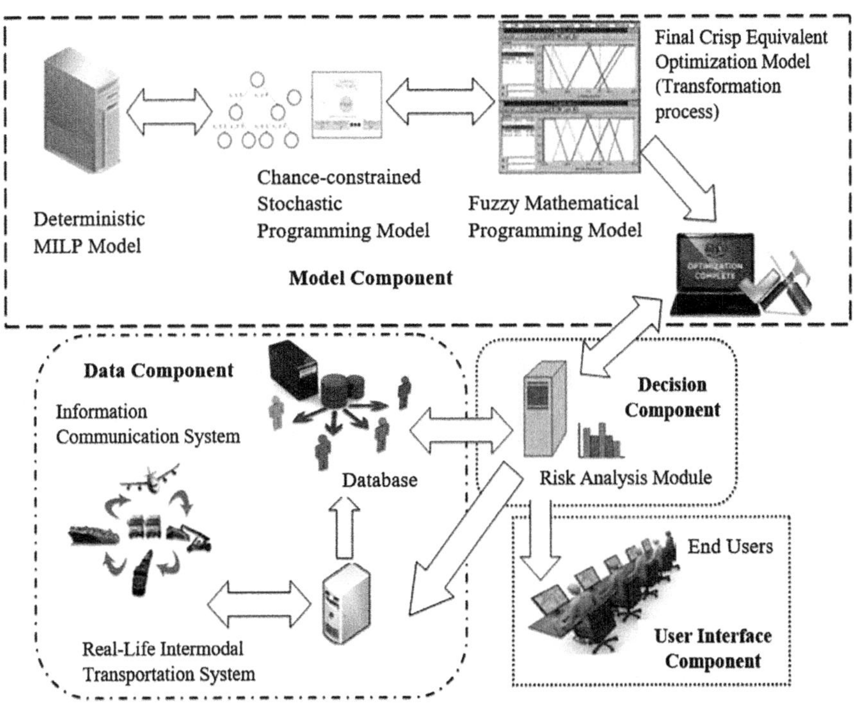


**Fig. 2** Framework and components of the proposed decision support system for intermodal fleet management

their solution approaches for transforming the fuzzy-stochastic models into the crisp equivalent form. Briefly, while developing the deterministic MIP model, there are four substantial model development phases. In detail, multi-mode load planning or intermodal freight transportation planning model is first formulated afterwards extended by incorporating the fleet sizing and composition, loaded and empty vehicle movements, vehicle inventories on the logistics network and fleet expansion/reduction decisions. The internet integration of MIP model was accomplished by connections between LINGO optimization software and *ASP.NET* by using DLL libraries in LINGO 15.0 (Lindo Systems Inc. 2016). Moreover, MVC (Model View Controller) was also used as a software architectural pattern to implement the web-based user interface.

2. *Decision component*: This component is able to present various alternative solutions to the company experts for freight and fleet planning activities under different uncertainty levels and risk attitudes. This component is also needed for obtaining just cost-oriented, customer-focused or environmental friendly fleet plans and also compromise plans for these conflicting objectives. Furthermore, it also contains risk analysis which can generate risky and risk-free fleet plans in accordance with different feasibility degrees of model constraints under various stochastic freight demand scenarios.
3. *Data component*: This component incorporates a database and information communication technology which collect, store and process the crisp and imprecise data as inputs of the mathematical programming model. This component also maintains necessary subjective information (pessimistic, most likely and optimistic values of fuzzy inputs) from the company experts, external information from the intermodal transportation environment and organizational information from the logistics company's database and data warehouses.
4. *User interface*: This part is required for direct contract with end users. In other words, effective and efficient usage and operation of a DSS by the users can be achieved by this component. The web-based user interface of the proposed DSS is designed by using .Net and C# programming languages on Microsoft Visual Studio which is an integrated development environment. For graphical reports and documents, DevExpress which builds highly configurable and well-designed user interface controls is utilized. The graphical user interface of the proposed DSS is able to visualize freight/fleet plans, load/fleet allocations, monthly business volumes and trips of Ro-Ro vessels/block trains on a web-based platform.

## 4.2 Development of a Fuzzy-Stochastic Mixed-Integer Programming Model

Variability in freight demands, transit times and disruptions can be specified as the main sources of uncertainty in intermodal transportation systems. Therefore, other major research challenge is to overcome these uncertainties such as stochasticity, dynamism and fuzziness related to the elements of an intermodal transportation system (stochastic transport demands, fuzzy transit times, mode capacities etc.). Thus, monthly import/export freight demands of the customers cannot be exactly known and may be considered as stochastic parameters since they have randomness in nature. This paper ensures that total amounts of freights transferred by different modes of transport should meet the random freight demands which are represented by independent continuous random variables with cumulative distributions. The amounts of import/export freight demands of the customers are assumed to fit a normal distribution. Finally, construction of a chance-constrained stochastic program is performed by making use of some private functions in LINGO 15.0 optimization software so as to identify random variables, declare probability distribution functions, set sample size, assign probabilistic constraints to chance constraint sets. Additionally, due to the epistemic uncertainty which is resulted from unavailability or incompleteness of the historical and insufficient data, transportation cost parameters, capacities of marine and rail transports, transit times, number of working days at each month, resource availabilities, $CO_2$ emission values of different transport modes etc. can also be stated as fuzzy parameters and described by possibility distributions. In other words, these parameters can be represented by triangular fuzzy numbers in which pessimistic, most likely and optimistic values are specified by company experts. Due to this fact that, we proposed a holistic optimization model by formulating a fuzzy and stochastic mixed-integer program for the examined multi-objective, multi-stage, multi-resource, multi-period intermodal fleet planning problem with strategic and tactical level decisions. Afterwards, different types of uncertainties, i.e., randomness and fuzziness, can also be taken into account by the proposed model concurrently. Because, real-life intermodal fleet planning problems may involve different types of uncertainties simultaneously. For this reason, the joint usage of stochastic programming and fuzzy mathematical programming techniques is required to solve this kind of problems. In this study, chance-constrained stochastic programming approach is employed in order to handle the uncertainty of freight demands. Furthermore, a popular fuzzy mathematical programming approach, i.e., interactive resolution method of Jimenez et al. (2007) is selected from the literature so as to deal with the fuzzy model parameters. In detail, this resolution method is used to transform the proposed fuzzy model into the crisp equivalent form. The most important reason for choosing this method is that one can achieve in a computationally efficient way of solving possibilistic programming problems with its linearity and without any increment in the objective functions and inequality constraints. In other words, it preserves linearity and model dimension will not increase after the transformation process. Furthermore, this method is able to provide feasible solutions in various degrees of feasibility. Thus, it can present $\alpha$-acceptable fuzzy efficient solutions which make possible to handle various uncertainty levels. In addition, the fuzzy efficient decision is made iteratively with the participation of the decision makers. Moreover, the decision maker may also have extra information about the risk of violation related to the model constraints. In general, the bigger the feasibility degree is, the worst the objective function value will be. In addition to minimize the overall transportation costs, the proposed model also aims at optimizing total transit time and environmental impact resulted from $CO_2$ emitted by different modes of transport. It is targeted to achieve maximal customer satisfaction or service level by minimizing the total transit times. Alternatively, the amount of total $CO_2$ emissions resulting from the usage of different transportation modes is also minimized to provide an environmental conscious fleet and freight plans. In order to handle these conflicting objectives simultaneously and obtain compromise solutions, a weighted additive fuzzy goal programming approach is applied. The constraints of the proposed fuzzy-stochastic model can be listed as follows:

- Multi-mode load planning constraints
  - Stochastic freight demand satisfaction constraints
  - Maritime transportation capacity constraints and balance equations
  - Railway transportation capacity and conservation of flow constraints
- Fleet sizing and composition constraints
  - Constraints for relationship between load plans and loaded vehicle movements
  - Constraints related to fleet sizing and composition of multiple resources
  - Logical constraints for relationship between loaded truck movements and fleet sizes
- Empty vehicle repositioning constraints
  - Constraints for empty flows resulted from the imbalance of import/export load quantities within the same time period
  - Constraints for empty flows resulted from the difference between loaded vehicle movements in the successive time periods
- Constraints of vehicle inventory on the intermodal logistics network
  - Periodic inventory constraints of different resources
  - Constraints for relationship between vehicle inventory levels and fleet sizes
  - Constraints for idle vehicles awaiting at hubs, seaports, rail stations etc.
  - Constraints of new vehicle purchasing and sales of second-hand vehicles

To the best of our knowledge, there is no a similar fleet planning model in the literature which is studied through this paper. The mathematical nomenclature and formulation of the proposed fuzzy-stochastic model can be accessed by the link in Appendix 1. Some of the dataset of a real-life application in an international logistics company can be accessed at the link in Appendix 2. The simplified LINGO 15.0 code of the proposed model can also be retrieved by the link in Appendix 3. For more details on the mathematical model development refer to the following references (Baykasoğlu et al. 2015; Dudaklı et al. 2015; Baykasoğlu and Subulan 2016, 2017).

## 4.3 Data Modelling and Database Integration

In the data component of the proposed DSS, Oracle Database which is an object-relational database management system is utilized. The schema of database consists of four main tables named as "Upload", "Entities", "Enums" and "Run information" and shown by Fig. 3. All of the relavent data for each month in intermodal fleet planning is defined on "Entities" table. For representing the data types in "Entities" table, "Enum" table is constructed.

All of the model outputs for the fleet and freight plans such as the objective function values, lower bounds of these objectives, load and fleet sizing quantities etc. were included in "Run information" table. Finally, all of the input data related to the specific solution requests (Model type: deterministic, fuzzy, stochastic or fuzzy-stochastic; number of samples and probability values for stochastic program, feasibility degree-$\alpha$ for fuzzy mathematical program etc.) were involved by "Upload" table.

The unified modeling language (UML) class diagrams for the database and all inputs-outputs of the proposed DSS are also depicted in Fig. 4. As mentioned previously, LINGO codes of the proposed mathematical models were run over C# by making use its own of DLL libraries and scripts. The LINGO scripts would be terminated when the termination criteria such as maximum run time, threshold for optimality gap etc. were satisfied. Then, the model outputs were obtained on different Excel spreadsheets. In order to provide graphs and output tables dynamically into these spreadsheets, an Excel macro was written. These Excel spreadsheets were also stored as a BLOB (Binary large object) data type on the database. The Excel macros were run over C# via Microsoft Office Interop Excel library. For the connection among Oracle Database and C#, object-oriented matching was utilized instead of SQL queries. When the hardware requirements of the DSS were evaluated, Windows server 2012 R2 64-bit was prepared for solution of the fuzzy-stochastic program since LINGO 15.0 has a 64-bit operating

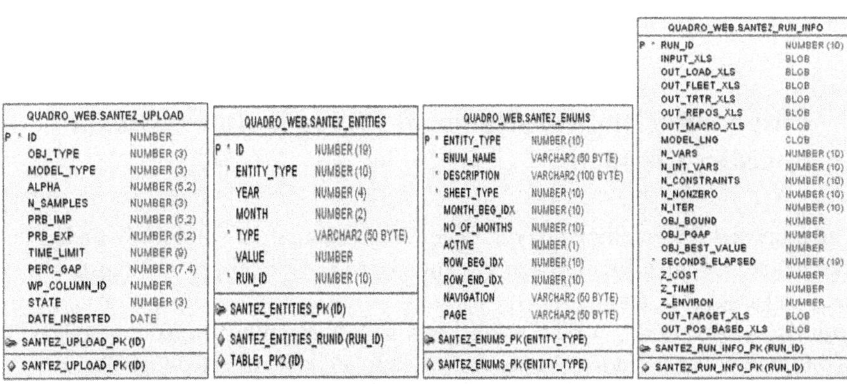


**Fig. 3** Names of the user tables in Oracle Database

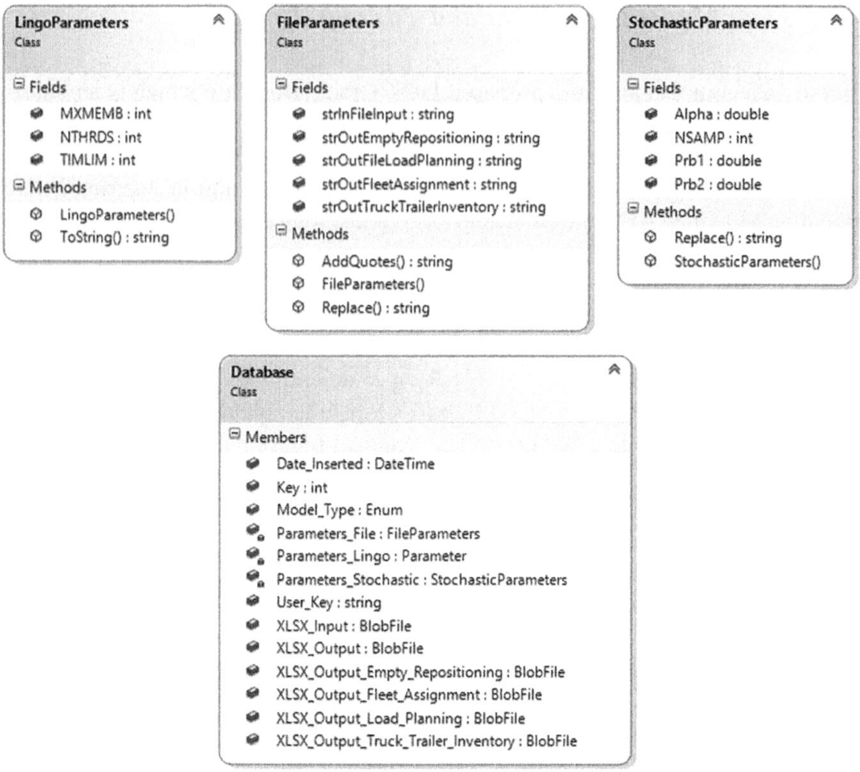


**Fig. 4** UML class diagrams for database, inputs and outputs of the DSS

system. The organized virtual machine has a RAM of 8 GB, 2.4 GHz Core i5 processor and 80 GB hard disk. Actually, fuzzy-stochastic programming model required approximately 8 GB memory whereas deterministic model needed only 4 GB memory.

## 5 Case Study: Implementation of DSS in an International Logistics Company

The proposed DSS methodology has been implemented for solving the underlying intermodal fleet management problems of a large-sized international logistics company in Turkey. The company is the fastest growing Logistics Corporation in Europe with over 100,000 m$^2$ of own warehouses and 1200 multinational employees. In addition, it is listed among the pioneering suppliers of integrated logistics services in both Turkey and Europe with its distribution centers with more than 450,000 m$^2$ indoor areas in Turkey and a fleet of over 4000 trucks, trailers and containers. The company provides intermodal transportation services and intercontinental logistics

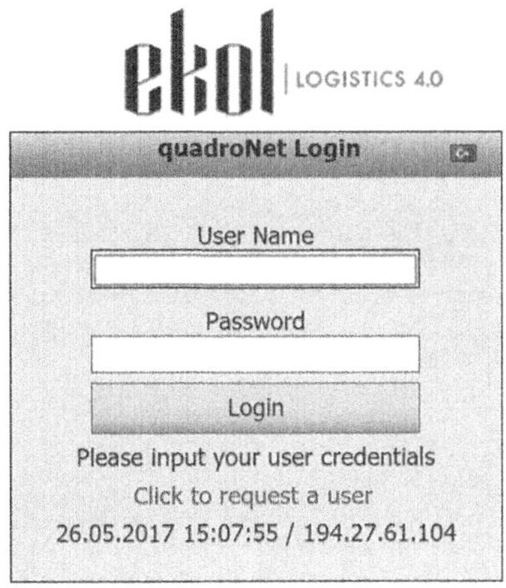


**Fig. 5** User login screen of the proposed "FILOPT" DSS

solutions by carrying thousands of containers full of materials, finished products, industrial and consumer goods between Turkey and Europe (Ekol Logistics Inc. 2017). The users can access FILOPT DSS via an internet browser. In detail, it is possible to login the system by entering user name and password over a website portal hosted at quadronet.ekol.com. The user login screen designed for FILOPT DSS is depicted in Fig. 5.

## *5.1 User Interface for Visualization of Different Fleet Management Modules*

The proposed FILOPT DSS has multiple fleet management modules which correspond to different intermodal fleet planning problems under responsibility of distinct decision makers. In other words, FILOPT has a modular structure that it contains number of modules accessible via one login to the proposed DSS. Actually, these modules are classified into seven categories namely optimization, load planning, fleet planning, cost planning, empty flow planning, transit time planning and environmental impacts according to the logistics company's user requirements and system functionalities. The web interface of the general menu for FILOPT is depicted in Fig. 6. At the lower left corner of this menu, the basic modules of FILOPT are listed.

## *5.2 Fleet Optimizer: Optimization Module*

The first and most vital module can be called as fleet optimizer. Actually, this module belongs to the model component of the proposed DSS. The screen details of

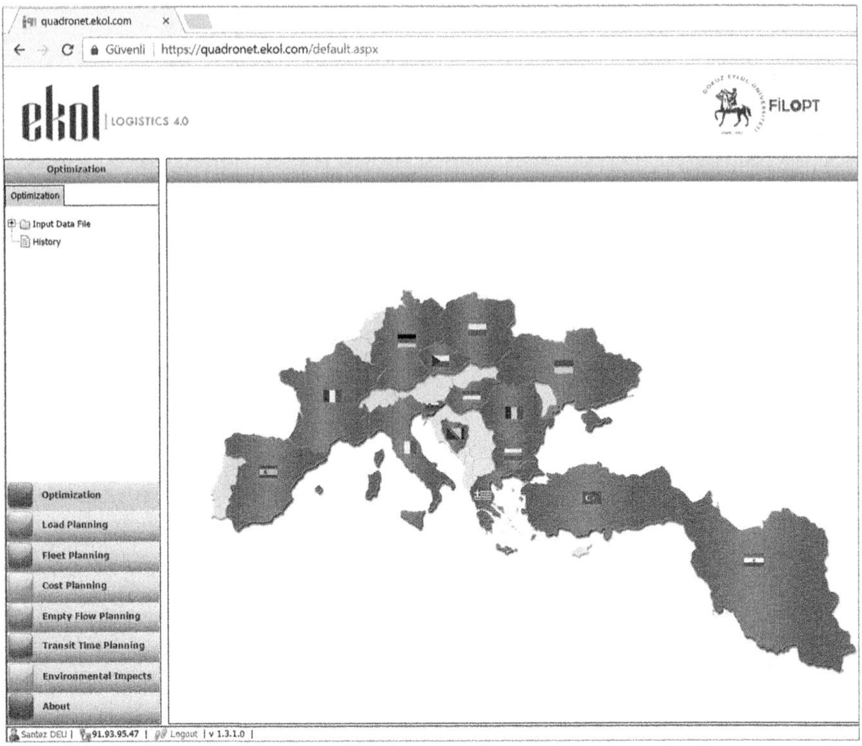


**Fig. 6** The general menu and web interface of the proposed DSS with modular structure

the optimization module can be seen as shown by Fig. 7. First of all, the users should upload the current fleet management data by choosing Excel spreadsheets. On the other hand, these data may change over time. Therefore, due to the dynamic changes in fleet management data such as the variation in freight demands, re-optimization is needed and the models which are background of FILOPT should be rerun. After uploading the data file, the user is able to select the model type according to his/her preferences and some uncertainty conditions. For instance, the user may disregard the uncertain data and thus can run the deterministic mixed-integer program.

Alternatively, the user may only take into account stochastic freight demands by excluding fuzzy input data. In this case, chance-constrained stochastic program should be run. Before running the model, number of samples and probabilities for both import/export freight demands should be determined. Otherwise, by running the fuzzy-stochastic programming model, one can handle both stochastic and fuzzy parameters while generating the fleet plans. Before running fuzzy or fuzzy-stochastic mathematical program, risk level which is related to the feasibility degree of fuzzy constraints should be specified. By making use of this module, the user can also adjust the desired percentage of optimality gap and maximum running time of the model.

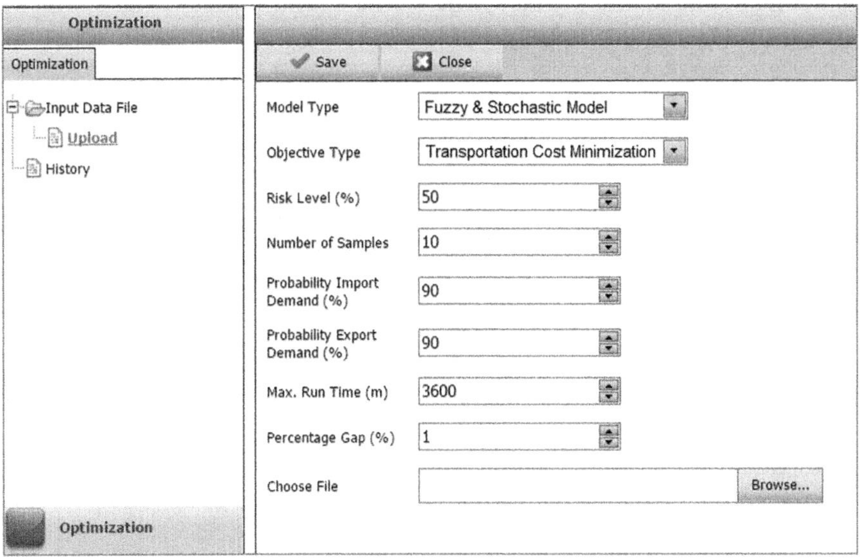


**Fig. 7** Optimization module: details and settings options of the fuzzy-stochastic mathematical program

## 5.3 Multi-Mode Load Planning Module

In this module, effective transportation mode selection and periodic load allocation to different transportation services are completed. In detail, composition of direct road freight transport, combined marine-road transport, intermodal transportation and outsourcing option is determined for importation and exportation operations simultaneously. The objective is to find the least costly, less time consuming and environmental friendly load plans on the intermodal logistics network. The proposed mathematical programming model which only incorporates multi-mode load planning decisions of this module can be found in (Baykasoğlu and Subulan 2016). First of all, aggregated freight demands or business volume of different customer zones are evaluated in Fig. 8. It is clear that there is no significant difference between importation/exportation operations.

Additionally, deviation of the monthly import/export freight demands of the customers can be seen in Fig. 9. After evaluation of the business volume, the optimization model was run and it was stated in Fig. 10 that a large portion of the customer freight demands should be met by using intermodal transportation where the railway transportation services are involved.

In fact, it is the most preferred transport option and also presents the most appropriate way of freighting in terms of economic, customer satisfaction and environmental issues. After transportation mode and transportation service selection, import and export loads are allocated to the relevant seaports and rail stations as

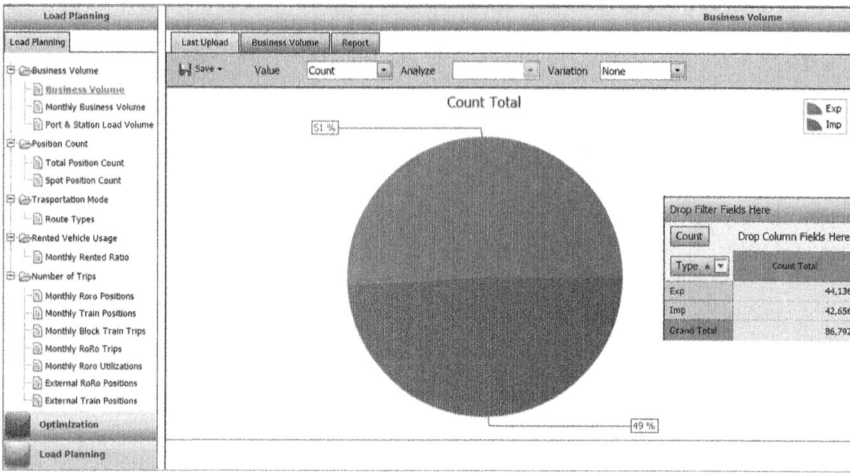


**Fig. 8** Percentages of the annual import/export freight demands of the customer zones

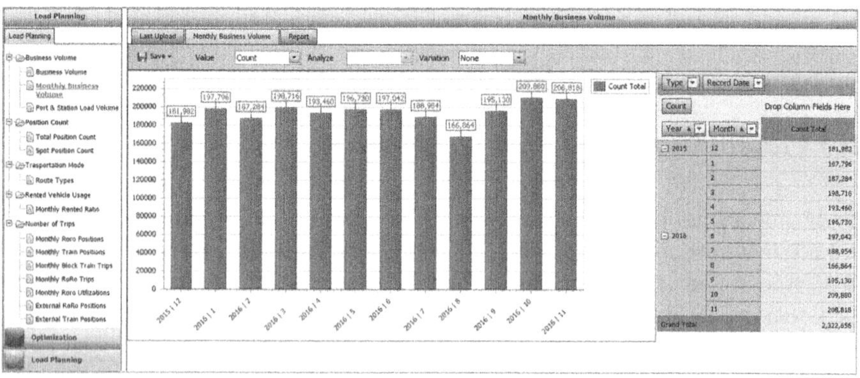


**Fig. 9** Variation of the monthly import/export freight demands or total business volume

depicted in Fig. 11 for transferring by trains and Ro-Ro vessels. Actually, the loads allocated to rail stations are relatively larger quantities than the loads assigned to the seaports. This means that intermodal freight transportation which includes road-marine-railway transportation services was largely utilized than combined marine-road freight transportation. It should also be mentioned here that allocated export loads are generally higher than the imported ones. The allocation of these loads to the company resources and outsourcing alternatives is also so vital under capacity limitations and transit time goals. Figures 12 and 13 present the total load allocations to the company resources such as Ro-Ro vessels, trucks and trailers etc. and an external outsourced transportation alternative, respectively. The variation of monthly usage rate of outsourcing option for import/export can also be

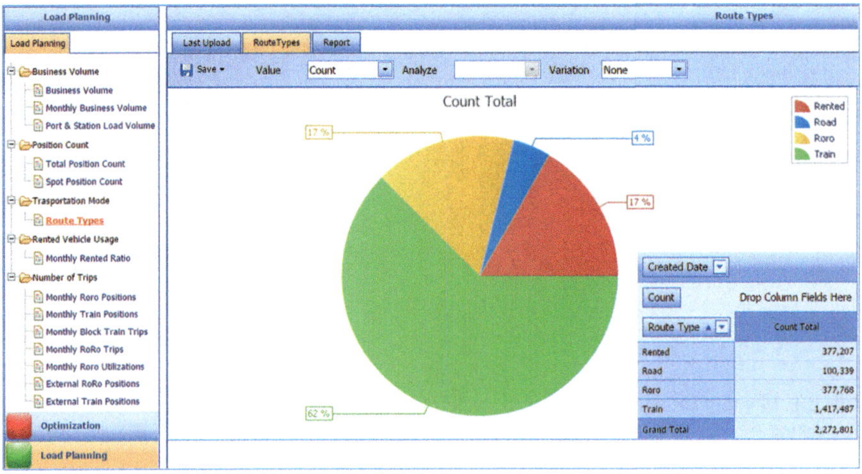


**Fig. 10** Usage rates of different transportation modes or route types

**Fig. 11** Monthly import and export load allocations to rail stations and seaports

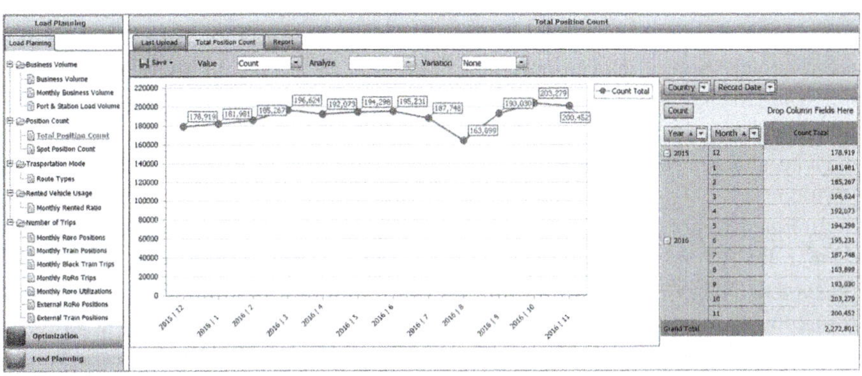


**Fig. 12** Monthly loaded truck/trailer positions among different locations on the intermodal logistics network

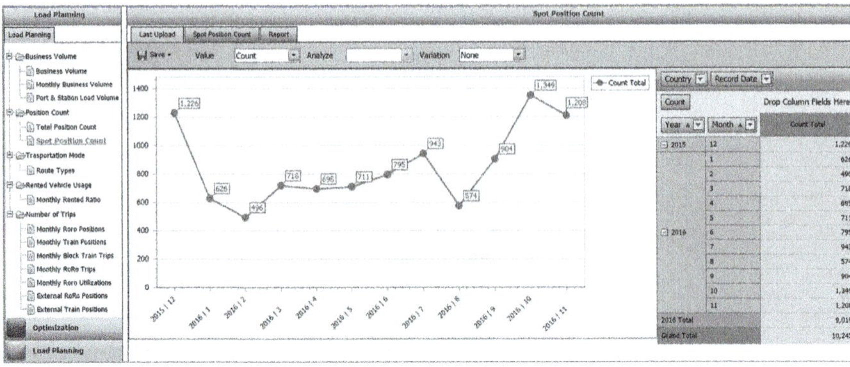


**Fig. 13** Monthly variation of the loaded truck/trailer positions performed by an external outsourcing option

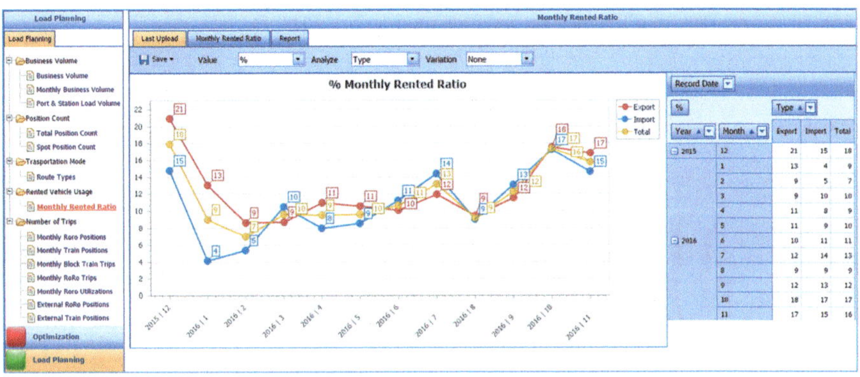


**Fig. 14** Monthly variation of the outsourcing rate for import/export transportation operations

demonstrated as in Fig. 14. In fact, amounts of the loads shipped by an outsourcing option will increase when the transit time objective is optimized.

It is clearly seen from Fig. 14 that outsourcing quantities in freight exportation are relatively higher than importation operations at the beginning of the planning horizon. On the other hand, usage ratios of outsourcing for import operations will increase dramatically after a quarter of the year. Then, these outsourcing amounts may be almost the same for import/export operations to the end of the planning period.

In Fig. 15, monthly import and export loaded positions performed by company-owned Ro-Ro vessels and other logistics service providers Ro-Ro vessels are presented. Due to the cost disadvantage of an external Ro-Ro vessel, it is rarely used for exportation operations. The detailed monthly Ro-Ro cruises among the different seaports of Europe and Turkey are given in the following Fig. 16.

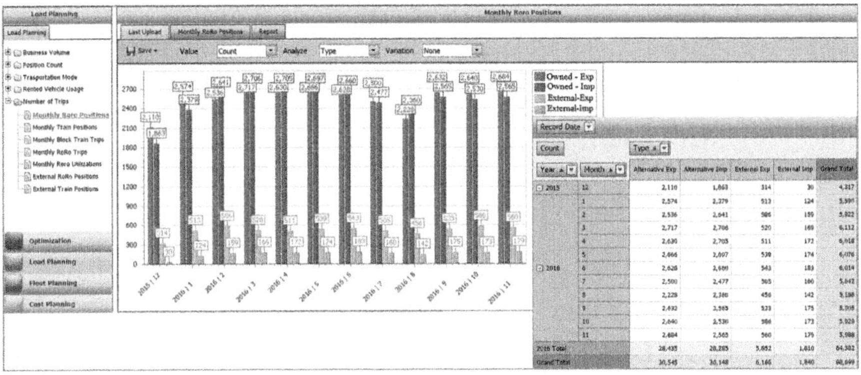

**Fig. 15** Monthly import/export loaded trailer positions by company-owned Ro-Ro vessels

**Fig. 16** Number of monthly cruises by company-owned Ro-Ro vessels on the maritime transportation network

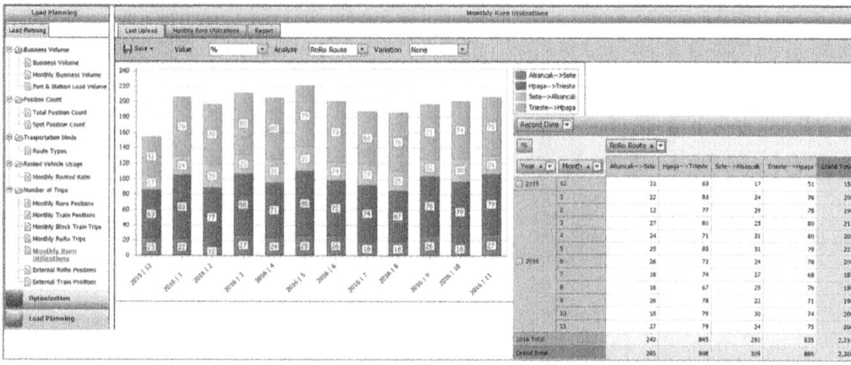


**Fig. 17** Monthly utilization rates of the company-owned Ro-Ro vessels operated between different seaports

In addition to these loaded positions and cruise numbers of the Ro-Ro vessels, utilization rates of the company owned Ro-Ro vessels on each marine link at each month are determined as in Fig. 17. These rates are computed by dividing the total number of monthly load units shipped by Ro-Ro vessels into the multiplication of maximum number of monthly cruises and capacity of a Ro-Ro vessel. Since some agreements on marine tariffs and weekly fixed schedules, there are lower/upper bounds available on these rates.

Finally, import/export loaded positions by the other logistics service providers' Ro-Ro vessels are shown by Fig. 18. Similarly, monthly import and export loaded train positions performed by block train and public train services are presented in Fig. 19. In fact, block train services have fixed timing schedules and full-than-train load characteristic, i.e., chartering the whole capacity of a block train. Moreover, some of the loads can be shipped by public trains which have less-than-train load property and flexible time schedules. In other words, loads of the different logistics companies can be shipped simultaneously by public trains and do not require chartering the whole train capacity by the logistics company. The detailed monthly block train trips among the different rail stations in Europe are given in the following Fig. 20. There is a balance between the import and export block train services at each month which fulfills the fixed scheduling limitations of block trains. Moreover, loading/unloading operations for some freights at the transshipment rail stations are also taken into account while calculating the number of block trains and loaded positions by public trains. Finally, import/export loaded positions by public trains on different rail lines are shown by Fig. 21. It should be noted here that load planning in intermodal transportation may become more complex than unimodal transportation systems due to the large amounts of freights, truck and trailer fleet sizes, multiple time periods and high level of uncertainty. Besides, load plans for all of the transportation modes should be optimized in an integrated manner instead of optimizing each mode separately. As highlighted by Baykasoğlu et al. (2015), load or intermodal freight transportation planning is one of the most vital problem since

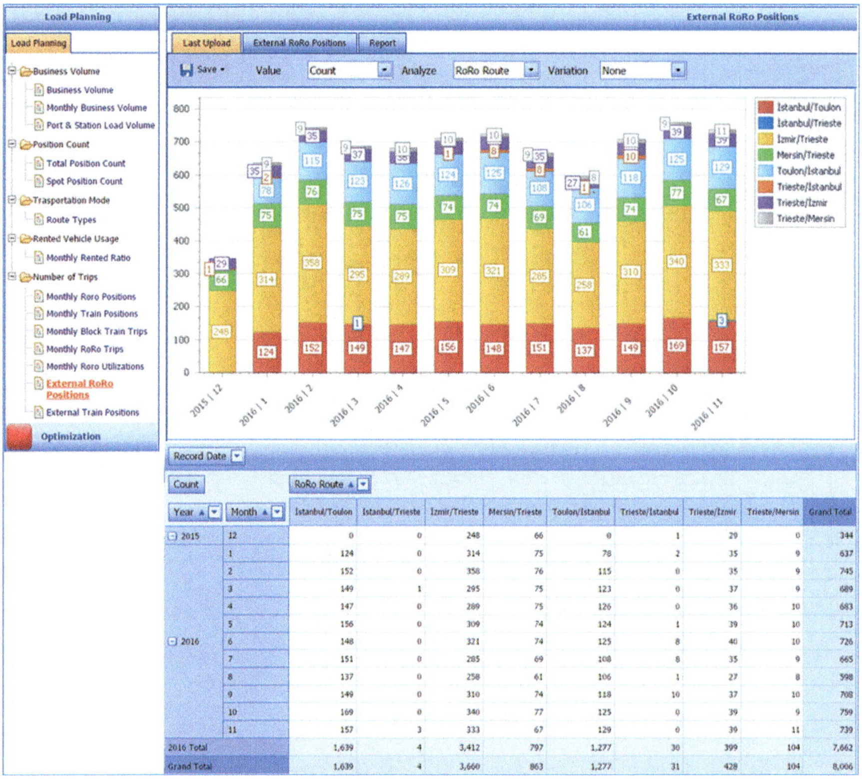

Fig. 18 Number of monthly cruises by other logistics service providers' Ro-Ro vessels

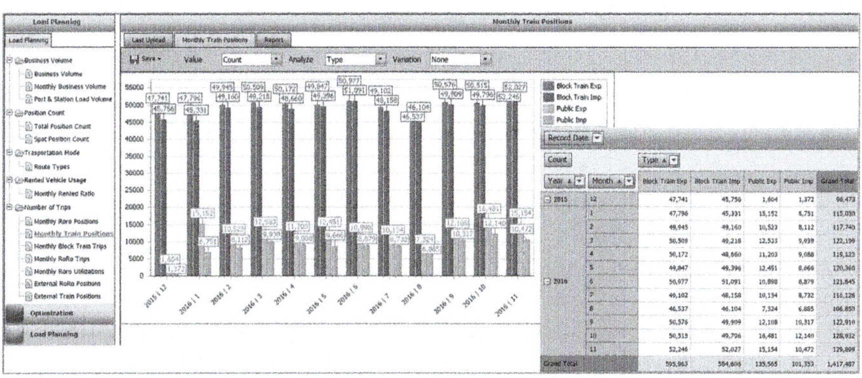

Fig. 19 Monthly import/export loaded trailer positions by block and public train services

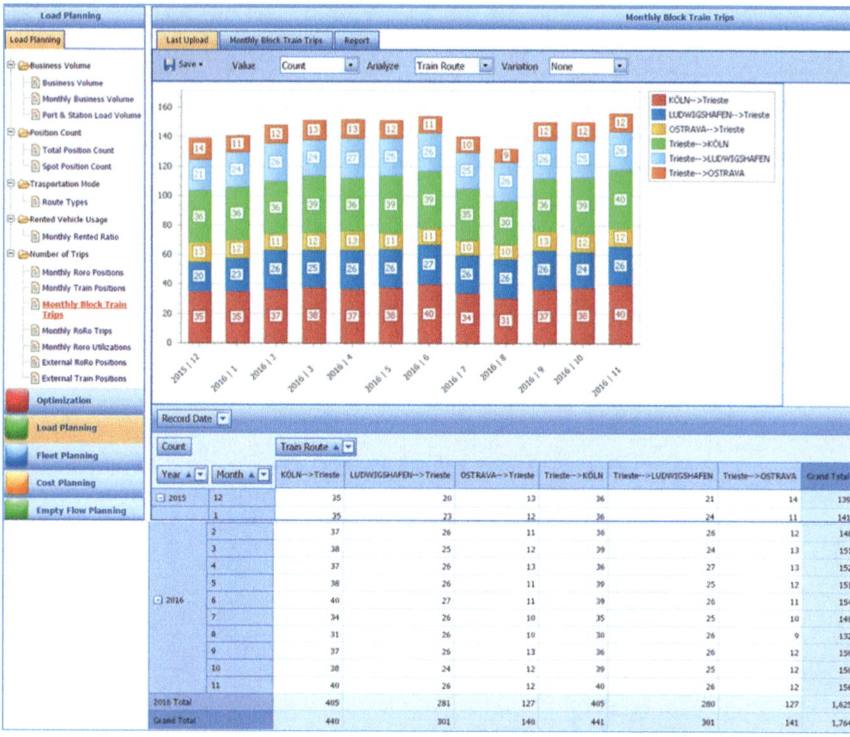


**Fig. 20** Number of monthly trips for block trains among different railway stations

influencing other planning decisions directly. Because, outputs of load planning are used as inputs by all of the other fleet planning issues.

## 5.4 *Fleet Planning Module*

In this module, both truck and trailer fleet size optimization are carried out. Fleet size of vehicles is determined for each month by considering loaded vehicle positions, empty vehicle repositions, transit times, duration of a month in hours and vehicle availabilities (List et al. 2003). In detail, quantities of loaded vehicle positions can be obtained from the solution of multi-mode load planning. In addition to the loaded vehicle movements, road-marine-rail transit times between the origin and destination pairs, monthly duration, number of working days in each month, and percent of time for vehicle availabilities are needed so as to determine appropriate fleet size. The minimum number of required trucks at arcs $(i,j)$ and $(j,i)$ can be calculated by the following mathematical formulas (1)–(2) (List et al. 2003;

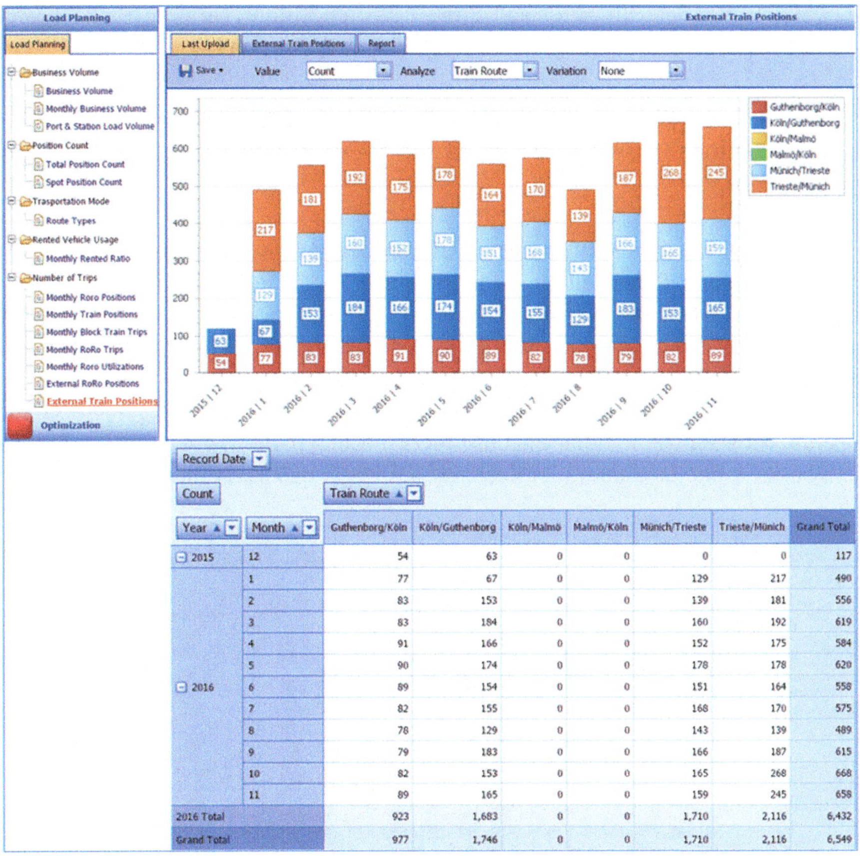


| Year | Month | Guthenborg/Köln | Köln/Guthenborg | Köln/Malmö | Malmö/Köln | Münich/Trieste | Trieste/Münich | Grand Total |
|---|---|---|---|---|---|---|---|---|
| 2015 | 12 | 54 | 63 | 0 | 0 | 0 | 0 | 117 |
| 2016 | 1 | 77 | 67 | 0 | 0 | 129 | 217 | 490 |
| | 2 | 83 | 153 | 0 | 0 | 139 | 181 | 556 |
| | 3 | 83 | 184 | 0 | 0 | 160 | 192 | 619 |
| | 4 | 91 | 166 | 0 | 0 | 152 | 175 | 584 |
| | 5 | 90 | 174 | 0 | 0 | 178 | 178 | 620 |
| | 6 | 89 | 154 | 0 | 0 | 151 | 164 | 558 |
| | 7 | 82 | 155 | 0 | 0 | 168 | 170 | 575 |
| | 8 | 78 | 129 | 0 | 0 | 143 | 139 | 489 |
| | 9 | 79 | 183 | 0 | 0 | 166 | 187 | 615 |
| | 10 | 82 | 153 | 0 | 0 | 165 | 268 | 668 |
| | 11 | 89 | 165 | 0 | 0 | 159 | 245 | 658 |
| 2016 Total | | 923 | 1,683 | 0 | 0 | 1,710 | 2,116 | 6,432 |
| Grand Total | | 977 | 1,746 | 0 | 0 | 1,710 | 2,116 | 6,549 |

**Fig. 21** Monthly import/export loaded positions by public trains on different rail lines

Baykasoğlu et al. 2015; Baykasoğlu and Subulan 2017):

$$Truck_{ij} + Truck_{ji} \geq 2 \times \left[ \frac{\max \left\{ Time_{ij} \times Position_{ij}, Time_{ji} \times Position_{ji} \right\}}{Period\ Length \times Truck\ Availability} \right]$$
$$= \frac{Time_{ij} \times Position_{ij} + Time_{ji} \times Position_{ji} + \left| Time_{ij} \times Position_{ij} - Time_{ji} \times Position_{ji} \right|}{Period\ Length \times Truck\ Availability} \tag{1}$$

$$Period\ Length = Number\ of\ working\ days \times duration\ (in\ hours) \tag{2}$$

The required number of empty truck repositions between the nodes $i$ and $j$ can be determined by the following formulas (3)–(4):

$$Reposition_{ji} \geq Position_{ij} - Position_{ji} - Truck_{ij} \tag{3}$$

$$Reposition_{ij} \geq Position_{ji} - Position_{ij} - Truck_{ji} \tag{4}$$

Finally, after fleet allocation and empty truck repositions, periodic truck inventories at the end of the planning period can be formulated as in Eq. (5).

$$\begin{cases} Inventory_i = Truck_{ij} - Position_{ij} + Position_{ji} + Reposition_{ji} - Reposition_{ij} \\ Inventory_j = Truck_{ji} - Position_{ji} + Position_{ij} + Reposition_{ij} - Reposition_{ji} \end{cases} \tag{5}$$

Moreover, fixed ownership costs of European/Turkish plates trucks and trailers are also included. Composition of the truck fleet is also determined by considering the different types of identification plates such as European and Turkish plates vehicles. The monthly trailer and truck fleet size requirements are also depicted in Figs. 22 and 23.

The fleet sizing results have shown that most of the European plates trucks should operate between rail stations and customers in Europe. Furthermore, it can also be emphasized that the total trailer fleet size has larger quantities than truck fleet size.

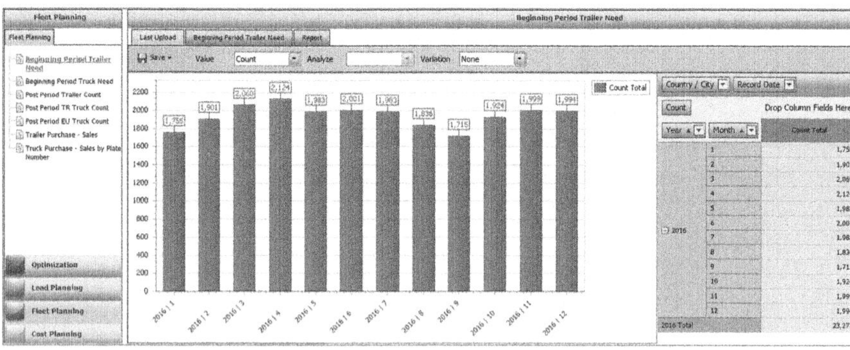


**Fig. 22** Monthly trailer fleet size requirements on the overall intermodal logistics network

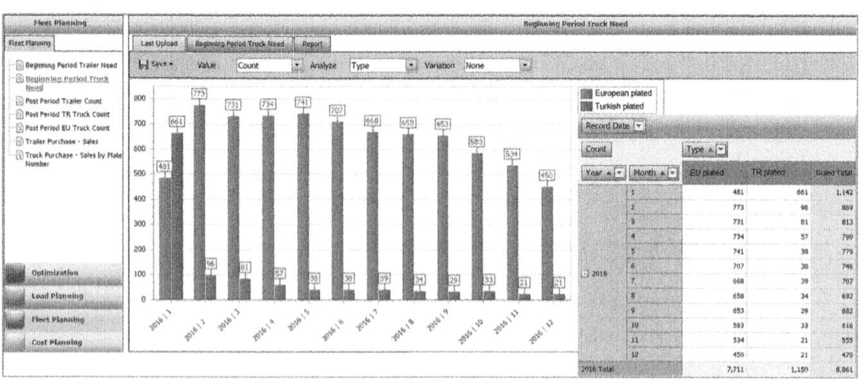


**Fig. 23** The monthly truck fleet size requirements with different types of plates on the road network

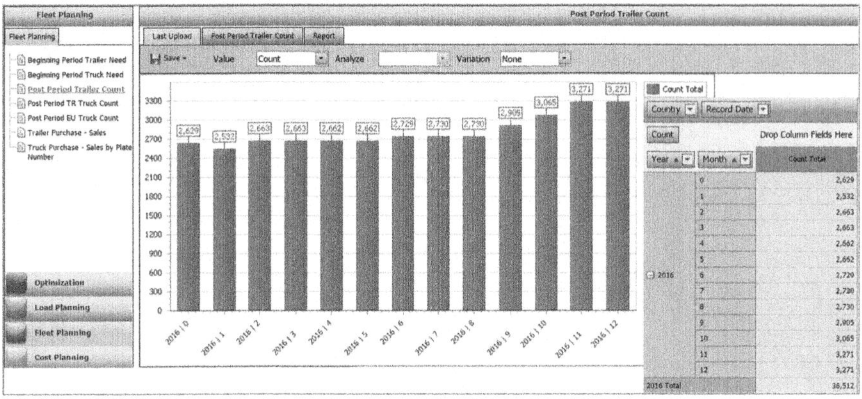


**Fig. 24** Periodic trailer inventories at the seaports in Turkey and all of the order countries in Europe

The monthly variations of the truck and trailer fleet size requirements have similar characteristics.

The large amount of truck fleet size composed of European plates trucks. Finally, most of the trailers should be allocated to seaports in Turkey rather than customers in Europe. In addition to fleet sizing and composition, this module also covers the periodic truck/trailer inventories on the intermodal logistics network, idle trucks/trailers awaiting at different locations and fleet expansion/reduction decisions. The periodic trailer inventory levels on the different locations at the end each month are demonstrated by Fig. 24.

In order to compute the truck inventories of any given node at the end of each time period, initial truck inventory levels, outgoing loaded truck positions, incoming loaded truck positions, internal/external empty truck repositions, new truck purchases and sales of second-hand trucks are taken into account. The periodic Turkish and European plates truck inventories on the different locations at the end of each month are shown by Figs. 25 and 26, respectively. The truck inventory levels at the customer zones and rail stations are comparatively higher than seaports. Opposite to the unstable Turkish plates truck inventories, European plates truck inventory levels have stable characteristic at the customer zones, seaports and rail stations. Most of the European plates truck inventories will accumulate at the customer zones and seaports to the end of the planning horizon. Moreover, it is also possible to calculate the idle trucks/trailers awaiting at any given location by subtracting the fleet size requirement of this node from its beginning inventory level. In fact, truck sales tend to increase so as to balance the purchasing costs and decrease the total transportation costs. The amounts of trailer purchases and sales can be seen from the following Fig. 27. Trailer purchases will increase to the end of planning horizon. According to Fig. 28, vast amounts of Turkish plates trucks should be sold at the first time period.

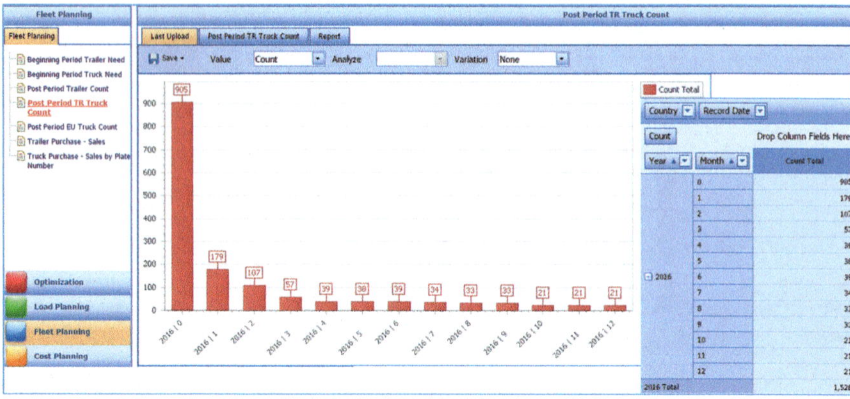


Fig. 25 Periodic Turkish plates truck inventories at different nodes on the intermodal logistics network

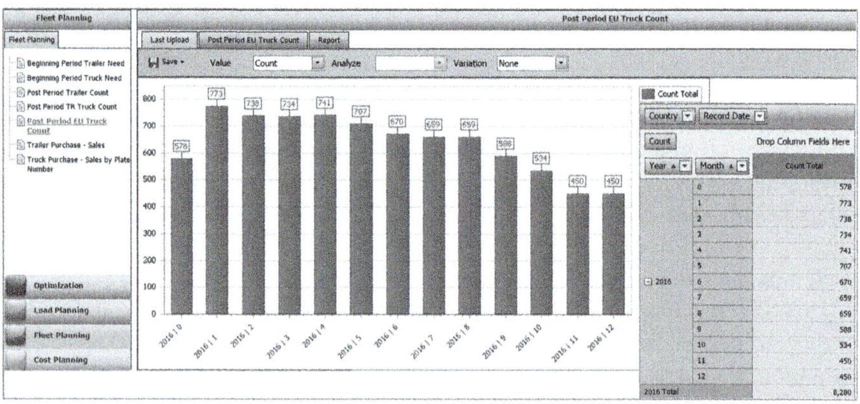


Fig. 26 Periodic European plates truck inventories at different nodes on the intermodal logistics network

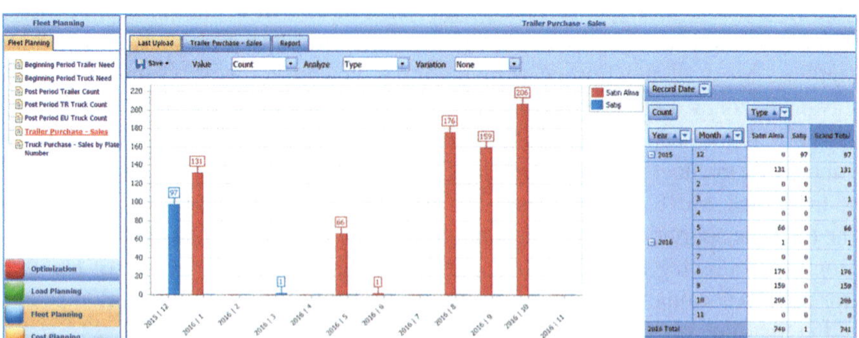


Fig. 27 The amounts of monthly trailer purchases and sales

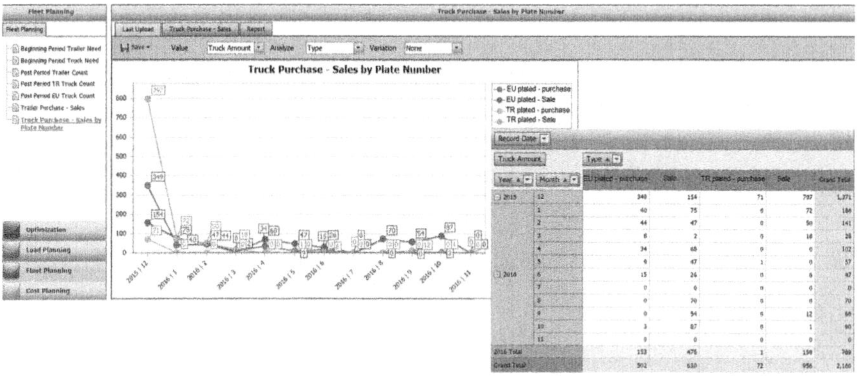

Fig. 28 The amounts of monthly Turkish and European plates truck purchases and sales

After this month, sales/purchasing amounts of both European and Turkish plates trucks will be in steady state. Moreover, purchasing/sales of trucks cannot be performed simultaneously within the same month for any given location. Figure 28 presents just total ones.

## 5.5 *Transportation Cost Planning Module*

In order to improve the economic performance of the logistics companies, transportation cost planning module was constructed. In detail, this module involves a function of total transportation costs including costs of road, marine, railway transports and outsourcing expenses. The road freight transportation cost is divided into three categories, i.e., costs for loaded trucks, empty trucks and trucks with empty trailers. The empty vehicle repositioning costs are defined for both trucks and trailers. Moreover, fixed ownership costs of European/Turkish plates trucks and trailers are taken into account for both active and idle resources. Marine transport costs consist of cost of per transportation unit by company-owned or other logistics service providers' Ro-Ro vessels. Rail transport costs are defined based on per block train trip and cost of per transportation unit by public trains. Finally, purchasing costs for new trucks/trailers and revenues obtained from the sales of second hand trucks/trailers are also included so as to determine periodic fleet expansion and reduction decisions.

The variation of total monthly transportation costs is shown by Fig. 29. Additionally, comparisons of cost components are made as in Fig. 30 which shows that fuel oil, marine and railway transportation costs constitute large portion of the overall annual transportation costs.

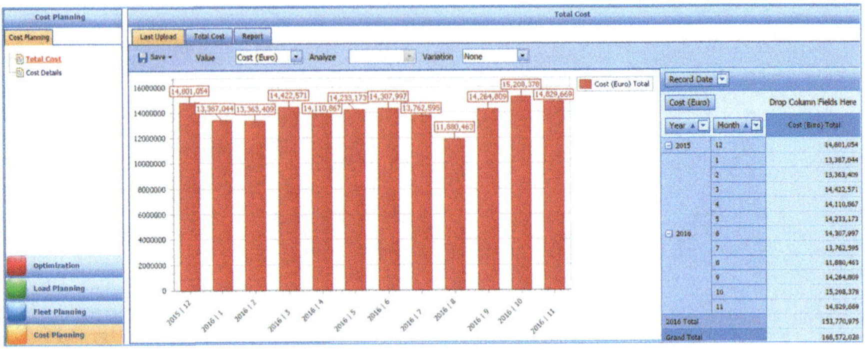


Fig. 29 The variation of monthly overall transportation costs

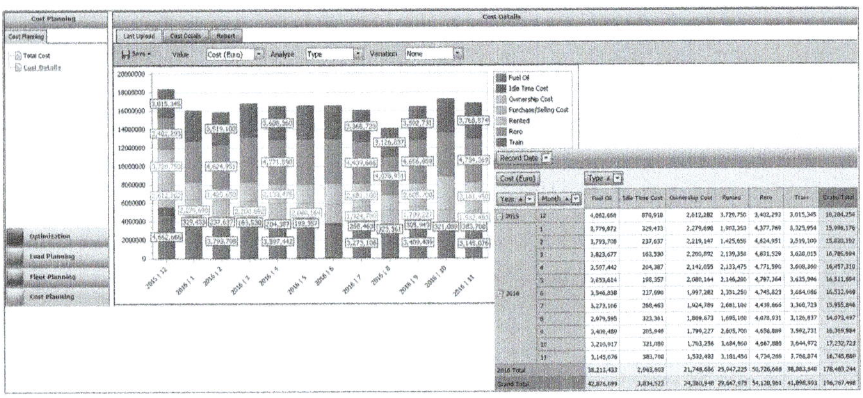


Fig. 30 Comparison of the detailed transportation costs components

## 5.6 *Empty Truck/Trailer Repositioning Planning Module*

There are two different sources of empty flows namely internal and external empty vehicle repositions. Due to the imbalance of import/export loaded vehicle movements within the same month, a need for internal empty vehicle repositioning occurs. An appropriate fleet size and allocation can provide minimal internal empty vehicle repositions. These repositions can be performed between a pair of nodes where loaded vehicle movements are available among these nodes. On the other hand, due to the imbalance between export vehicle fleet size and vehicle inventories in the consecutive time periods, a need for external vehicle repositioning may arise. In detail, difference between outgoing vehicle fleet size from a node and the vehicle inventory of this node at the beginning of the month can be satisfied by external empty vehicle repositions. Opposite to internal empty vehicle repositioning, external empty repositions can be performed between any given nodes where loaded vehicle movements may not be possible.

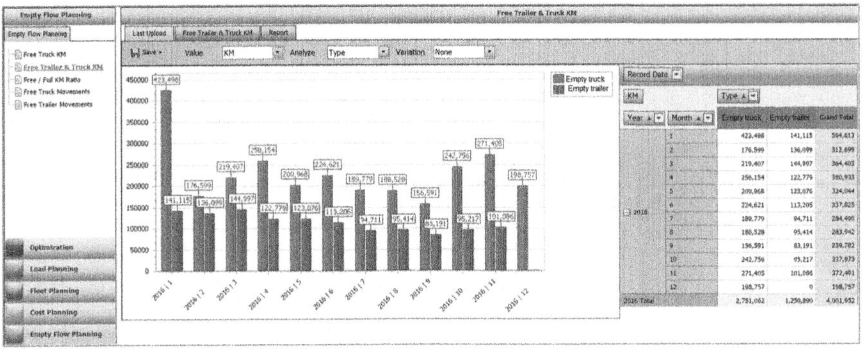


Fig. 31 Comparison of the empty truck and trailer repositions

In Fig. 31, total amount of mileages (km) by empty truck and trailer repositions are presented. It is obvious that amounts of empty truck repositions are relatively higher than the empty trailer repositions. Moreover, majority of the empty trailer repositions are performed by European plates trucks within Europe. Due to the large travel distances, there is no empty flows by Turkish plates trucks between Europe and Turkey.

European plates empty trucks are generally repositioned between the railway stations and final destinations or customers. It should also be noted here that there are no empty truck or trailer repositions among the seaports themselves in Turkey. Furthermore, most of the empty truck/trailer repositions will occur among the customers themselves. Besides the large amounts of empty trailer repositions by European plates trucks, some empty trailers are also repositioned by Turkish plates trucks at the first quarter of the year. According to the company experts' opinions, amount of empty vehicle repositions is at acceptable and stabile level when compared to loaded vehicle positions as it is seen by Fig. 32. In addition

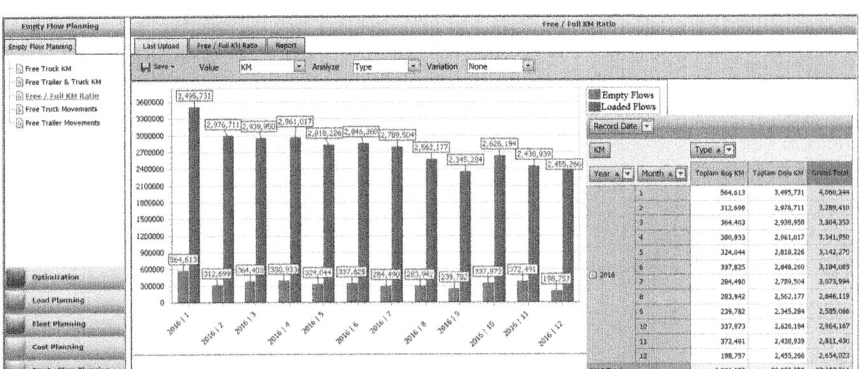


Fig. 32 Comparison of the monthly loaded positions and empty repositions

Fig. 33 The amounts of monthly empty truck repositions

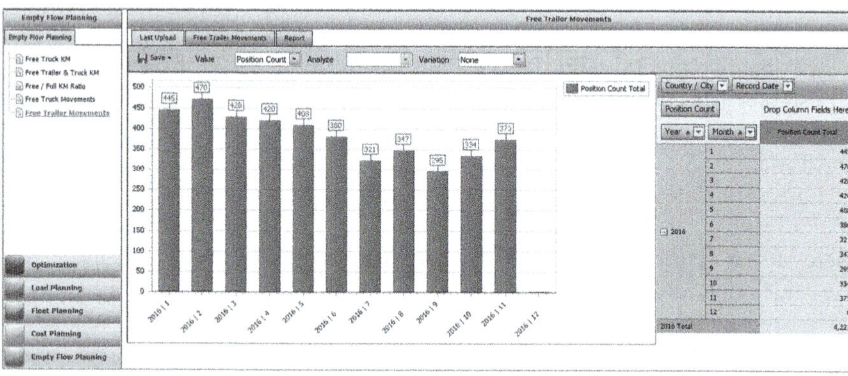


Fig. 34 The amounts of monthly empty trailer repositions

to mileages of empty flows, amounts of empty flows are also given in Figs. 33 and 34 in terms of reposition numbers for empty trucks and trailers, respectively. While calculating the internal vehicle repositions, load/freight plans, loaded vehicle positions and relevant fleet sizes are utilized. The relationship between loaded vehicle positions and fleet sizes is also taken into consideration. On the other hand, vehicle fleet sizes and periodic vehicle inventory levels are used to compute external empty vehicle repositions. The relationship between fleet sizes and vehicle inventory levels is also considered. For the transit time goal, repositioning times are included in the objective function. Finally, cabotage limitations within Europe are also handled for performing empty vehicle repositions by European and Turkish plates trucks.

## 5.7 Transit Time Planning Module

In order to provide on time delivery of loads to the destinations and satisfaction of customer freight demands as soon as possible, minimization of total transit time including loading/unloading times, travel times, times of consolidation operations, customs clearance times and awaiting times at the terminals is also carried out. Transit time objective was formulated by incorporating travel times of loaded vehicle movements, combined marine and road freight transit times, intermodal transit times, transfer times of outsourced services and empty vehicle repositioning times. The variations of total monthly transit times are given in Fig. 35. It is obvious that transit time values are steady throughout the planning horizon.

Moreover, the components of the transit time objective are compared in Fig. 36. Intermodal transportation has the largest transit times and is the most low-speed

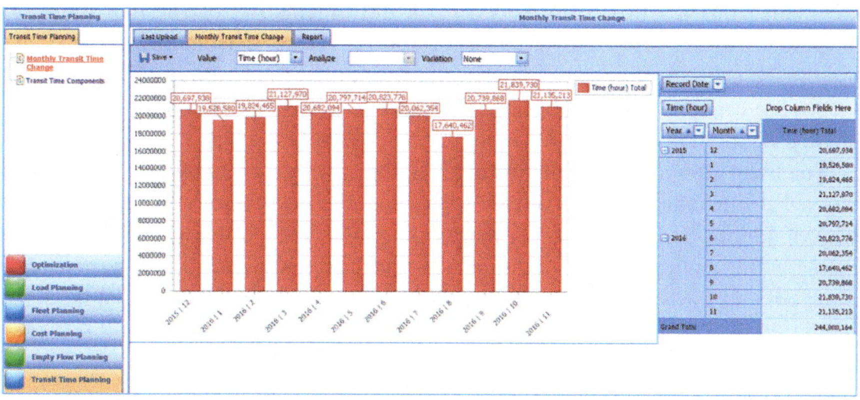


**Fig. 35** The variation of total monthly transit times

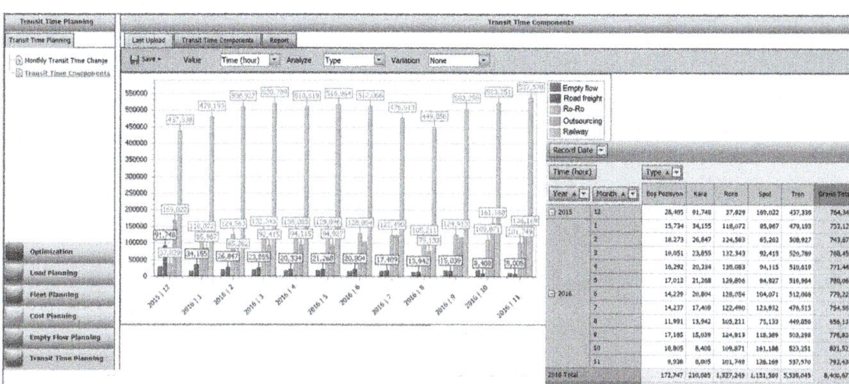


**Fig. 36** Comparison of the transit time components

transportation mode. The combined marine-road freight transportation and outsourcing option are the other alternatives which may cause significant increases in transit times. The most important reason is that highly utilization of maritime logistics services into these alternatives. It should also be noted here that only minor increases may exist in transit time when the empty flows are included.

## 5.8 *Environmental Impact Assessment Module*

Nowadays, most of the logistics companies are committed to produce environmentally conscious freight and fleet plans so as to reduce the environmental damage of transportation operations. In this module, minimization of total environmental impact or total $CO_2$ emissions by different transportation modes is intended. For this reason, an environmental objective function is formulated by including the $CO_2$ emissions resulted from loaded truck movements by road freight transport, maritime logistics services, rail transportation services, outsourcing services and empty vehicle repositions. Similarly, variations of the total monthly $CO_2$ emission values are first presented in Fig. 37. It should be noted here that there is a balance among the months in terms of the environmental impact values.

The main components of environmental objective are given by Fig. 38. Computational results have shown that rail transport is the most environmentally friendly way of freight transportation. Furthermore, empty vehicle repositioning does not increase the environmental pollution significantly. However, direct road freight transport, marine transport and outsourcing services may have substantial influence on the total $CO_2$ emission. Because, road and marine transports have the largest carbon dioxide, hydrocarbon, particle and nitrogen emission values.

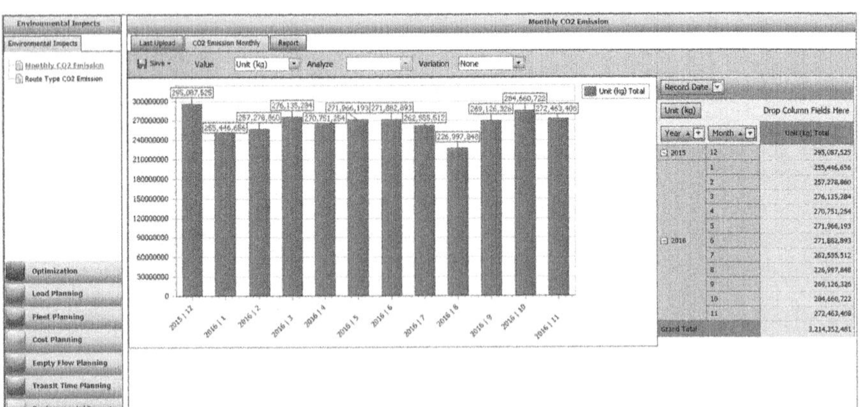


**Fig. 37** The variation of total monthly $CO_2$ emission values

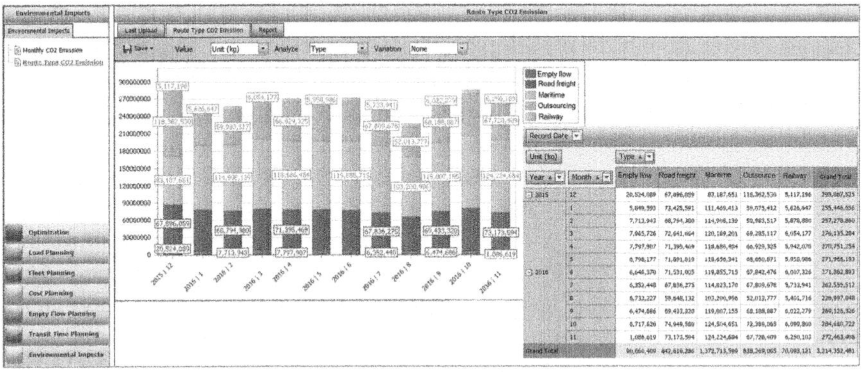


**Fig. 38** Comparison of the total environmental impact components

# 6 Concluding Remarks and Future Research Directions

In this paper, a web-based DSS with modular structure is developed for an integrated multi-objective, multi-mode, multi-stage and multi-period intermodal fleet planning problem. The proposed DSS consists of four main components. The model component includes a fuzzy-stochastic mixed-integer programming model since the real-life fleet management systems may have different uncertainty categories simultaneously. Moreover, load planning, fleet sizing/composition, fleet allocation, empty vehicle repositioning, active and inactive vehicle inventories on the transportation network, fleet expansion/reduction decisions constitute different modules of the developed optimization model. In the data component, user tables in Oracle database and UML class diagrams are first constructed for the database, solver options, input data and model outputs. The fuzzy-stochastic mathematical programming model is solved through a Windows server 2012 R2 64-bit where the LINGO codes are run over C# by making use its own of DLL libraries and scripts. Then, because of the re-optimization is needed under different objectives, uncertainty levels and dynamic changes in data, Excel macros are written for generating graphs and tables dynamically from Excel spreadsheets related to the model outputs. The Excel macros were run over C# via Microsoft Office Interop Excel library and an object-oriented matching was utilized for connection among Oracle database and C# programs. In the decision component, it is able to produce various freight and fleet plans with respect to some sustainability objectives, i.e., cost minimization, customer satisfaction maximization and environmental impact minimization under different uncertainty levels and risk attitudes. It is also possible to generate compromise solutions to these conflicting objectives via a fuzzy multi-objective optimization method. While generating solutions in accordance with different risk levels or feasibility degrees of fuzzy model constraints, an interactive resolution method was applied. Besides, chance-constrained stochastic programming approach was also employed for handling probabilistic freight demands of the customers. Finally, a web-based user interface component of the proposed DSS is designed by using .Net and C# programs on an integrated development environment namely Microsoft Visual Studio. Furthermore, DevExpress which aims to build highly configurable and well-designed user interface controls is used for presenting graphical reports and documents. The proposed DSS is implemented on a real-life case study of a large-sized international logistics company. When single and multi-objective optimization results are compared in deterministic environments, empty truck and trailer repositions will be higher than stochastic and fuzzy optimization environments. Furthermore, it is expressed by the company experts that compromise solution contains acceptable level of external empty flows. Moreover, deterministic and fuzzy fleet planning cases have the largest internal empty repositions. Fortunately, compromise solution has the minimal level of internal empty truck flows which make the solution more applicable. After developing such an optimization-based DSS, results of the proposed fuzzy-stochastic multi-objective fleet planning model are compared with the company's realized performance measures on transportation cost, transit time and $CO_2$ emission values. The computational results have shown that the logistics company could achieve approximately more than 13.2% of total transportation cost over the realized cost objective if it had used the proposed optimization-based DSS. In other words, usage of such an optimization-based DSS gives the opportunity of decreasing the annual transportation costs down nearly €168572028 with 70.2% and 62.2% satisfaction degrees on total transit time and $CO_2$ emission objectives, respectively. In fact, there will be approximately 10.7% and 8.9% improvements on the total transit time and environmental pollution objectives. It is also possible to provide approximately 15.3% fleet size reduction for trucks and trailers. In detail, the truck and trailer fleet size of the company can be decreased down to almost 750 trucks and 2000 trailers, respectively. This will also provide decreases in ownership and inventory costs of the available vehicles.

The proposed DSS mainly focused on the strategic and tactical fleet planning problems. It can be extended by adding some operational level decisions such as truck routing, scheduling of trains and Ro-Ro vessels and load consolidation at the terminals/hubs. Furthermore, the proposed DSS is the lack of some other essential fleet planning problems such as container fleet sizing and empty repositioning, fleet replacement decisions for trucks/trailers and coordination/collaboration among the different actors of the intermodal transportation network.

In the future, model component can be extended by adding some operational fleet planning decisions such as truck routing and scheduling of block trains and Ro-Ro vessels. In addition, developing a cloud-based distributed DSS which is also accessible by mobile devices is scheduled as a future work to facilitate the decision making process for planning and design of such a complex fleet management decisions.

**Acknowledgement** This paper is supported by Ministry of Science, Industry & Technology of Turkey in the scope of SAN-TEZ project No: 0617.STZ.2014.

## Appendix 1 Mathematical Nomenclature and Model Formulations

Supplementary mathematical formulations associated with the proposed fuzzy-stochastic programming model can be found, in the online version, at http://web.deu.edu.tr/baykasoglu/Appendix_A.pdf

## Appendix 2 Some of the Dataset for Real-Life Application

Some of the dataset of the real-life application in an international logistics company can be accessed at the following link: http://web.deu.edu.tr/baykasoglu/Appendix_B.pdf

## Appendix 3 LINGO Code of the Proposed Model

The reader can refer to the following link for the LINGO 15.0 code of the proposed fuzzy-stochastic mathematical programming model: http://web.deu.edu.tr/baykasoglu/Appendix_C.pdf

## References

Andersson, T. (2005). *Decision support tools for dynamic fleet management – applications in airline planning and ambulance logistics*. Dissertation No. 942, Department of Science and Technology Linköpings Universitet, Norrköping.

Andersson, T., & Värbrand, P. (2007). Decision support tools for ambulance dispatch and relocation. *Journal of the Operational Research Society, 58*(2), 195–201.

Arnold, P., Peeters, D., & Thomas, I. (2004). Modelling a rail/road intermodal transportation system. *Transportation Research Part E, 40*(3), 255–270.

Assadipour, G., Ke, G. Y., & Verma, M. (2015). Planning and managing intermodal transportation of hazardous materials with capacity selection and congestion. *Transportation Research Part E, 76*, 45–57.

Avramovich, D., Cook, T. M., Langston, G. D., & Sutherland, F. (1982). A decision support system for fleet management: A linear programming approach. *Interfaces, 12*(3), 1–9.

Bandeira, D. L., Becker, J. L., & Borenstein, D. (2009). A DSS for integrated distribution of empty and full containers. *Decision Support Systems, 47*(4), 383–397.

Basnet, C., Foulds, L., & Igbaria, M. (1996). FleetManager: A microcomputer-based decision support system for vehicle routing. *Decision Support Systems, 16*, 195–207.

Bauer, J., Bektaş, T., & Crainic, T. G. (2010). Minimizing greenhouse gas emissions in intermodal freight transport: An application to rail service design. *Journal of the Operational Research Society, 61*, 530–542.

Baykasoğlu, A., & Subulan, K. (2016). A multi-objective sustainable load planning model for intermodal transportation networks with a real-life application. *Transportation Research Part E, 95*, 207–247.

Baykasoğlu, A., & Subulan, K. (2017). A new intermodal fleet planning model via hybrid fuzzy-stochastic mathematical programming method. In *Abstract book of international workshop on mathematical methods in engineering* (p. 123). Istanbul.

Baykasoğlu, A., Subulan, K., Dudaklı, N., Taşan, A. S., Kaplan, M. C., & Turan, M. (2015). A new mixed-integer programming model for strategic and tactical fleet planning problems in intermodal transportation networks with a real life application. In *Proceeding book of XIII: International logistics and supply chain congress* (pp. 637–648). Izmir.

Bhattacharya, A., Kumar, S. A., Tiwari, M. K., & Talluri, S. (2014). An intermodal freight transport system for optimal supply chain logistics. *Transportation Research Part C, 38*, 73–84.

Bierwirth, C., Kirschstein, T., & Meisel, F. (2012). On transport service selection in intermodal rail/road distribution networks. *BuR-Business Research, 5*(2), 198–219.

Bruns, F., & Knust, S. (2012). Optimized load planning of trains in intermodal transportation. *OR Spectrum, 34*(3), 511–533.

Caris, A., Macharis, C., & Janssens, G. K. (2008). Planning problems in intermodal freight transport: Accomplishments and prospects. *Transportation Planning and Technology, 31*(3), 277–302.

Caris, A., Macharis, C., & Janssens, G. K. (2013). Decision support in intermodal transport: A new research agenda. *Computers in Industry, 64*(2), 105–112.

Chang, T. S. (2008). Best routes selection in international intermodal networks. *Computers & Operations Research, 35*(9), 2877–2891.

Cho, J. H., Kim, H. S., & Choi, H. R. (2012). An intermodal transport network planning algorithm using dynamic programming—A case study: From Busan to Rotterdam in intermodal freight routing. *Applied Intelligence, 36*(3), 529–541.

Couillard, J. (1993). A decision support system for vehicle fleet planning. *Decision Support Systems, 9*(2), 149–159.

Dejax, P. J., & Crainic, T. G. (1987). Survey paper-a review of empty flows and fleet management models in freight transportation. *Transportation Science, 21*(4), 227–248.

Demir, E., Burgholzer, W., Hrusovsky, M., Arıkan, E., Jammernegg, W., & Woensel, T. V. (2016). A green intermodal service network design problem with travel time uncertainty. *Transportation Research Part B, 93*, 789–807.

Dudaklı, N., Baykasoğlu, A., Subulan, K., Tasan, A. S., Kaplan, M. C., & Turan, M. (2015). A mathematical model proposal for fleet planning problem of a real-life intermodal transportation network. In *Abstract book international conference on operations research*. Vienna.

Ekol Logistics Inc. (2017). Retrieved from https://quadronet.ekol.com/login.aspx on 6th June.

El-Din, M. M. M., Ghali, N. I., Sadek, A., & Abouzeid, A. A. (2015). Decision support system for airlines fleet capacity management. *International Journal of Computer Applications, 109*(16), 1–8.

Ezabadi, M. G., & Vergara, H. A. (2016). Decomposition approach for integrated intermodal logistics network design. *Transportation Research Part E, 89*, 53–69.

Fagerholt, K. (2004). A computer-based decision support system for vessel fleet scheduling – experience and future research. *Decision Support Systems, 37*, 35–47.

Fagerholt, K., Christiansen, M., Hvattum, L. M., Johnsen, T. A. V., & Vabø, T. J. (2010). A decision support methodology for strategic planning in maritime transportation. *Omega, 38*(6), 465–474.

Fagerholt, K., & Lindstad, H. (2007). TurboRouter: An interactive optimisation-based decision support system for ship routing and scheduling. *Maritime Economics & Logistics, 9*(3), 214–233.

Febbraro, A. D., Sacco, N., & Saeednia, M. (2016). An agent-based framework for cooperative planning of intermodal freight transport chains. *Transportation Research Part C, 64*, 72–85.

Garcia, J., Florez, J. E., Torralba, A., Borrajo, D., Lopez, C. L., Garcia-Olaya, A., & Saenz, J. (2013). Combining linear programming and automated planning to solve intermodal transportation problems. *European Journal of Operational Research, 227*(1), 216–226.

Guelat, J., Florian, M., & Crainic, T. G. (1990). A multimode multiproduct network assignment model for strategic planning of freight flows. *Transportation Science, 24*(1), 25–39.

Grzybowska, H., & Barceló, J. (2012). Decision support system for real-time urban freight management. *Procedia – Social and Behavioral Sciences, 39*, 712–725.

Herrera, M., Agrell, P. J., Manrique-de-Lara-Peñate, C., & Trujillo, L. (2017). Vessel capacity restrictions in the fleet deployment problem: An application to the Panama Canal. *Annals of Operations Research, 253*(2), 845–869.

Hill, A., & Böse, J. W. (2017). A decision support system for improved resource planning and truck routing at logistic nodes. *Information Technology and Management, 18*(3), 241–251.

Imai, A., & Rivera, F. (2001). Strategic fleet size planning for maritime refrigerated containers. *Maritime Policy & Management, 28*(4), 361–374.

Inghels, D., Dullaert, W., & Vigo, D. (2016). A service network design model for multimodal municipal solid waste transport. *European Journal of Operational Research, 254*(1), 68–79.

Jang, W., Noble, J., Klein, C., Nemmers, C., Yu, Z., & Zhang, Y. (2008). *A decision support system for optimal depot and fleet management.* A report from The Midwest Transportation Consortium Iowa State University.

Jimenez, M., Arenas, M., Bilbao, A., & Rodriguez, M. V. (2007). Linear programming with fuzzy parameters: An interactive method resolution. *European Journal of Operational Research, 177*(3), 1599–1609.

Kalinina, M., Olsson, L., & Larsson, A. (2013). A multi objective chance constrained programming model for intermodal logistics with uncertain time. *International Journal of Computer Science Issues, 10*(6), 35–44.

Kek, A. G. H., Cheu, R. L., Meng, Q., & Fung, C. H. (2009). A decision support system for vehicle relocation operations in carsharing systems. *Transportation Research Part E, 45*(1), 149–158.

Li, L., Negenborn, R. R., & Schutter, B. D. (2015). Intermodal freight transport planning – A receding horizon control approach. *Transportation Research Part C, 60*, 77–95.

Lin, C. C., & Lin, S. W. (2016). Two-stage approach to the intermodal terminal location problem. *Computers & Operations Research, 67*, 113–119.

Lindo Systems Inc. (2016). Retrieved from http://www.lindo.com/downloads/PDF/LINGO.pdf

List, G. F., Wood, B., Nozick, L. K., Turnquist, M. A., Jones, D. A., Kjeldgaard, E. A., & Lawton, C. R. (2003). Robust optimization for fleet planning under uncertainty. *Transportation Research Part E, 39*(3), 209–227.

Macharis, C., & Bontekoning, Y. M. (2004). Opportunities for OR in intermodal freight transport research: A review. *European Journal of Operational Research, 153*(2), 400–416.

Macharis, C., Caris, A., Jourquin, B., & Pekin, E. (2011). A decision support framework for intermodal transport policy. *European Transport Research Review, 3*(4), 167–178.

Martínez-López, A., Sobrino, P. C., González, M. C., & Trujillo, L. (2018). Optimization of a container vessel fleet and its propulsion plant to articulate sustainable intermodal chains versus road transport. *Transportation Research Part D, 59*, 134–147.

Matsatsinis, N. F. (2004). Towards a decision support system for the ready concrete distribution system: A case of a Greek company. *European Journal of Operational Research, 152*, 487–499.

Meisel, F., Kirschstein, T., & Bierwirth, C. (2013). Integrated production and intermodal transportation planning in large scale production–distribution networks. *Transportation Research Part E, 60*, 62–78.

Müller, D., & Tierney, K. (2017). Decision support and data visualization for liner shipping fleet repositioning. *Information Technology and Management, 18*(3), 203–221.

Polo, A., Robol, F., Nardin, C., Marchesi, S., Zorer, A., Zappini, L., Viani, F., Massa, A. (2014). Decision support system for fleet management based on TETRA terminals geolocation. *IEEE 8th European conference on antennas and propagation.*

Puetmann, C., & Stadtler, H. (2010). A collaborative planning approach for intermodal freight transportation. *OR Spectrum, 32*(3), 809–830.

Repede, J. F. (1994). Developing and validating a decision support system for locating emergency medical vehicles in Louisville, Kentucky. *European Journal of Operational Research, 75*(3), 567–581.

Resat, H. G., & Turkay, M. (2015). Design and operation of intermodal transportation network in the Marmara region of Turkey. *Transportation Research Part E, 83*, 16–33.
Ruiz, R., Maroto, C., & Alcaraz, J. (2004). A decision support system for a real vehicle routing problem. *European Journal of Operational Research, 153*(3), 593–606.
Saidane, C. (2007). *Decision support system for the management of an Army's tracked and wheeled vehicle fleet*, Master Thesis, Naval Postgraduate School Monterey, California.
Santos, L., Countinho-Rodrigues, J., & Antunes, C. H. (2011). A web spatial decision support system for vehicle routing using Google Maps. *Decision Support Systems, 51*(1), 1–9.
Schorpp, S. (2011). *Dynamic fleet management for international truck transportation*. Germany: Springer.
SteadieSeifi, M., Dellaert, N. P., Nuijten, W., Van Woensel, T., & Raoufi, R. (2014). Multimodal freight transportation planning: A literature review. *European Journal of Operational Research, 233*(1), 1–15.
Verma, M., & Verter, V. (2010). A lead-time based approach for planning rail–truck intermodal transportation of dangerous goods. *European Journal of Operational Research, 202*(3), 696–706.
Verma, M., Verter, V., & Zufferey, N. (2012). A bi-objective model for planning and managing rail-truck intermodal transportation of hazardous materials. *Transportation Research Part E, 48*(1), 132–149.
Wang, X., & Meng, Q. (2017). Discrete intermodal freight transportation network design with route choice behavior of intermodal operators. *Transportation Research Part B, 95*, 76–104.
Wayne, R. (1988). *A decision support system for military vehicle fleet management*, Thesis (M. Eng. Sc.), Australian Defence Force Academy, pp. 1–162.
Yang, X., Low, J. M. W., & Tang, L. C. (2011). Analysis of intermodal freight from China to Indian Ocean: A goal programming approach. *Journal of Transport Geography, 19*(4), 515–527.
Zografos, K. G., & Androutsopoulos, K. N. (2008). A decision support system for integrated hazardous materials routing and emergency response decisions. *Transportation Research Part C, 16*(6), 684–703.

# Integrated Production Scheduling and Distribution Planning with Time Windows

**Saadettin Erhan Kesen and Tolga Bektaş**

## 1 Introduction

Severe competition in global markets coupled with advances in computer, communication and transportation technologies (e.g., mobile technology, internet, same-day delivery, route navigation) require companies to improve on and invest in their supply chain. The overall performance of a supply chain depends, to a large extent, on a better integration of its traditional functions. Thomas and Griffin (1996) define these functions as being procurement, production, and distribution. Production and outbound distribution (i.e., delivery of finished products) are the two key interdependent operations that need to be coordinated in order to reach a desired service level and to reduce the overall system cost (Steiner and Zhang 2011). Traditionally, these functions have been optimized separately and sequentially, whereby an intermediate inventory phase is used as a buffer. Such approaches lead to sub-optimal decisions resulting in unnecessary inventory that degrades the system effectiveness (Pundoor and Chen 2005). Many companies operating a just-in-time production system recognize the need to establish a closer link between their production and outbound distribution operations in order to lower inventory levels and the associated costs.

S. E. Kesen (✉)
Industrial Engineering Department, Faculty of Engineering and Natural Sciences, Konya Technical University, Konya, Turkey
e-mail: skesen@selcuk.edu.tr

T. Bektaş
Centre for Operational Research, Management Science and Information Systems (CORMSIS), Southampton Business School, University of Southampton, Southampton, UK
e-mail: T.Bektas@soton.ac.uk

 
T. Paksoy et al. (eds.), *Lean and Green Supply Chain Management*, International Series in Operations Research & Management Science 273,
https://doi.org/10.1007/978-3-319-97511-5_8

There are many other applications where production and distribution decisions must be linked. Time sensitive or perishable products, make-to-order items, newspaper printing, and catering services are examples to such applications. In particular, ready-mix concrete (Garcia et al. 2002, 2004) cannot be put in stock. Some adhesive chemicals used in plywood panel have limited life span and must be delivered to the user before the adhesive solidifies and becomes unusable (Geismar et al. 2008, 2011). In make-to-order business, every product is custom-made and must be delivered to customer in very short lead times (Li et al. 2005; Stecke and Zhao 2007; Rasti-Barzoki and Hejazi 2013). Catering services (Chen and Vairaktarakis 2005; Farahani et al. 2012) must deliver the foods to the customers shortly after they are prepared. Similarly, fashion apparels (Chen and Pundoor 2006) have a short period of shelf life and may become obsolete unless production and delivery operations are closely linked. Another example is newspaper printing (Hurter and Van Buer 1996; Van Buer et al. 1999), which cannot start before midnight on a given day and must be printed and delivered to home delivery subscribers no later than 6:00 am in the same day. Popular grocery brands enable their customers to place their orders online and book delivery slots. A staff in the store then collects the ordered items from shelves, following which they are delivered within the time windows requested by each customer. For the relevant literature and state-of-the-art surveys on the topic, the reader is referred to Chen (2004, 2010).

From a practical point of view, resources carrying out similar tasks in parallel can be seen in many industries and hence are quite common. Examples include machines at a production site, multi-processor computers, tills in a bank branch, runways in an airport, rail tracks or platforms at a train station, emergency doctors/nurses at a hospital and so forth. In each of these examples, there are multiple resources with the same (or similar) processing capabilities used for incoming tasks/jobs which have similar processing requirements, and which need to be used to process the jobs prior to delivery. Recently, tardiness as a due-date related performance measure has been commonly used in the scheduling literature as due-date is contractually negotiated and failing to deliver product to customer later than this pre-defined date results in a penalty. However, this objective does not account for potential costs or penalties arising from early deliveries due to potential holding costs, thefts, product deterioration, insurance and so on. Readers are referred to review paper by Baker and Scudder (1990) for detailed earliness and tardiness literature. This chapter describes an Integrated Production and Outbound Distribution Scheduling (IPODS) problem, which involves finding a detailed schedule at an individual order level in a two-stage supply chain environment, where joint decisions of production and distribution operations are involved. In particular, a set of given jobs (also referred as orders), each of which is destined to a unique customer, has to be processed on identical parallel machines in a single facility and subsequently delivered to associated customer locations by a fleet of limited number of vehicles within the time windows requested by each customer. From an operational effectiveness point of view, it would be ideal to consolidate orders as much as possible in one shipment but this may imply an increase in the earliness or tardiness. Since the number of machines and the number of vehicles are limited, the joint schedule can imply early or late deliveries. If an order arrives at its customer later than the requested time window, the company will potentially loose customer goodwill and prestige. On the other hand, if an order is delivered to the customer earlier than the time window, then the vehicle will have idle wait, affecting driver welfare and increasing certain costs (e.g., driver wage, vehicles' operational cost, etc.). The objective of the problem is therefore the minimization of total earliness and tardiness of all orders. The joint schedule includes information on the two main sub-problems, namely machine scheduling and vehicle routing. Scheduling decisions include machine assignment and job ordering on each machine. Routing decisions involve the dispatch time of each vehicle from the main depot, the orders to be served on each trip and their sequence, and the arrival time of vehicle to customers.

The contributions of the study can be stated as follows. First, we study, for what we believe to be the first time, the joint problem described above. In particular, we study two variants, depending on whether a vehicle can wait idle at a customer site or not with the assumption that lower time bound is soft. Contrary to the existing literature where delivery operations are generally assumed to be direct, we handle the distribution part as a vehicle routing problem with time windows and limited number of vehicles. Such problems are more challenging but deserve to be studied more than any other kind of IPODS problems due to their practicality (Chen and Vairaktarakis 2005). We present computational results on the general case of the problem to illustrate the performance of the formulations and to show the impact of problem parameters on the solution. The remainder of the chapter is organized as follows. The next section presents an overview of the relevant literature. Section 3 introduces the problem definition and describes the variants. Mathematical formulations are described in Sect. 4. Computational experiments are presented in Sect. 5. The chapter concludes in Sect. 6.

## 2 An Overview of the Relevant Literature

The existing literature on integrated production and outbound distribution scheduling can be divided into two categories with respect to the way in which delivery is performed. In the first category, a simple delivery method is assumed, such as direct shipping of an order to a customer after its completion (see Potts 1980; Garcia and Lozano 2005; Wang and Lee 2005; Ullrich 2012), batch delivery to a single customer (see Hall and Potts 2005; Averbakh and Xue 2007; Chen and Vairaktarakis 2005; Ng and Lu 2012), batch delivery to multiple customers with direct shipping (see Gao et al. 2015; Hall and Potts 2003; Li and Vairaktarakis 2007; Chen and Lee 2008; Averbakh and Baysan 2012). In the second category, there exist a limited number of papers that take the consolidation aspect into account, i.e., deliveries are made using routes where multiple customers are visited.

In Table 1, we list a tabulated and a chronological summary of the studies in the second category, as they are of more relevance to our study. It can be observed from the table that the number of studies on such problems have started to increase in recent years. Problems with general order size are given more attention than the ones with equal order size despite the additional complexity.

**Table 1** A chronological and a tabulated summary of the IPODS literature considering routing decisions

| Paper | Order size | Vehicle type | Vehicle number | Machine configuration | Objective | Solution method |
|---|---|---|---|---|---|---|
| Van Buer et al. (1999) | General | Homogenous | Sufficient | Single | Cost | Exact/heuristic |
| Chang and Lee (2004) | General | Homogenous | Sufficient | Single | Service | Heuristic |
| Chen and Vairaktarakis (2005) | Equal | Homogenous | Sufficient | Single/parallel | Cost/service | Exact/heuristic |
| Li et al. (2005) | Equal | Homogenous | Limited | Single | Service | Exact/heuristic |
| Li and Vairaktarakis (2007) | Equal | Homogenous | Sufficient | Bundling | Cost | Heuristic |
| Armstrong et al. (2008) | General | Homogenous | Limited | Single | Service | Exact |
| Geismar et al. (2008) | General | Homogenous | Limited | Single | Service | Exact/heuristic |
| Geismar et al. (2011) | General | Homogenous | Limited | Single | Cost | Heuristic |
| Farahani et al. (2012) | Equal | Homogenous | Limited | Parallel | Cost | Exact/heuristic |
| Condotta et al. (2013) | General | Homogenous | Limited | Parallel | Service | Exact/heuristic |
| Ullrich (2013) | Equal | Heterogeneous | Limited | Single | Service | Exact/heuristic |
| Hajiaghaei-Keshteli et al. (2014) | General | Heterogeneous | Limited | Single | Cost/service | Heuristic |
| Lee et al. (2014) | General | Heterogeneous | Limited | Parallel | Cost | Exact/heuristic |
| Low et al. (2014) | General | Heterogeneous | Sufficient | Single | Cost | Heuristic |
| Viergutz and Knust (2014) | General | Homogenous | Limited | Single | Service | Exact/heuristic |
| Kang et al. (2015) | General | Homogenous | Limited | – | Cost | Exact/heuristic |
| Karaoğlan and Kesen (2017) | General | Homogenous | Limited | Single | Service | Exact |
| *Our study* | General | Homogenous | Limited | Parallel | Service | Exact |

Homogeneous vehicles are used more than heterogeneous, and with preference on limited vehicle number. Single machine operation is the most frequently studied machining environment, followed by parallel machines. The objective is often modeled as the minimization of system cost generally including transportation costs and maximization of customer service level, depending on whether customer due date or time windows are included. As for solution techniques, heuristics are more commonly applied but exact solution techniques are also presented.

Below, we review the studies listed in Table 1 in greater detail.

Van Buer et al. (1999) describe a mathematical formulation and a solution method for a problem where production and distribution activities must be coordinated in the newspaper industry. They find that re-using the trucks helps in reducing the cost. Chang and Lee (2004) study the three different scenarios where jobs require varying amount of space during delivery, and present a worst case analysis. Chen and Vairaktarakis (2005) consider a production problem under single and parallel machine environment, with two customer service levels based on average and maximum shipment times where the objective is to minimize the total distribution cost. Through numerical results, they show that jointly solving the production and distribution problem results in a better system performance as compared to solving them sequentially. Li et al. (2005) develop a polynomial time algorithm for an IPODS problem where the number of customer is fixed. For more special cases, they present dynamic programming algorithms. The study of Li and Vairaktarakis (2007) differs from the others since it models the production environment as bundling operations in which two components processed on different machines form the end-product. They present polynomial-time heuristics and approximation schemes. Armstrong et al. (2008) study the problem with a single transporter and a fixed sequence of customers with time windows. They solve the problem with a branch and bound algorithm, in which a subset of customers from the given customer sequence is chosen to maximize the total satisfied demand. Viergutz and Knust (2014) extend the problem of Armstrong et al. (2008) by considering the delays of the production start and varying production and distribution sequences.

Geismar et al. (2008) study a problem in which a subset of customer orders are produced and transported in a given sequence. They propose a lower bounding scheme and a two-phase heuristic. In the first phase of the heuristic, a genetic or a memetic algorithm is used to choose a local optimal solution and then customer subsets are ordered by using the Gilmore–Gomory algorithm. Karaoğlan and Kesen (2017) address the same problem defined in the study of Geismar et al. (2008) and develop branch and cut algorithm, which provides better performance than that of Geismar et al. (2008). Geismar et al. (2011) extend the study of Geismar et al. (2008) by taking pool-point delivery into consideration. Customers are defined as pool points requiring multiple trucks to satisfy the total demand. Using real catering industry data, Farahani et al. (2012) study a problem involving food with a high spoilage rate, where the aim is to reduce the time interval between production and transportation. They propose an iterative scheme to coordinate production scheduling and distribution. They propose an iterative scheme to coordinate production scheduling and transportation. Condotta et al. (2013) study a problem in which each job with a certain release date is processed on a single machine prior to delivery. Delivery is then performed by a fleet of homogeneous vehicles with limited loading capacity. The objective is to minimize the maximum lateness. They model the problem as a mixed integer programming formulation and propose different methods to find lower bounds. A tabu search algorithm is developed to produce partial solutions in the production phase. Each partial solution is turned to an integrated solution through an optimal polynomial-time transportation schedule. Ullrich (2013) investigates a problem in which jobs/orders are produced under a parallel machine environment and distributed to customers with a fleet of vehicles with different loading capacities. Each vehicle is allowed to make multiple trips. Ullrich (2013) describes a formulation and a genetic algorithm to solve the problem, and concludes that an integrated decision is better than sequential separate decisions with respect to total tardiness value. Hajiaghaei-Keshteli et al. (2014) look at the synchronization of rail transportation. They aim to find a production schedule and an order allocation in rail transportation while optimizing service level at minimum cost. To tackle the problem, they propose a heuristic and two metaheuristic procedures. They also optimize level of parameters through the use of Taguchi experimental design technique. Lee et al. (2014) investigate the coordinated decisions of production and distribution of nuclear medicine. The half-life of a particular radioactive substance used in diagnostic treatment of many cancer types is 110 min, which is quite short. Therefore, the production and transportation process must be coordinated to deliver the medicine to patients (or end medical-users) before the half-life. They propose a mixed-integer linear programming formulation and develop large neighborhood search algorithm. Low et al. (2014) study a production scheduling and delivery problem where different items are processed in a distribution center and subsequently delivered to retailers. Each retailer may require a different type of product. The problem is to determine the production and customer visiting sequence so as to minimize the transportation cost. Two adaptive genetic algorithms are tested on a wide range of test instances. Kang et al. (2015) formulate a mixed integer programming formulation for the integrated production and transportation problem in semiconductor manufacturing where different outsourcing processes involving circuit probing testing, integrated circuit assembly, and final testing need to be coordinated by wafer fabricator. Clustered by their types, jobs must be processed in these outsourcing factories. They then propose Genetic Algorithm as the problem becomes complicated with increasing problem size. Results show the good performance of algorithm.

The problem introduced in this chapter differs from those in the existing studied described above in several aspects. First, earliness and tardiness objective function is studied for the first time. Second, the problem that we study assumes a limited number of vehicles, each of which can only be used once in the planning horizon. Third, we break away from the literature by studying variants and by allowing idle waiting in the tours.

## 3 Problem Definition

The problem we consider here can formally be defined as follows: A set $\{1, \ldots, J\}$ of jobs arrive at a single facility, where each job requires processing by a single machine in a single machine environment with $M$ identical machines. We assume that each job $j \in \{1, \ldots, J\}$ is non-splittable and characterized by a nonnegative processing time $p_j$, load size $q_j$ and time windows $[a_j, b_j]$ for delivery. We further assume that all jobs are ready at time 0 and preemption is not permitted, which means that once started, processing cannot be interrupted until its completion. Following the machining operation, jobs are kept in the facility for temporary storage. A set $\{1, \ldots, K\}$ of homogeneous vehicles is available to distribute the jobs to respective customers, each of which has a load capacity of $Q$ units and is allowed to be used only once in the planning cycle. A service time of $s_j$ units is needed at each customer for delivery and also at the main facility (depot) for loading the vehicles. The routing part of the problem is modelled on a graph with $\{0, 1, \ldots, J\}$ as the set of nodes where node 0 denotes the facility and the rest are customers. Each job corresponds to a unique customer, and vice versa, for which reason we use the terms jobs and customers interchangeably, as there is one-to-one correspondence between the two. The travel time $t_{ij}$ between customers $i$ and $j$ is assumed to be a constant. Since the number of vehicles is limited and each customer order needs to be delivered, tours are allowed to visit multiple customers.

The problem consists of assigning each order to one of parallel machines, finding the order sequence in each machine, and determining the sequence of customers to visit for each vehicle. At each customer location, there are three possible cases: (i) vehicle arrives at customer location $j$ between time $a_j$ and $b_j$, which means that the order is delivered on time (ii) vehicle arrives at location $j$ earlier than time $a_j$, resulting in earliness, and (iii) vehicle arrives at location $j$ later than time $b_j$, resulting in tardiness. The problem minimizes the total earliness and tardiness of all customer orders. Figure 1 illustrates a feasible solution to an instance of the problem with nine jobs, three identical machines and three homogeneous vehicles.

Inspired by the scheduling literature (Graham et al. 1979), Chen (2010) suggests the use of a five-field $\alpha|\beta|\pi|\delta|\gamma$ notation to represent and classify IPODS problems. In this representation, $\alpha$, $\beta$ and $\gamma$ fields are borrowed from production scheduling and denote the machine environment, restrictions and constraints on the orders, and the objective function of the problem, respectively. Additionally, $\pi$ describes the delivery characteristics, including the number of vehicles, capacities and delivery methods, and $\delta$ represents the customer (or job) number. Based on this notation, we can represent the problem under consideration as $P_m|[a_j, b_j]|V(v, Q), routing|n|\sum(E_j + T_j)$, where $P_m$ indicates that there are $m$ identical parallel machines, but only one is needed to process each job, $[a_j, b_j]$ is the delivery time window requested by each customer in which the order is expected to be delivered, $V(v, Q)$ denotes the vehicles each with capacity $Q$, where $Q$ might be unlimited and that there is a limited number $v$ of vehicles available. Furthermore, *routing* means that orders of different customers can be delivered in

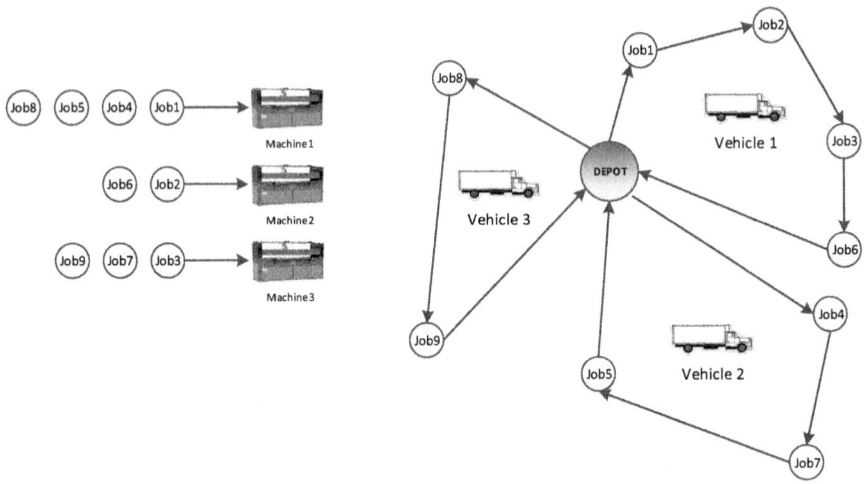


**Fig. 1** Solution to a sample instance with nine jobs, three machines, and three vehicles

the same shipment, $n$ simply states the number customers, meaning that every order belongs to a different customer. Finally, $\sum(E_j + T_j)$ is the objective function to be minimized representing the total earliness and tardiness of all orders.

## 3.1 Notation

We now formally present the parameters and decision variables which will be used to formulate the problem as follows:

**Parameters**

$q_j$ Demand of job $j$ $(j = 1, \ldots, J)$
$p_j$ Processing time of job $j$ $(j = 1, \ldots, J)$ for all the $q_j$ units
$s_j$ Service time at node $j$ $(j = 0, 1, \ldots, J$; index 0 indicates the facility)
$t_{ij}$ Travel time from customer $i$ to customer $j$ $(i, j = 0, 1, \ldots, J; i \neq j)$
$Q$ The capacity of vehicles
$K$ The number of vehicles available
$a_j$ Lower bound of the delivery time window of job $j$
$b_j$ Upper bound of the delivery time window of job $j$
$M_1 = \sum_{j=1}^{J} p_j - \min p_j + \max p_j$
$M_2 = \sum_{j=1}^{J} p_j - \min p_j$
$M_3$ A sufficiently large number

**Decision Variables**

$C_j$ Machine completion time of job $j$ ($j = 1, \ldots, J$)
$E_j$ Earliness of job $j$ ($j = 1, \ldots, J$)
$T_j$ Tardiness of job $j$ ($j = 1, \ldots, J$)
$F_{ij}$ Amount of load on a vehicle just after leaving the destination of job $i$ to destination of job $j$ ($i,j = 0, \ldots, j; i \neq j$)
$Y_j$ Arrival time of vehicle at customer $j$ ($j = 1, \ldots, J$)
$Y_0^k$ Service starting time at the depot ($k = 1, \ldots, K$)
$V_{ij}$ Arrival time at customer $i$ for which the subsequent visit is at node $j$ ($i,j = 0, \ldots, J$)
$Z_{ij}$ 1 If job $i$ directly precedes job $j$ on a machine, 0 otherwise ($i,j = 1, \ldots, J; i \neq j$)
$W_{mj}$ 1 If job $j$ is the first to be processed on machine $m$, 0 otherwise ($m = 1, \ldots, M; j = 1, \ldots, J$)
$Z_{j,J+1}$ 1 If job $j$ is the last to be processed on a machine, 0 otherwise ($j = 1, \ldots, J$), where job $J + 1$ is a dummy last job
$X_{ij}^k$ 1 If vehicle $k$ goes directly from customer $i$ to customer $j$, 0 otherwise ($i,j = 0, \ldots, J; i \neq j; k = 1, \ldots, K$)

## 3.2 *Two Variants of the Problem*

We now differentiate between two variants of the problem, depending on the assumptions or restrictions made on the time of arrivals at customer nodes. In order to illustrate the differences among the variants, we present a simple example with four customers, each with their own time window as shown in Fig. 2 and a single vehicle. For the sake of simplicity, we assume that jobs are processed in advance and are ready to be serviced at time 0. All service times at the depot and at customer locations are assumed to be 0. An optimal order of visit for this instance has been found as 1, 2, 3, 4 where the travel time between consecutive customers is equal to 10 min. It should be noted that the solution presented here remains optimal for both cases. In Case 1, the time windows are soft, meaning that the vehicle is allowed to start the service to customers earlier than the prescribed lower time bound. In this case, the time spent between consecutive customer visits can be longer than the actual travel time between them. For example, in Case 1, the vehicle arrives at customer 3 at time 35, while the arrival time at customer 2 was at time 20. The difference between the two arrival times is more than the actual travel time of 10 units between these customers due to an idle time of 5 min after servicing customer 2. In Case 1, the only penalty incurred is due to the early arrival at customer 1, which is 5 min. In Case 2, the vehicle is not allowed to wait idle between customers. In this case, Fig. 2 shows that the arrival times between consecutive customers are the same and equal to the travel time 10. Due to the "no idle-wait" restriction, an earliness of 5 min at customers 1 and 3, and an earliness of 10 min at customer 4 are incurred. Hence, a total of 20 min of earliness penalty occurs in Case 2.

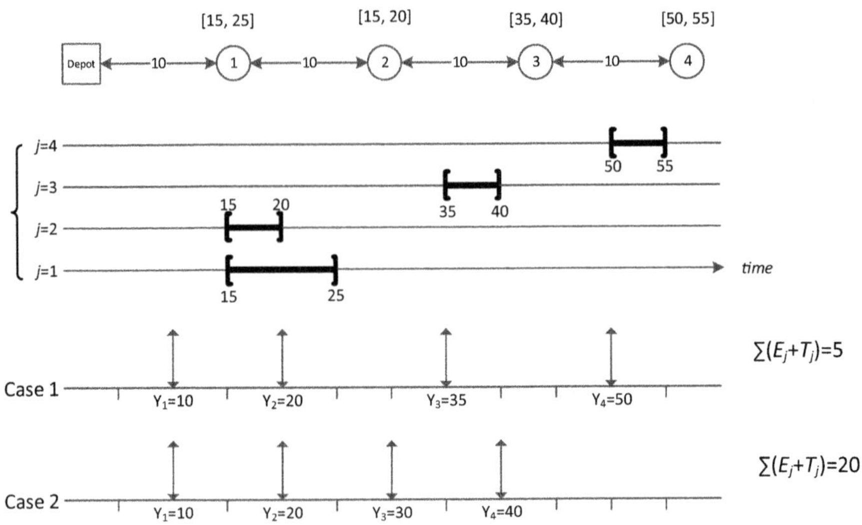


**Fig. 2** Objective function based on each case

# 4 Formulations

In this section, we present a mixed integer linear programming formulations for both cases described above. The formulations include, as a sub problem, a parallel machine scheduling problem, for which we base ourselves on the work of Biskup et al. (2008), since it requires fewer binary variables than traditional models. All formulations presented in the following sections use the notation shown in Sect. 3.1, where additional notation is introduced in sections where needed.

## *4.1 Case 1: Soft Time Window with Idle Wait*

We present a formulation based on nodes for this case. In this strategy, a delivery arriving before lower time bound is accepted but travel time between any consecutive customers can be loose. In other words, if vehicle $k$ visits customer $j$ just after customer $i$, then the arrival time at customer $j$ can be later than or equal to the sum of the arrival time at customer $i$, service time at customer $i$ and the travel time between customers $i$ and $j$. The formulation is given as follows:

$$\textbf{NB1}\ \text{Minimize} \sum_{j=1}^{J} (E_j + T_j) \tag{1}$$

Subject to

$$\sum_{j=1}^{J} W_{mj} \leq 1 \qquad m = 1, \ldots, M \tag{2}$$

$$\sum_{i=1,\ i \neq j}^{J+1} Z_{ji} = 1 \qquad j = 1, \ldots, J \tag{3}$$

$$\sum_{m=1}^{M} W_{mj} + \sum_{i=1,\ i \neq j}^{J} Z_{ij} = 1 \qquad j = 1, \ldots, J \tag{4}$$

$$C_j \geq W_{mj} p_j \qquad j = 1, \ldots, J; m = 1, \ldots, M \tag{5}$$

$$C_i - C_j + M_1 Z_{ij} + \left(M_1 - p_i - p_j\right) Z_{ji} \leq M_1 - p_j \qquad i, j = 1, \ldots, J; i \neq j \tag{6}$$

$$\sum_{j=1}^{J} X_{0j}^{k} \leq 1 \qquad k = 1, \ldots, K \tag{7}$$

$$\sum_{k=1}^{K} \sum_{i=0, i \neq j}^{J} X_{ij}^{k} = 1 \qquad j = 1, \ldots, J \tag{8}$$

$$\sum_{h=0, h \neq i}^{J} X_{hi}^{k} = \sum_{j=0, i \neq j}^{J} X_{ij}^{k} \qquad i = 0, \ldots, J; k = 1, \ldots, K \tag{9}$$

$$\sum_{h=0, h \neq i}^{J} F_{hi} - \sum_{j=0, i \neq j}^{J} F_{ij} = q_i \qquad i = 1, \ldots, J \tag{10}$$

$$q_j \sum_{k=1}^{K} X_{ij}^{k} \leq F_{ij} \qquad i, j = 0, \ldots, J; i \neq j \tag{11}$$

$$(Q - q_i) \sum_{k=1}^{K} X_{ij}^{k} \geq F_{ij} \qquad i, j = 0, \ldots, J; i \neq j \tag{12}$$

$$Y_0^k \geq C_i - M_2\left(1 - X_{ij}^k\right) \qquad i, j = 0, \ldots, J; i \neq j; k = 1, \ldots, K \tag{13}$$

$$Y_0^k \geq C_j - M_2\left(1 - X_{ij}^k\right) \qquad i, j = 0, \ldots, J; i \neq j; k = 1, \ldots, K \tag{14}$$

$$Y_i - Y_j + s_i + t_{ij} \leq M_3\left(1 - \sum_{k=1}^{K} X_{ij}^k\right) \qquad i, j = 1, \ldots, J; i \neq j \tag{15}$$

$$Y_0^k - Y_j + s_0 + t_{0j} \leq M_3\left(1 - X_{0j}^k\right) \qquad j = 1, \ldots, J; k = 1, \ldots, K \tag{16}$$

$$E_j \geq a_j - Y_j \qquad j = 1, \ldots, J \tag{17}$$

$$T_j \geq Y_j - b_j \qquad j = 1, \ldots, J \tag{18}$$

$$E_j \geq 0 \qquad j = 1, \ldots, J \tag{19}$$

$$T_j \geq 0 \qquad j = 1, \ldots, J \tag{20}$$

$$C_j \geq 0 \qquad j = 1, \ldots, J \tag{21}$$

$$Z_{ij} \in \{0, 1\} \qquad i, j = 1, \ldots, J; i \neq j \tag{22}$$

$$W_{mj} \in \{0, 1\} \qquad m = 1, \ldots, M; j = 1, \ldots, J \tag{23}$$

$$X_{ij}^k \in \{0, 1\} \qquad i, j = 0, \ldots, J; i \neq j; k = 1, \ldots, K \tag{24}$$

$$F_{ij} \geq 0 \qquad i, j = 0, \ldots, J; i \neq j \tag{25}$$

$$Y_0^k \geq 0 \qquad k = 1, \ldots, K \tag{26}$$

$$Y_j \geq 0 \qquad j = 1, \ldots, J. \tag{27}$$

The objective function (1) models the total earliness and tardiness of all jobs. Constraint (2) ensures that at most one job can be the first to be processed on any machine. Constraints (3) and (4) indicate that a job can either be the first on any machine or should follow any other job. Constraint (5) implies that the completion time of the first job on a machine must be greater than or equal to its processing time. Constraint (6) guarantees that if job $j$ is preceded directly by job $i$ then the completion time of job $j$ must be greater than or equal to the completion time of job $i$ plus processing time of job $j$. Constraint (7) models the limitation that the number of tours starting from the depot must be less than or equal to the available number of vehicles. Constraints (8) and (9) are the degree constraints; in particular, constraint (8) states that each customer must be visited exactly once by any vehicle and constraint (9) ensures that the number of arrivals to and the number of departures from each customer location are equal and are made by the same vehicle. Balance flow of vehicle load between customers is ensured through constraint (10). Constraints (11) and (12) limit the total amount of load carried on a vehicle by its capacity. These constraints are based on a single-commodity flow formulation described by Gavish and Graves (1978). There are two reasons as to why we have opted to use such constraints in this formulation, as opposed to others such as those based on the Miller et al. (1960). The first reason is that there is no significant computational advantage in using the latter over the former, as recent evidence by Roberti and Toth (2012) suggests, but the former set of constraints are more amenable to bespoke approaches such as decomposition (Gavish and Graves 1978). The second reason is that such constraints have recently been shown to be useful in modeling variants of the problem where load needs to be explicitly calculated, or to impose various side constraints, such as balancing of workload.

We additional note here that it is possible to reformulate constraints (10–12) as follows,

$$\sum_{k=1}^{K} \sum_{h=0, i \neq j}^{J} F_{hi}^{k} - \sum_{k=1}^{K} \sum_{j=0, i \neq j}^{J} F_{ij}^{k} = q_i \qquad i = 1, \ldots, J \tag{28}$$

$$q_j X_{ij}^{k} \leq F_{ij}^{k} \qquad i, j = 0, \ldots, J; i \neq j; k = 1, \ldots, K \tag{29}$$

$$(Q - q_i) X_{ij}^{k} \geq F_{ij}^{k} \qquad i, j = 0, \ldots, J; i \neq j; k = 1, \ldots, K, \tag{30}$$

where variables $F_{ij}^{k}$ are used to explicitly represent the load on vehicle $k$ when traversing arc $(i,j)$. Preliminary computational results did not yield any benefits by using the alternative representation above. However, one advantage of constraints (28–30) is that they can be used to model an extension of the problem with a heterogeneous fleet in which the load capacity of vehicle $k$ is $Q_k$ units, which would feature in constraint (30). Constraints (13) and (14) guarantee that if vehicle $k$ travels from customer $i$ to customer $j$ (i.e., $X_{ij}^{k} = 1$), then the arrival time of vehicle $k$ at the depot must be greater than both the completion time of job $i$ and completion time of job $j$. Constraint (15) ensures that the arrival time at customer $j$ must be later than the arrival time plus the service time of customer $i$ and the travel time from customer $i$ to customer $j$ if vehicle $k$ directly travels from former to the latter. In a similar fashion, constraint (16) ensures that the arrival time at customer $j$ must be later than the arrival time at the depot, plus the service time at the depot, and the travel time from the depot to customer $j$ if vehicle $k$ directly travels from the depot to customer $j$. Constraint (17) simply states that earliness can only occur if the arrival time at customer $j$ is earlier than the lower time bound $a_j$. Likewise, constraint (18) ensures that tardiness of customer $j$ is the difference between the arrival time at its destination and upper time bound $b_j$. Constraints (19–27) represent the integrality and non-negativity restrictions on the variables.

## 4.2 Case 2: Soft Time Window with No Idle Wait

Similar to Case 1, the no idle wait restriction also allows arriving at a customer $j$ earlier than lower time bound $a_j$, but the travel time spent between successive customers should not include any idle wait. In other words, if customer $j$ is visited just after customer $i$ in any tour, then the arrival time at customer $j$ must be equal to the sum of the arrival time at customer $i$, the service time at customer $i$, and the travel time between customers $i$ and $j$.

With the new requirement, the following constraints are required to model NB2:

$$Y_j - Y_i - s_i - t_{ij} \le M_3 \left(1 - \sum_{k=1}^{K} X_{ij}^k\right) \qquad i, j = 1, \ldots, J; i \ne j \qquad (31)$$

$$Y_j - Y_0^k - s_0 - t_{0j} \le M_3 \left(1 - X_{0j}^k\right) \qquad i, j = 1, \ldots, J; k = 1, \ldots, K \quad (32)$$

Constraint (31) together with constraint (15) will guarantee that if there is a direct travel from customer $i$ to customer $j$, then the arrival time at customer $j$ will be equal to the sum of the arrival time at customer $i$, the service time at customer $i$, and travel time between customer $i$ and customer $j$ (i.e., $Y_j = Y_i + s_i + t_{ij}$). Based on the same idea, constraint (32) together with constraint (16) will ensure that if vehicle $k$ visits customer $j$ just after the depot (i.e., $X_{0j}^k = 1$), the arrival time at customer $j$ will be equal to the sum of the arrival time at the depot, service time at the depot, and the travel time from depot to customer $j$ (i.e., $Y_j = Y_0^k + s_0 + t_{0j}$). NB2 is modelled as follows:

**NB2:** Minimize (1) subject to (2–27), (31–32).

# 5 Computational Experiments

This section presents the computational experimentation conducted to test the performance of the formulations and to draw some computational and managerial insights to the problem.

## 5.1 Instance Generation

In this section, we explain how to generate our test instances, which are randomly produced in a similar way as described by Ullrich (2013). The processing time $p_j$ of each job $j$ is drawn from a discrete uniform distribution $UNIF(1, \rho)$ with parameters 1 and $\rho$. Processing times can therefore be regarded as integer numbers drawn from the interval $[1, \rho]$, with each number having an equal likelihood of being chosen.

When generating the travel time matrix, we first randomly assign the $x$ and $y$ coordinates of each customer location on a two-dimensional plane. We then calculate the Euclidian distance between each pair of nodes $(i,j)$ such that $i < j$, which is set to be the travel time $t_{ij}$. We assume that the travel time matrix is symmetric (i.e., $t_{ij} = t_{ji}\ \forall\ i, j = 0, \ldots, J$) and $t_{ii} = 0$. We generate the customer locations in such a way that the maximum traveling time between any pair of customers does not exceed $\lfloor \rho(K/M) \rfloor$, where function $\lfloor * \rfloor$ denotes the largest integer value less than or equal to $*$. For an effective management of both production and distribution, it is necessary to balance the machine and vehicle capacities. If production capacities are abundant as compared to vehicle capacities, then the production scheduling problem becomes trivial and the integrated problem boils down to the routing problem. In practice, though, it is common that both the production and distribution operations are performed with limited and well-fitted capacities. We balance the two different capacities by using the machines ($M$), the vehicles ($K$), and the processing times ($\rho$) to generate traveling times. The service times $s_j$ are also set based on processing times and are generated as $s_j \sim UNIF(1, \lfloor \lambda\rho \rfloor)$ for all $j = 0, \ldots, J$.

Although the mathematical models we have given in Sect. 4 consider that each job has different demand size and that vehicles are capacitated, we conduct our computational analysis by assuming unit demands (i.e., $q_j = 1, \forall\, j = 1, \ldots, J$) and vehicles have sufficient capacity ($q_j = 1$ and $Q \geq \sum_{j=1}^{J} q_j$) so as to remove the effects of these parameters on the resulting solution.

In generating the test instances, the number of jobs is chosen between 5 and 10 in increments of one and the number of machines and vehicles as equal to 1, 2 or 3, producing a total $6 \times 3 \times 3 = 54$ of test instances.

When generating the time windows we follow the procedure whose details are described as follows. In this case, one would expect that the delivery time of any customer $j$ must be at least as much as the sum of the processing time, the service time at the depot, and the travel time from the depot to the customer, which also corresponds to the lower time bound $a_j$, generated as $a_j = p_j + s_0 + t_{0j} + \pi_j$ where $\pi_j \sim UNIF(0, \lfloor \delta_1 \rho J/(M + K) \rfloor)$. The parameter $\delta_1$ controls the tightness of the lower bound. The smaller the value of $\delta_1$, the tighter the time for production and distribution. The term $J/(M + K)$ tells us that the number of customers $J$, the number of machines $M$, and the number of vehicles $K$ can be changed with a minor effect on the tightness of the test instances. Parameter $\rho$ is used as a multiplier to convert the term into time units. The upper bound $b_j$ is calculated as $a_j + \xi_j$ where $\xi_j \sim UNIF(0, \lfloor \delta_2 \rho \rfloor)$. All instances are generated in C++ by using the embedded random number stream and with the following parameter settings: $\rho = 100$, $\lambda = 0.2$, $\delta_1 = 1$, and $\delta_2 = 0.5$.

## 5.2 *Computational Results*

In this section, we present results of computational experience with the formulations for both cases. All experiments are conducted on a server with 1 GB RAM and Intel Xeon 2.6 GHz processor, and the models are solved by CPLEX 12.5. We allow a maximum time limit of 3 h (10,800 s) for each instance.

### 5.2.1 Results for the General Problem

In this section, we report results of the node-based formulation for both cases. The results are presented in Table 2, where the first four columns present the name and the parameters of each instance. For each case, *UB* stands for the upper bound, which is either the optimal value or the best value found within 3 h. Column *Gap(%)* represents the percentage gap between *UB* and *LB*, where LB is the lower bound found at the end of the time limit by CPLEX, and is calculated as $(UB - LB)/(UB) \times 100$. If *Gap(%)* = 0 is for a particular instance, this indicates that an optimal solution has been identified. Finally, *CPU(sec.)* is the computational time in seconds. A ‘3 h’ value in *CPU(sec.)* column for some instances indicates that an optimal solution cannot be found in 3 h. For Case 1, *NB*1 has an average gap of 34.56% over the 54 instances. For Case 2, an average gap of 33.8% is found for *NB*2 formulation. Average computational times of the formulations for Cases 1 and 2 are conflicting, however. While average computational time for *NB*1 is 4948 s, that of *NB*2 is 5162 s. The number of instances solved to optimality can be regarded as an additional performance criterion for both models, which both find optimal solutions in 30 out of 54 instances for Case 1 and Case 2.

If results are examined in terms of parameters, it is clearly seen that both formulations are able to optimally solve instances with six or seven jobs in very short CPU times. For larger number of jobs, the solution time increases dramatically, meaning that the number of jobs plays crucial role on problem complexity. In contrast, when the number of machines and vehicles increase, the solution time decreases. In fact, these parameters reflect two sets of resources and with higher

**Table 2** Comparative results of the node-based formulation for Cases 1 and 2

| Parameters | | | | Case-1: soft a(i), with idle time | | | Case-2: soft a(i), with no-idle time | | |
|---|---|---|---|---|---|---|---|---|---|
| Instance | J | M | K | UB | Gap (%) | CPU (s.) | UB | Gap (%) | CPU (s) |
| P1 | 5 | 1 | 1 | 148 | 0 | 3 | 148 | 0 | 4 |
| P2 | 5 | 1 | 2 | 616 | 0 | 14 | 616 | 0 | 27 |
| P3 | 5 | 1 | 3 | 158 | 0 | 3 | 158 | 0 | 4 |
| P4 | 5 | 2 | 1 | 121 | 0 | 0 | 124 | 0 | 0 |
| P5 | 5 | 2 | 2 | 86 | 0 | 0 | 86 | 0 | 0 |
| P6 | 5 | 2 | 3 | 0 | 0 | 0 | 0 | 0 | 0 |
| P7 | 5 | 3 | 1 | 139 | 0 | 0 | 139 | 0 | 0 |
| P8 | 5 | 3 | 2 | 1 | 0 | 0 | 1 | 0 | 0 |
| P9 | 5 | 3 | 3 | 17 | 0 | 0 | 17 | 0 | 0 |
| P10 | 6 | 1 | 1 | 310 | 0 | 40 | 310 | 0 | 112 |
| P11 | 6 | 1 | 2 | 616 | 0 | 271 | 616 | 0 | 704 |
| P12 | 6 | 1 | 3 | 293 | 0 | 54 | 311 | 0 | 140 |
| P13 | 6 | 2 | 1 | 265 | 0 | 30 | 265 | 0 | 23 |
| P14 | 6 | 2 | 2 | 60 | 0 | 3 | 60 | 0 | 5 |
| P15 | 6 | 2 | 3 | 16 | 0 | 2 | 16 | 0 | 3 |
| P16 | 6 | 3 | 1 | 84 | 0 | 0 | 84 | 0 | 1 |
| P17 | 6 | 3 | 2 | 75 | 0 | 6 | 75 | 0 | 11 |
| P18 | 6 | 3 | 3 | 0 | 0 | 0 | 2 | 0 | 1 |
| P19 | 7 | 1 | 1 | 1274 | 67.3 | 3 h | 1274 | 59.6 | 3 h |
| P20 | 7 | 1 | 2 | 659 | 32.6 | 3 h | 659 | 30.3 | 3 h |
| P21 | 7 | 1 | 3 | 386 | 59.2 | 3 h | 395 | 84.4 | 3 h |
| P22 | 7 | 2 | 1 | 162 | 0 | 103 | 173 | 0 | 23 |
| P23 | 7 | 2 | 2 | 107 | 0 | 301 | 107 | 0 | 299 |
| P24 | 7 | 2 | 3 | 37 | 0 | 3 | 65 | 0 | 18 |
| P25 | 7 | 3 | 1 | 246 | 0 | 86 | 257 | 0 | 38 |
| P26 | 7 | 3 | 2 | 77 | 0 | 122 | 77 | 0 | 218 |
| P27 | 7 | 3 | 3 | 3 | 0 | 2 | 3 | 0 | 4 |
| P28 | 8 | 1 | 1 | 2810 | 95.3 | 3 h | 2810 | 92.2 | 3 h |
| P29 | 8 | 1 | 2 | 960 | 81.3 | 3 h | 1014 | 76.7 | 3 h |
| P30 | 8 | 1 | 3 | 1123 | 97.2 | 3 h | 1075 | 96.5 | 3 h |
| P31 | 8 | 2 | 1 | 258 | 68.7 | 3 h | 258 | 88.8 | 3 h |
| P32 | 8 | 2 | 2 | 72 | 0 | 55 | 87 | 0 | 40 |
| P33 | 8 | 2 | 3 | 145 | 0 | 3542 | 154 | 0 | 10, 521 |
| P34 | 8 | 3 | 1 | 155 | 0 | 3074 | 155 | 0 | 6996 |
| P35 | 8 | 3 | 2 | 21 | 0 | 12 | 23 | 0 | 4 |
| P36 | 8 | 3 | 3 | 20 | 0 | 10 | 35 | 0 | 11 |
| P37 | 9 | 1 | 1 | 2205 | 90.2 | 3 h | 2205 | 80.7 | 3 h |
| P38 | 9 | 1 | 2 | 1256 | 98 | 3 h | 1256 | 95.8 | 3 h |
| P39 | 9 | 1 | 3 | 1125 | 77.6 | 3 h | 1125 | 79.4 | 3 h |
| P40 | 9 | 2 | 1 | 558 | 69.7 | 3 h | 558 | 73.3 | 3 h |

(continued)

**Table 2** (continued)

| Parameters | | | | Case-1: soft a(i), with idle time | | | Case-2: soft a(i), with no-idle time | | |
|---|---|---|---|---|---|---|---|---|---|
| Instance | J | M | K | UB | Gap (%) | CPU (s.) | UB | Gap (%) | CPU (s) |
| P41 | 9 | 2 | 2 | 565 | 96.1 | 3 h | 565 | 69.7 | 3 h |
| P42 | 9 | 2 | 3 | 387 | 84.1 | 3 h | 397 | 83.3 | 3 h |
| P43 | 9 | 3 | 1 | 316 | 25 | 3 h | 316 | 40.4 | 3 h |
| P44 | 9 | 3 | 2 | 179 | 79.8 | 3 h | 179 | 73.9 | 3 h |
| P45 | 9 | 3 | 3 | 8 | 0 | 229 | 26 | 0 | 340 |
| P46 | 10 | 1 | 1 | 2230 | 92.6 | 3 h | 2230 | 81.2 | 3 h |
| P47 | 10 | 1 | 2 | 1594 | 97.2 | 3 h | 1594 | 94.3 | 3 h |
| P48 | 10 | 1 | 3 | 2479 | 94.7 | 3 h | 2635 | 96.3 | 3 h |
| P49 | 10 | 2 | 1 | 760 | 96.7 | 3 h | 748 | 85 | 3 h |
| P50 | 10 | 2 | 2 | 374 | 69.3 | 3 h | 405 | 70.3 | 3 h |
| P51 | 10 | 2 | 3 | 416 | 97.7 | 3 h | 326 | 97.2 | 3 h |
| P52 | 10 | 3 | 1 | 704 | 62.6 | 3 h | 703 | 62.1 | 3 h |
| P53 | 10 | 3 | 2 | 236 | 48.7 | 3 h | 254 | 53.5 | 3 h |
| P54 | 10 | 3 | 3 | 172 | 84.7 | 3 h | 209 | 60.1 | 3 h |
| **Average** | | | | | **34.56** | **4948** | | **33.8** | **5162** |
| **# Of solved ins optimality** | | | | | | **30/54** | | | **30/54** |

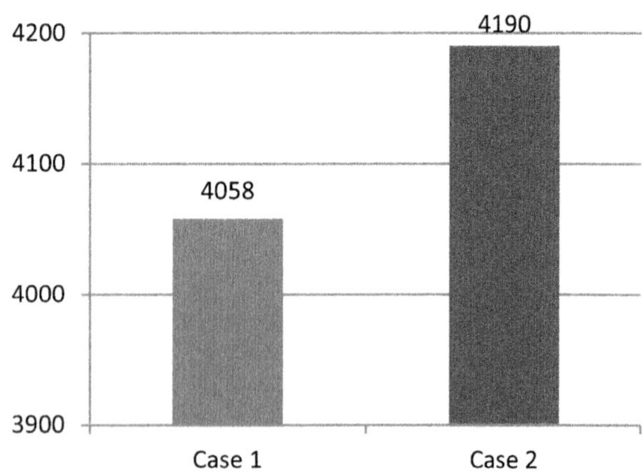


**Fig. 3** Sum of the objective function values of the 30 optimally solved instances in each case

number of machines and jobs, the complexity of delivering the customer orders within the given time windows is reduced.

We now look at 30 optimally solved instances out of the 54 for each case in terms of the objective function values and report the total objective function values for each case separately in Fig. 3.

As Fig. 3 shows, Case 1 results in a total earliness/tardiness equal to 4058. Case 2 produces higher objective value 4190 than Case 1, which is expected, as idle time is not allowed, resulting in an increase in the objective function value. This is a potentially useful managerial insight that may help decision maker to understand the economics of the problem and prefer to use one case over another depending on the particular conditions.

## 6 Conclusions

In the literature, two important phases of a supply chain, production and distribution stages, were handled separately. But in many of the real applications involving ready-mix concrete, newspaper printing, same day delivery, online shopping provided by food retailers, catering services, perishable products require integration of production and distribution decisions, whereby eliminating the need for intermediate inventory phase. Therefore production and distribution schedules are made in a coordinated manner. This chapter introduced, formulated and studied an integrated production scheduling and transportation problem that is characterized by identical parallel machines, a set of jobs (customer orders), each with a destined customer. Two cases of the problem were examined depending on the time spent in traveling between customers (i.e., idle wait or no-idle wait) under the assumption that the lower time bound is soft. We described node-based formulations for both cases. Computational results suggested that the formulation developed for Case 2 provided slightly better performance than that of Case 1 in terms of relative gap between upper bound and lower bound given by CPLEX. However, the average solution time for Case 1 was lower than that of Case 2. As regards the number of optimally solved instances, both formulations were able to solve 30 instances over 54. Among the optimally solved instances, formulation developed for Case 1 had lower objective function value than that of Case 2 since Case 1 is less restrictive than Case 2, allowing delivery even before the lover time bound and idle waiting after the arrival time. A final remark based on our analysis is that both models were able to find optimal solutions to instances with six or seven jobs in given time limit, but the solution time increased dramatically for larger number of jobs. This indicates the significant role that the number of jobs plays on problem complexity. In contrast, the solution time is reduced when the number of machines and vehicles increase. For further research on the problem, we suggest the design of bespoke exact and heuristic solution algorithms to be able to solve instances that are larger in size as compared to the ones considered in this study.

**Acknowledgement** This research was partially supported by The Scientific and Technological Research Council of Turkey (TUBITAK). Whilst this research was being carried out, the first author was a Visiting Fellow at the Southampton Business School, the support of which is greatly acknowledged. Authors are also thankful to the two anonymous referees whose comments greatly contributed to the content of the study.

# References


Armstrong, R. S., Gao, S., & Lei, L. (2008). A zero-inventory production and distribution problem with a fixed customer sequence. *Annals of Operations Research, 159*, 395–414.

Averbakh, I., & Baysan, M. (2012). Semi-online two-level supply chain scheduling problems. *Journal of Scheduling, 15*, 381–390.

Averbakh, I., & Xue, Z. (2007). On-line supply chain scheduling problems with preemptions. *European Journal of Operational Research, 181*, 500–504.

Baker, K. R., & Scudder, G. D. (1990). Sequencing with earliness and tardiness penalties: A review. *Operations Research, 38*, 22–36.

Biskup, D., Herrman, J., & Gupta, J. N. D. (2008). Scheduling identical parallel machines to minimize total tardiness. *International Journal of Production Economics, 115*, 134–142.

Chang, Y.-C., & Lee, C.-Y. (2004). Machine scheduling with job delivery coordination. *European Journal of Operational Research, 158*, 470–487.

Chen, Z.-L. (2004). Integrated production and distribution operations – Taxonomy, models, and review. In D. Simchi-Levi, S. D. Wu, & Z.-J. Shen (Eds.), *Handbook of quantitative supply chain analysis* (pp. 711–740). Boston: Kluwer Academic.

Chen, Z.-L. (2010). Integrated production and outbound distribution scheduling: Review and extensions. *Operations Research, 58*, 130–148.

Chen, B., & Lee, C.-Y. (2008). Logistics scheduling with batching and transportation. *European Journal of Operational Research, 189*, 871–876.

Chen, Z.-L., & Pundoor, G. (2006). Order assignment and scheduling in a supply chain. *Operations Research, 54*, 555–572.

Chen, Z.-L., & Vairaktarakis, G. L. (2005). Integrated scheduling of production and distribution operations. *Management Science, 51*, 614–628.

Condotta, A., Knust, S., Meier, D., & Shakhlevich, N. V. (2013). Tabu search and lower bounds for a combined production-transportation problem. *Computers and Operations Research, 40*, 886–900.

Farahani, P., Grunow, M., & Günther, H.-O. (2012). Integrated production and distribution planning for perishable food products. *Flexible Services and Manufacturing Journal, 24*, 28–51.

Gao, S., Qi, L., & Lei, L. (2015). Integrated batch production and distribution scheduling with limited vehicle capacity. *International Journal of Production Economics, 160*, 13–25.

Garcia, J. M., & Lozano, S. (2005). Production and delivery scheduling problem with time windows. *Computers and Industrial Engineering, 48*, 733–742.

Garcia, J. M., Lozano, S., & Canca, D. (2004). Coordinated scheduling of production and delivery from multiple plants. *Robotics and Computer-Integrated Manufacturing, 20*, 191–198.

Garcia, J. M., Lozano, S., Smith, K., Kwok, T., & Villa, G. (2002). Coordinated scheduling of production and delivery from multiple plants and with time windows using genetic algorithms. In *Proceedings of the 9th international conference on neuronal information processing (ICONP'02)* (vol. 3, pp. 1153–1158).

Gavish, B., & Graves, S. C. (1978). The travelling salesman problem and related problems (Working paper OR-078-78). Cambridge, MA: Operations Research Center, Massachusetts Institute of Technology.

Geismar, H. N., Dawande, M., & Sriskandarajah, C. (2011). Pool-point distribution of zero-inventory products. *Production and Operations Management, 20*, 737–753.

Geismar, H. N., Laporte, G., Lei, L., & Srinkandarajah, C. (2008). The integrated production and transportation scheduling problem for a product with a short lifespan. *INFORMS Journal on Computing, 20*, 21–33.

Graham, R. L., Lawler, E. L., Lenstra, J. K., & Rinnooy Kan, A. H. G. (1979). Optimization and approximation in deterministic sequencing and scheduling: A survey. *Annals of Discrete Mathematics, 4*, 287–326.

Hajiaghaei-Keshteli, M., Aminnayeri, M., & Fatemi Ghomi, S. M. T. (2014). Integrated scheduling of production and rail transportation. *Computers and Industrial Engineering, 74*, 240–256.

Hall, N. G., & Potts, C. N. (2003). Supply chain scheduling: Batching and delivery. *Operations Research, 51*, 566–584.
Hall, N. G., & Potts, C. N. (2005). The coordination of scheduling and batch deliveries. *Annals of Operations Research, 135*, 41–64.
Hurter, A. P., & Van Buer, M. G. (1996). The newspaper production/distribution problem. *Journal of Business Logistics, 17*, 85–107.
Kang, H.-Y., Pearn, W. L., Chung, I.-P., & Lee, A. H. I. (2015). An enhanced model for the integrated production and transportation problem in a multiple vehicles environment. *Soft Computing, 20*(4), 1415–1435.
Karaoğlan, İ., & Kesen, S. E. (2017). The coordinated production and transportation scheduling problem with a time-sensitive product: A branch-and-cut algorithm. *International Journal of Production Research, 55*, 536–557.
Lee, J., Kim, B.-I., Johnson, A. L., & Lee, K. (2014). The nuclear medicine production and delivery problem. *European Journal of Operational Research, 236*, 461–472.
Li, K. P., Ganesan, V. K., & Sivakumar, A. I. (2005). Synchronized scheduling of assembly and multi-destination air-transportation in a consumer electronics supply chain. *International Journal of Production Research, 43*, 2671–2685.
Li, C.-L., & Vairaktarakis, G. (2007). Coordinating production and distribution of jobs with bundling operations. *IIE Transactions, 39*, 203–215.
Li, C.-L., Vairaktarakis, G., & Lee, C.-Y. (2005). Machine scheduling with deliveries to multiple customer locations. *European Journal of Operational Research, 164*, 39–51.
Low, C., Chang, C.-M., Li, R.-K., & Huang, C.-L. (2014). Coordination of production scheduling and delivery problems with heterogeneous fleet. *International Journal of Production Economics, 153*, 139–148.
Miller, C. E., Tucker, A. W., & Zemlin, R. A. (1960). Integer programming formulations and travelling salesman problems. *Journal of the Association for Computing Machinery, 7*, 326–329.
Ng, C. T., & Lu, L. (2012). On-line integrated production and outbound distribution scheduling to minimize the maximum delivery completion time. *Journal of Scheduling, 15*, 391–398.
Potts, C. N. (1980). Analysis of a heuristic for one machine sequencing with release dates and delivery times. *Operations Research, 28*, 1436–1441.
Pundoor, G., & Chen, Z.-L. (2005). Scheduling a production-distribution system to optimize the tradeoff between delivery tardiness and total distribution cost. *Naval Research Logistics, 52*, 571–589.
Rasti-Barzoki, M., & Hejazi, S. R. (2013). Minimizing the weighted number of tardy jobs with due date assignment and capacity-constrained deliveries for multiple customers in supply chain. *European Journal of Operational Research, 228*, 345–357.
Roberti, R., & Toth, P. (2012). Models and algorithms for the asymmetric traveling salesman problem: An experimental comparison. *EURO Journal on Transportation and Logistics, 1*, 113–133.
Stecke, K. E., & Zhao, X. (2007). Production and transportation integration for a make-to-order manufacturing company with a commit-to-delivery business mode. *Manufacturing and Service Operations Management, 9*, 206–224.
Steiner, G., & Zhang, R. (2011). Minimizing the weighted number of tardy jobs with due date assignment and capacity-constrained deliveries. *Annals of Operations Research, 191*, 171–181.
Thomas, D. J., & Griffin, P. M. (1996). Coordinated supply chain management. *European Journal of Operational Research, 94*, 1–15.
Ullrich, C. A. (2012). Supply chain scheduling: Makespan reduction potential. *International Journal of Logistics: Research and Applications, 15*, 323–336.
Ullrich, C. A. (2013). Integrated machine scheduling and vehicle routing with time windows. *European Journal of Operational Research, 227*, 152–165.
Van Buer, M. G., Woodruff, D. L., & Olson, R. T. (1999). Solving the medium newspaper production/distribution problem. *European Journal of Operational Research, 115*, 237–253.

Viergutz, C., & Knust, S. (2014). Integrated production and distribution scheduling with lifespan constraints. *Annals of Operations Research, 213*, 293–318.

Wang, H., & Lee, C.-Y. (2005). Production and transport logistics scheduling with two transport mode choices. *Naval Research Logistics, 52*, 796–809.

# Achieving Shojinka by Integrated Balancing of Multiple Straight Lines with Resource Dependent Task Times

**Yakup Atasagun, Yakup Kara, and Gözde Can Atasagun**

## 1 Introduction

Toyota Production System (TPS), which is also called as "lean manufacturing" or "lean production", is known as one of the most flexible, efficient and productive production systems. TPS has emerged with the philosophy of Just-in-Time (JIT) in 1950s and pioneered the continuous development model. Several techniques such as level production, one piece flow, kanban system, shojinka etc. are used in accordance with the main purpose of JIT, which is producing the necessary products in the necessary quantities at the necessary time and eliminating all kinds of waste in the production environment (Monden 1993).

The concept of Shojinka, which was originally an important element of TPS, is easily to increase or decrease the number of workers in a production facility when the demand rate is increased or decreased. In a production facility, different types of products may be produced on different lines. The fluctuations in demands of products will probably require adding workers to some lines and removing from others (Monden 1993; Gökçen et al. 2010). In TPS, gaining the flexibility to adjust the number of workers in a workshop due to demand changes is called *shojinka*. Shojinka is one of the main components of JIT philosophy in the field of reducing workforce wastes. Monden (1993) states that three factors are prerequisite in order to realize shojinka: "proper design of machinery layout", "multi-functional workers" and "continuous evaluation of standard operation routine". TPS adopts

Y. Atasagun (✉) · Y. Kara · G. Can Atasagun
Department of Industrial Engineering, Faculty of Engineering and Natural Sciences, Konya Technical University, Konya, Turkey
e-mail: yatasagun@selcuk.edu.tr



T. Paksoy et al. (eds.), *Lean and Green Supply Chain Management*, International Series in Operations Research & Management Science 273,
https://doi.org/10.1007/978-3-319-97511-5_9

U-shaped lines as the most proper machinery layout for realizing the concept of shojinka. U-shaped lines have more advantages compared to straight lines such as high communication between workers, easy problem solving, convenience for workers to be transformed into multi-functional workers and potential to utilize less number of stations (Miltenburg and Wijngaard 1994; Atasagun and Kara 2016).

On the other hand, for a given planning period, the number of workers required on a U-shaped line may be fractional. Since it is not possible to allocate a fractional number of workers, either the demand will not be met or at least one worker will be idle for a proportion of his/her available time. In order to make this idle time productive and eliminate the workforce waste, it is decided to combine several U-shaped lines into a single integrated line in Toyota. This way, a worker can perform operations from two or more neighbour U-shaped lines, and idle times can be eliminated or reduced (Monden 1993; Gökçen et al. 2010).

Even though JIT production system adopts U-shaped line layout, traditional straight lines can be combined and balanced in an integrated manner to obtain the advantages of shojinka. It should be noted that traditional straight assembly lines are still one of the most important elements and an important fact of today's production systems. Therefore, if applicable, a company can combine its multiple straight assembly lines and obtain many advantages of shojinka more or less (Gökçen et al. 2010).

In the case of multiple straight assembly lines are combined, the problem, which is called as Integrated Balancing of Multiple Straight Lines (MSLB), arises. Essentially, MSLB is a generalized version of a well-known problem, simple assembly line balancing (SALB) which was first studied by Salveson (1955). Numerous studies have been published on SALB up to date. The literature on SALB is not practical to present here but interested researchers can refer to the review studies of Baybars (1986), Ghosh and Gagnon (1989), Erel and Sarin (1998), Becker and Scholl (2006), Scholl and Becker (2006) and Battaïa and Dolgui (2013).

The pioneering study on MSLB was performed by Gökçen et al. (2006). Gökçen et al. (2006) suggested that more than one assembly line can be located parallel and they can be balanced in an integrated manner. They proposed a binary integer mathematical model and a heuristic procedure for minimizing total number of stations, assuming that common stations can be utilized between two adjacent lines. The experimental study performed by Gökçen et al. (2006) revealed that integrated balancing approach utilized less number of stations for 65 out of 95 (68.4%) test problems compared to the independent balances of the lines.

In another study on MLSB, Gökçen et al. (2010) improved the work of Gökçen et al. (2006) considering not only parallel connectivity of the lines but also different connectivity opportunities such as consecutive and perpendicular connectivity. Gökçen et al. (2010) stated that if assembly lines are tightly related to each other and some supplier–customer relationships exist between these assembly lines then they may be connected to each other with consecutive or perpendicular connectivity.

The consecutive connectivity type can appear between an upstream line (supplier) and its downstream line (customer). If the output of an upstream line is the main input (part) of a downstream line, then assembly line managers may desire to locate these lines consecutively and close to each other (Gökçen et al. 2010). In this case, mentioned two lines can be connected to each other with a consecutive connectivity by utilizing a common station which includes tasks from the end of the upstream line and from the beginning of the downstream line.

The outputs of an upstream line may not always be the main part for a downstream line. In other words, the outputs of the upstream line may not be processed throughout most of the tasks on the downstream line. The outputs may be components that are attached to the main parts processed on the downstream line. This attachment can be performed at a stage (task) of assembly process. In this case, assembly line managers probably desire to locate the upstream line to the nearest point of use so as to minimise material handling from the upstream line to the downstream line. If such a location exists in a production system, two lines can be connected to each other with perpendicular connectivity (Gökçen et al. 2010).

One of the most important decision problems in assembly lines is assembly line balancing (ALB). ALB is the problem of assigning task to stations in such a way that precedence relations among tasks are not violated, sum of the processing times of the tasks in a station does not exceed cycle time and a performance measure is optimized. This performance measure is usually minimization of the number of utilized stations for a given cycle time. Cycle time is the time interval between two completed products. ALB problems with the objective of minimizing the number of stations is categorized as Type-I. Conversely, cycle time may also be minimized for a given number of stations in order to maximize the output. This type of ALB problems are known as Type-II.

In addition, the basic assumption in most of the studies of ALB literature is that every task's time is fixed. However, in practice, different resource alternatives such as equipment or assistant worker may be available to process a task with different times. The problem in this case is to assign tasks and resources to stations that minimize total cost (Kara et al. 2011). Faaland et al. (1992) defined mentioned problem as Resource Dependent Assembly Line Balancing (RDALB). There are several studies in ALB literature dealing with resource dependency such as Pinto et al. (1983), Bukchin and Tzur (2000), Bukchin and Rubinovitz (2002) and Jayaswal and Agarwal (2014). Additionally, Corominas et al. (2008), Moon et al. (2009) and Corominas et al. (2011) implemented some resource restrictions to the ALB problem.

Kara et al. (2011) generalized the RDALB problem for straight and U-shaped assembly lines and proposed binary integer formulations considering both resource dependency and resource restrictions with practice oriented assumptions.

Kara and Atasagun (2013) adapted the RDALB approach of Kara et al. (2011) to the concept of parallel assembly lines and they obtained promising results in the case of parallel lines are balanced in an integrated manner instead of independent balancing.

In this chapter, RDALB approach proposed by Kara et al. (2011) is adapted to the problem of MSLB which was developed by Gökçen et al. (2010). The problem is called as integrated balancing of multiple straight assembly lines with resource dependent task times (RDMSLB). A binary integer mathematical model is presented with the objective of minimizing total equipment and operating costs of assembly lines for RDMSLB. An experimental analysis is also conducted to emphasize the advantages of the integrated balancing concept, compared to the situation in which the lines are balanced independent from each other. The remainder of the chapter is structured as follows. Notations and the proposed mathematical model are explained in Sect. 2. The proposed model is validated on an illustrative example in Sect. 3. Design and results of the experimental study are presented in Sect. 4. Finally, conclusion and some future research suggestions are given in Sect. 5.

## 2 Mathematical Model

In this section we propose binary integer programming formulations for RDMSLB by adhering to the assumptions of Kara et al. (2011) for RDALB and the assumptions of Gökçen et al. (2010) for the concept of MSLB.

### 2.1 Assumptions

The assumptions of the proposed model are as follows:

- There are more than one assembly lines located close to each other with parallel, consecutive and perpendicular connectivity options.
- Connectivity options of each line pair is predetermined and known.
- Each assembly line in the facility is single model.
- The precedence relationships among tasks of each line are known.
- The processing time of a task is deterministic, but depends on the resources (equipment type and assistant) allocated to perform the task. The processing time of a task is independent of the station to which the task is assigned.
- Some tasks cannot be completed by only one worker. If such a task is assigned to a station, an assistant should be assigned to this station as well.
- The processing times of some tasks can be reduced by performing these tasks with the assistance of an assistant.
- Some tasks should be performed using particular equipment. There may be alternate equipment types for a task. Some tasks can be performed with an equipment type or without equipment. Each equipment type is specified with a cost.

- At most one equipment type can be allocated to perform a task. An equipment type can be used to perform more than one task in a station. In the case of two tasks from different lines require the same equipment type in a common station; it is assumed that those tasks can share the related equipment type.
- Two or more equipment types can be assigned to a station for performing different tasks.
- There is sufficient number of workers required to operate stations. But, the amounts of other resources (equipment types and assistants) are limited.
- Employment cost of an assistant is independent of the station to which the assistant is assigned.
- Utilization costs of the stations are assumed to be equal.
- No work-in-process inventory is allowed between workstations.
- It can also be worked each side of any line.
- At most one common station can be utilized for a consecutive line pair, including tasks from the end of the upstream line and from the beginning of the downstream line.
- At most one common station can be utilized for a perpendicular line pair, including tasks from the end of the upstream line and the task of the downstream line to which outputs of the upstream line are input.
- Cycle times of the lines may differ. The cycle time for a common station among lines with different cycle times is the minimum of the cycle times of the lines that it spans.

## 2.2 Notation

The notations of the proposed model are as follows:

| | |
|---|---|
| *Indices* | |
| $h, g, k, l, m, n$: | Assembly line |
| $i, r, s, a, b, c, d$: | Task |
| $j$: | Station |
| $e$: | Equipment |
| *Parameters and sets* | |
| $t^0_{hie}$: | Completion time of task $i$ on line $h$ with equipment $e$ without assistant |
| $t^1_{hie}$: | Completion time of task $i$ on line $h$ with equipment $e$ with assistant |
| $N_h$: | Set of tasks on line $h$ |
| $H$: | Set of assembly lines |
| $E$: | Set of equipment |
| $E_{hi}$: | Set of equipment which can be used to process task $i$ on line $h$ |

(continued)

| | |
|---|---|
| $NE_e$: | Available number of equipment $e$ |
| $NA$: | Available number of assistants |
| $J$: | Set of stations |
| $CT_h$: | Cycle time of line $h$ |
| $PR_h$: | Set of precedence relations on line $h$ |
| $(r, s) \in PR_h$: | A precedence relation on line $h$; task $r$ is an immediate predecessor of task $s$ |
| $P^*_{hi}$: | Set of predecessors of task $i$ of line $h$ |
| $S^*_{hi}$: | Set of successors of task $i$ of line $h$ |
| $F$: | Set of disconnection relationships |
| $(h, g) \in F$: | A disconnection relationship; common stations between lines $h$ and $g$ are not allowed |
| $CC$: | Set of consecutive connectivity relationships |
| $(m, n) \in CC$: | A consecutive connectivity relationship; upstream line $m$ is connected to downstream line $n$ with a consecutive connectivity |
| $c$: | A task of upstream line $m$; $S*_{mc} = \emptyset$ |
| $d$: | A task of downstream line $n$; $P*_{mc} = \emptyset$ |
| $PC$: | Set of perpendicular connectivity relationships |
| $(k, l) \in PC$: | A perpendicular connectivity relationship; upstream line $k$ is connected to downstream line $l$ with a perpendicular connectivity |
| $a$: | The last task of the upstream line $k$ |
| $b$: | The task of downstream line $l$ to which outputs of the upstream line $k$ are input |
| $CW$: | Annual utilization cost of a station (worker + fixed costs) |
| $CA$: | Annual employment cost of an assistant |
| $c_e$: | Annual operating cost of equipment $e$ |
| $K_{max}$: | Maximum number of stations |
| $M$: | A big number |
| *Variables* | |
| $x_{hij}$: | 1, if task $i$ on line $h$ is assigned to station $j$; 0, otherwise |
| $p_{hije}$: | 1, if task $i$ on line $h$ is assigned to station $j$ with equipment $e$ without assistant; 0, otherwise |
| $q_{hije}$: | 1, if task $i$ on line $h$ is assigned to station $j$ with equipment $e$ with assistant; 0, otherwise |
| $z_{je}$: | 1, if equipment $e$ is assigned to station $j$; 0, otherwise |
| $u_j$: | 1, if station $j$ is utilized; 0, otherwise |
| $t_j$: | 1, if an assistant is assigned to station $j$; 0, otherwise |
| $y_{hj}$: | 1, if station $j$ is utilized on line $h$; 0, otherwise |
| $W_{(k,l)j}$: | 1, if station $j$ includes tasks of both line $k$ and line $l$ and it is a common station; 0, otherwise |
| $V_{(k,l)j}$: | 1, if station $j$ includes tasks of one of the line $k$ and line $l$ and it is not a common station; 0, otherwise |

## 2.3 Model

The objective function and constraints of the proposed model are as follows:

$$Min \sum_{j \in J} \left( CWu_j + CAt_j \right) + \sum_{j \in J} \sum_{e \in E} c_e z_{je} \tag{1}$$

$$\sum_{j \in J} x_{hij} = 1 \qquad \forall h \in H; \forall i \in N_h \tag{2}$$

$$\sum_{e \in E_{hi}} \left( p_{hije} + q_{hije} \right) = x_{hij} \qquad \forall h \in H; \forall i \in N_h; \forall j \in J \tag{3}$$

$$\begin{aligned} &\sum_{h \in H} \sum_{i \in N_h} \sum_{e \in E_{hi}} \left( t^0_{hie} p_{hije} + t^1_{hie} q_{hije} \right) \\ &\quad \leq CT_g y_{gj} + M \left( 1 - y_{gj} \right) \qquad \forall g \in H; \forall j \in J \end{aligned} \tag{4}$$

$$\sum_{j \in J} \left( K_{max} - j + 1 \right) \left( x_{hrj} - x_{hsj} \right) \geq 0 \qquad \forall h \in H; \forall (r, s) \in PR_h \tag{5}$$

$$\sum_{i \in N_h} x_{hij} - \| N_h \| \, y_{hj} \leq 0 \qquad \forall h \in H; \forall j \in J \tag{6}$$

$$\sum_{i \in N_h} x_{hij} - \| N_h \| \, y_{hj} \geq 1 - \| N_h \| \qquad \forall h \in H; \forall j \in J \tag{7}$$

$$\sum_{h \in H} y_{hj} - \| H \| \, u_j \leq 0 \qquad \forall j \in J \tag{8}$$

$$y_{hj} + y_{gj} \leq 1 \qquad \forall (h, g) \in F; \forall j \in J \tag{9}$$

$$y_{kj} + y_{lj} - 2W_{(k,l)j} - V_{(k,l)j} = 0 \qquad \forall (k, l) \in PC; \forall j \in J \tag{10}$$

$$\sum_{j \in J} W_{(k,l)j} \leq 1 \qquad \forall (k, l) \in PC \tag{11}$$

$$x_{kaj} + x_{lbj} - 2W_{(k,l)j} \geq 0 \qquad \forall (k, l) \in PC; \forall j \in J \tag{12}$$

$$\begin{aligned} &\sum_{j \in J} \left( K_{max} - j + 1 \right) \left( x_{mcj} - x_{ndj} \right) \geq 0 \qquad \forall (m, n) \in CC; \forall c \in N_m; S^*_{mc} \\ &\quad = \varnothing; \forall d \in N_n; P^*_{nd} = \varnothing \end{aligned} \tag{13}$$

$$\sum_{h \in H} \sum_{i \in N_h} \left( p_{hije} + q_{hije} \right) - Mz_{je} \leq 0 \qquad \forall e \in E; \forall j \in J \tag{14}$$

$$\sum_{j \in J} z_{je} \leq NE_e \qquad \forall e \in E \tag{15}$$

$$\sum_{h \in H} \sum_{i \in N_h} \sum_{e \in E_{hi}} q_{hije} - Mt_j \leq 0 \qquad \forall j \in J \tag{16}$$

$$\sum_{j \in J} t_j \leq NA \tag{17}$$

$$x_{hij}, p_{hije}, q_{hije}, z_{je}, u_j, t_j, y_{hj}, W_{(k,l)j}, V_{(k,l)j} \in \{0, 1\} \tag{18}$$

The objective function in Eq. (1) denotes the total cost associated with station utilization, assistant worker and equipment allocation. Equation (2) ensures that each task of each line is assigned to at least and at most one station. Equation (3) determines the resources (equipment type and assistant) allocated to a station. Equation (4) ensures that the workload of a station does not exceed the predetermined cycle time. This equation also ensures that the cycle time for a common station between two or more lines is the minimum of the cycle times of the lines that it spans. Precedence relationships among tasks are satisfied by the set of constraints given in Eq. (5). Equations (6) and (7) determines whether a station $j$ is opened for line $h$ or not. Equation (8) determines whether a station $j$ is utilized or not. Equation (9) is used for disconnecting two assembly lines which cannot be connected with common stations. For a given $(k, l)$ perpendicular connectivity relationship, Eq. (10) determines whether a station $j$ is a common station between lines $k$ and $l$, or not. Equation (11) ensures that at most one common station can be utilized between perpendicular connected lines $k$ and $l$. If a common station is utilized between perpendicular connected lines $k$ and $l$, Eq. (12) guarantees this station contains the last task ($a$) of the upstream line $k$ and the task ($b$) of the downstream line $l$ to which the outputs of upstream line $k$ are the input. Note that, if there are two or more tasks with no successors in the precedence diagram of the upstream line $k$, a dummy task which succeeds all remaining tasks of the upstream line $k$ with zero processing time should be added. For a given $(m, n)$ consecutive connectivity relationship, Eq. (13) guarantees a common station between lines $m$ and $n$ includes tasks from the end of the upstream line $m$ and from the beginning of the downstream line $n$. Equation (14) determines whether an equipment $e$ is allocated to station $j$ or not. Equation (15) restricts the allocated number of equipment $e$ by the available number of this equipment type. Equation (16) determines whether an assistant worker is allocated to station $j$ or not. Equation (17) ensures that the number of assistant workers assigned to stations does not exceed the available number of assistant workers. Finally, Equation (18) denotes that all variables in the model are binary variables.

## 3 Illustrative Example

The proposed mathematical model is validated on an illustrative problem in this section. The illustrative problem consists of four assembly lines each with 10 tasks. Information about task processing times in seconds and precedence relationships are given in Tables 1, 2, 3 and 4. Additionally, precedence diagrams of the lines are also given in Appendix.

Task time and precedence relationships information of Line 1 is given in Table 1. The $IP_i$ column of the table denotes the immediate predecessors of task $i$. It is seen in the table that task 1 of the Line 1 has a manual processing time of 21 s with one worker. Task 2 also has a manual processing time of 21 s with one worker but this task can alternatively be processed by Equipment 2 and 9 with processing times of 18 s and 13 s, respectively. Task 3 can be processed by one worker in 45 s manually but assistance of an assistant worker can reduce the task time to 27 s. Note that task 4 has three different processing alternatives each with assistant. This means that task 4 necessarily requires assistance of an assistant worker, and so on.

Similarly, task time and precedence relationships information of Lines 2, 3 and 4 are given in Tables 2, 3 and 4, respectively.

**Table 1** Task time and precedence relationships information of Line 1

| | | Task processing times | | | | | | | | | | | | | |
|---|---|---|---|---|---|---|---|---|---|---|---|---|---|---|---|
| | | | Equipment | | | | | | | | | | | | |
| Task ($i$) | $IP_i$ | Assist. | No | 1 | 2 | 3 | 4 | 5 | 6 | 7 | 8 | 9 | 10 | 11 | 12 |
| 1 | – | Yes | | | | | | | | | | | | | |
| | | No | 21 | | | | | | | | | | | | |
| 2 | – | Yes | | | | | | | | | | | | | |
| | | No | 21 | | 18 | | | | | | | 13 | | | |
| 3 | – | Yes | 27 | | | | | | | | | | | | |
| | | No | 45 | | | | | | | | | | | | |
| 4 | 3 | Yes | 29 | 15 | 25 | | | | | | | | | | |
| | | No | | | | | | | | | | | | | |
| 5 | 2, 3 | Yes | 36 | | | | | | | | | | | | |
| | | No | 46 | | | | | | | | | | | | |
| 6 | 5 | Yes | 38 | | | | | | | | | | | | |
| | | No | | | | | | | | | | | | | |
| 7 | 1, 4, 6 | Yes | | | | | | | | | | | | | |
| | | No | 49 | | | | | 27 | | | | 29 | 36 | | |
| 8 | 6 | Yes | | | | | | | | | | | | | |
| | | No | 20 | | | | | | | | | | | | |
| 9 | 7 | Yes | | | | | | | | | | | | | |
| | | No | 33 | | | | | 19 | | | | | 26 | | 22 |
| 10 | 8, 9 | Yes | | | | | | | | | | | | | |
| | | No | 42 | 21 | | | 34 | 24 | | | | | | | |

**Table 2** Task time and precedence relationships information of Line 2

| | | Task processing times | | | | | | | | | | | | | | |
|---|---|---|---|---|---|---|---|---|---|---|---|---|---|---|---|---|
| | | | Equipment | | | | | | | | | | | | | |
| Task ($i$) | $IP_i$ | Assist. | No | 1 | 2 | 3 | 4 | 5 | 6 | 7 | 8 | 9 | 10 | 11 | 12 |
| 1 | – | Yes | | | | | | | | | | | | | |
| | | No | 34 | | 30 | | | | | | | | 26 | | 23 |
| 2 | – | Yes | 48 | | | | | | | | | | | | |
| | | No | | | | | | | | | | | | | |
| 3 | – | Yes | | | | | | | | | | | | | |
| | | No | 13 | | | | | | | | | | | | |
| 4 | – | Yes | | | | | | | | | | | | | |
| | | No | 42 | | | | | | | | | | | | |
| 5 | 1, 2, 4 | Yes | | | | | | | | | | | | | |
| | | No | 22 | | | | | | | | | | | | |
| 6 | 5 | Yes | | | | | | | | | | | | | |
| | | No | 47 | | | | | | | | | | | | |
| 7 | 3, 5 | Yes | 14 | | | | | | | | | | | | |
| | | No | | | | | | | | | | | | | |
| 8 | 7 | Yes | | | | | | | | | | | | | |
| | | No | 16 | | | | 13 | | 8 | | 13 | | | | |
| 9 | 7 | Yes | 8 | | | | | | | | | | | | |
| | | No | 14 | | | | | | | | | | | | |
| 10 | 6, 8, 9 | Yes | | | | | | | | | | | | | |
| | | No | 21 | | | | | | | | | | 16 | 11 | 14 |

The connection relationships among lines for the illustrative problem are shown in Table 5.

Table 5 indicates that Lines 2 and 4 are connected to each other with a perpendicular connectivity. Note that, the last task of the line 2 is numbered as "10" and outputs of the upstream line 2 are input for the task 6 of the downstream line 4.

Available numbers and operating costs of different equipment types are given in Table 6.

Additional information about the illustrative example is as follows: utilization cost of a station ($CW$) is 100 cost units; employment cost of an assistant worker ($CA$) is 70 cost units and the number of available assistant workers ($NA$) is 10. The case "no equipment" is considered as an equipment type by labelling the equipment number "0". Cycle time is 180 s and it is equal for all of the lines. $K_{max}$ is selected as nine and the problem is solved using CPLEX solver V12.5 on an Intel Xeon E5-1650 (6 Core) 3.20 GHz processor and 16 GB RAM workstation.

The optimal solution of the illustrative problem is obtained within 2 h and 2 min and 17 s with the objective function value of 899 cost units. Figure 1 illustrates the optimal solution of the problem.

**Table 3** Task time and precedence relationships information of Line 3

| | | Task processing times | | | | | | | | | | | | | |
|---|---|---|---|---|---|---|---|---|---|---|---|---|---|---|---|
| | | | Equipment | | | | | | | | | | | | |
| Task ($i$) | $IP_i$ | Assist. | No | 1 | 2 | 3 | 4 | 5 | 6 | 7 | 8 | 9 | 10 | 11 | 12 |
| 1 | – | Yes | 14 | | | | | | 7 | 7 | | | | | |
| | | No | | | | | | | | | | | | | |
| 2 | – | Yes | | | | | | | | | | | | | |
| | | No | 18 | | | | | | | | | | | | |
| 3 | 1, 2 | Yes | 9 | | | | | | | | | | | | |
| | | No | 19 | | | | | | | | | | | | |
| 4 | 3 | Yes | | | | | | | | | | | | | |
| | | No | 22 | | | | | | | | 19 | | | | |
| 5 | 2 | Yes | | | | | | | | | | | | | |
| | | No | 15 | | | | | | | | | | | | |
| 6 | 4, 5 | Yes | 12 | | | | | | | | | | | | |
| | | No | 16 | | | | | | | | | | | | |
| 7 | 6 | Yes | | | | | | | | | | | | | |
| | | No | 27 | | | | | | | | | | | | |
| 8 | 6 | Yes | | | | | | | | | | | | | |
| | | No | 43 | | | | | | | 21 | 36 | | | 25 | |
| 9 | 1, 5 | Yes | | | | | | | | | | | | | |
| | | No | 24 | | | | | | | | | | | | |
| 10 | 7, 8, 9 | Yes | 20 | | | | | | | | | | | | |
| | | No | | | | | | | | | | | | | |

Figure 1 shows that, optimal solution of the illustrative problem consists of six stations, three assistant workers and four different types of equipment such as equipment 2, 5, 8 and 10. Stations I and II are common stations between parallel connected lines 1 and 3. Station I includes tasks 1, 2, 3, 4, 5 and 6 of line 1 and task 1 of line 2 while station II includes tasks 7, 8, 9 and 10 of line 1 and tasks 2, 3, 4, 5 and 6 of line 2. Station III is a common station between consecutive connected lines 3 and 4, and it includes tasks from the end of line 3 and from the beginning of line 4. Similarly, Station V is a common station between perpendicular connected lines 2 and 4. This station includes the last task of line 2 (task 10) and the task of line 4 to which outputs of line 2 are input (task 6). Finally, stations IV and VI are single line stations and they are utilized for lines 2 and 4, respectively.

It is also seen in Fig. 1 that, additional resources allocated to stations are as follows: equipment 2 and an assistant to the station I, equipment 5 to the station II, an assistant to each of stations III and IV, and finally equipment 8 and equipment 10 to the station V.

**Table 4** Task time and precedence relationships information of Line 4

| | | Task processing times | | | | | | | | | | | | | |
|---|---|---|---|---|---|---|---|---|---|---|---|---|---|---|---|
| | | | Equipment | | | | | | | | | | | | |
| Task ($i$) | $IP_i$ | Assist. | No | 1 | 2 | 3 | 4 | 5 | 6 | 7 | 8 | 9 | 10 | 11 | 12 |
| 1 | – | Yes | | | | | | | | | | | | | |
| | | No | 19 | | | | | | | | | | | | |
| 2 | 1 | Yes | 23 | | | | 18 | | | | | | | | |
| | | No | | | | | | | | | | | | | |
| 3 | 2 | Yes | | | | | | | | | | | | | |
| | | No | 22 | | 19 | | | | | | | | | | |
| 4 | – | Yes | 20 | | | | | | | | | | | | |
| | | No | | | | | | | | | | | | | |
| 5 | 3 | Yes | | | | | | | | | | | | | |
| | | No | 38 | | | | | | | | 33 | | | 21 | 25 |
| 6 | 5 | Yes | 31 | | | | | | | | | | | | |
| | | No | 40 | | | | 32 | | | | 35 | | | | |
| 7 | 6 | Yes | | | | | | | | | | | | | |
| | | No | 46 | | | | 37 | | | | | | | 25 | 29 |
| 8 | 6 | Yes | 12 | | | | | | | | | | | | |
| | | No | 16 | 8 | | 14 | | | 8 | | | | | | |
| 9 | 5 | Yes | | | | | | | | | | | | | |
| | | No | 18 | | | | | | | | | | | | |
| 10 | 4, 7, 8, 9 | Yes | 25 | | | | | | | | | | | | |
| | | No | 49 | | | | | | | | | | | | |

**Table 5** The connection relationships among lines

| | Line | | | |
|---|---|---|---|---|
| Line | 1 | 2 | 3 | 4 |
| 1 | – | None | Parallel | None |
| 2 | None | – | None | Perpendicular |
| 3 | Parallel | None | – | Consecutive |
| 4 | None | Perpendicular | Consecutive | – |

**Table 6** Available numbers and operating costs of equipment

| Equipment | Available number | Operating cost |
|---|---|---|
| 1 | 1 | 48 |
| 2 | 1 | 14 |
| 3 | 2 | 12 |
| 4 | 1 | 17 |
| 5 | 1 | 43 |
| 6 | 2 | 46 |
| 7 | 2 | 48 |
| 8 | 2 | 11 |
| 9 | 1 | 36 |
| 10 | 1 | 21 |
| 11 | 1 | 44 |
| 12 | 2 | 33 |

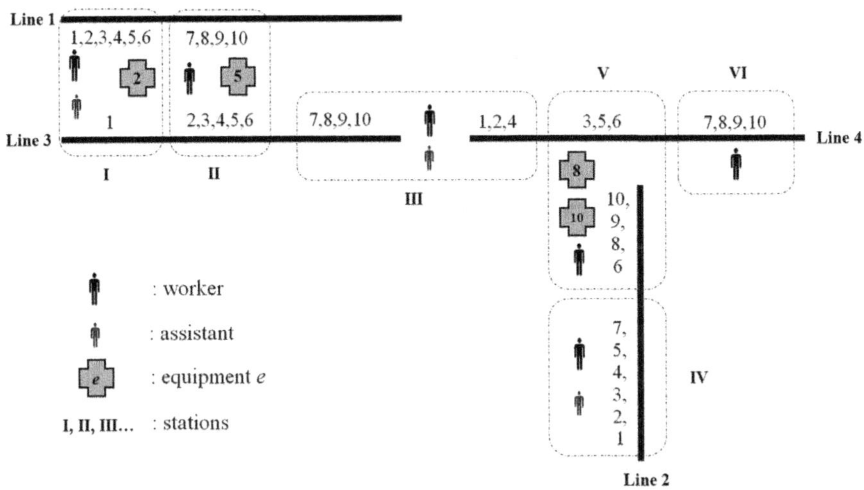


**Fig. 1** Optimal solution of the illustrative problem

# 4 Experimental Results

In this section an experimental study is conducted to compare integrated and independent balances of multiple straight lines in terms of total cost. A data set consists of 108 problems is generated and all those problems are solved using both integrated (RDMSLB) and independent (RDMSLB-*independent*) balancing approaches. Hence, a total of 216 solutions are obtained.

Assembly systems are generally characterised by three factors: problem size, strength of the precedence ordering relations among tasks, and the expected number of tasks per station (Aase et al. 2004; Kara et al. 2011). These factors can be defined by the number of tasks (*NT*), flexibility ratio (*FR*), and cycle time (*CT*) measures, respectively (Kara et al. 2011). *FR* is known as an indicator of the difficulty of the problem in ALB literature. A higher *FR* value indicates fewer precedence relations among tasks. This means more solution alternatives and a larger solution space. So, obtaining the optimal solution will be more difficult compared to an equivalent problem with a lower *FR* value.

It is assumed that there are four assembly lines in each test problem. Line 1 and Line 3 are located in parallel. Line 3 and Line 4 are connected to each other with a consecutive connectivity in which Lines 3 and 4 are upstream and downstream lines, respectively. Line 2 and Line 4 are connected to each other with a perpendicular connectivity in which outputs of Line 2 are inputs for Line 4. Finally, common stations are not allowed between line pairs (1, 2), (1, 4) and (2, 3).

Four levels of *NT* (20, 40, 60 and 80) are considered by assuming that each of the lines in the problem includes an equal number of tasks. For example, each line includes 5 tasks in a 20-task problem. Precedence relationships among tasks for each *NT* level are randomly generated and three different levels of *FR* (0.25, 0.50 and 0.75) are considered. In a *NT–FR* combination, it is assumed that all of the lines in the problem have precedence diagrams with the same level of *FR*. Hence, 12 assembly line systems are obtained.

In addition, it is assumed that all of the tasks in an assembly line system have a processing alternative without equipment (manual processing time). Processing times of the tasks without equipment are randomly generated following a discrete uniform distribution with $U(10, 50)$. The numbers of different equipment types are selected as 6, 12, 18 and 24 for *NT* levels of 20, 40, 60 and 80, respectively. Annual operating costs of the equipment are randomly generated following a discrete uniform distribution with $U(10, 50)$. Fifty percent of the tasks in each problem are considered to have processing alternatives with equipment while remaining 50% are assumed to be manual. The tasks with processing alternatives are also selected randomly. Processing times of the tasks with equipment are randomly generated in such a way that a task has a lesser processing time with equipment than its manual processing time. In the process of generating task times with equipment, it is assumed that task time reducing of a particular equipment type should be associated with the operating cost of that equipment type. For example, if the operating cost of equipment *e* is between 11 and 20, then this equipment reduces a task's manual processing time with a random percentage between 11 and 20. If the operating cost of equipment *e* is between 21 and 30 then this equipment reduces a task's manual processing time with a random percentage between 21 and 30, and so on. Percentages of task time reducing for equipment types are selected randomly in the related interval. In addition, it is assumed that a task can be processed using at most three different equipment alternatives. Available numbers of equipment types for all problems are restricted by 1 or 2 randomly.

Furthermore, 25% of the tasks in each problem are assumed to necessarily require assistance of an assistant worker. An additional 25% of the tasks can be processed by one worker but assistance of an assistant worker can reduce task processing time. Note that a task can be processed using both resource alternatives such as assistant worker and equipment. Percentages of task time reducing for assistant workers are randomly selected following a discrete uniform distribution with U(20, 50), for each related task. Available numbers of assistant workers are selected as 5, 10, 15 and 20 for *NT* levels of 20, 40, 60 and 80, respectively.

The process of generating resource dependent task times is repeated three times for each *NT* level. Each of these three task times is used together with each of *NT–FR* combination to construct a problem instance and hence 36 problem instances are obtained. For each problem instance, fixed station utilization and assistant worker costs are selected 100 and 70 cost units, respectively. Each problem instance is solved considering three levels of *CT* (180, 240 and 300) using the proposed mathematical model. That is, 108 solutions for RDMSLB are obtained. Additionally, 108 solutions for RDMSLB-*independent* are obtained using the proposed model by adding all of the line pairs to the set of disconnection relationships (*F*) and excluding the Eqs. (10–13). Note that, cycle times of the different assembly lines for each problem are assumed to be equal in the experimental study. All problems are solved using CPLEX solver V12.5 on an Intel Xeon E5-1650 (6 Core) 3.20 GHz processor and 16 GB RAM workstation. The CPU time limit of 3 h is used for all 216 solutions.

If the results of experimental study analysed, it is seen that 63 out of 108 (58.33%) RDMSLB solutions are optimal while 55 out of 108 (50.93%) RDMSLB-*independent* problems are optimally solved. In addition, all of the optimally solved 55 problems in RDMSLB-*independent* are also optimal in RDMSLB. Average of CPU times of optimally solved 63 RDMSLB problems is 21 min 54 s while average of CPU times of optimally solved 55 RDMSLB-*independent* problems is 20 min 45 s.

If the RDMSLB problems which have not been solved optimally are considered, it is seen that the average of optimality gaps is 14.99% while minimum and maximum of the optimality gaps are 0.55% and 28.83%, respectively. Similarly, for the RDMSLB-*independent* problems, minimum and maximum of the optimality gaps are 4.03% and 44.11%, respectively. The average of the optimality gaps for the RDMSLB-*independent* problems is 25.18%.

The effects of different *NT*, *FR* and *CT* values on the solution status are analysed in Tables 7, 8 and 9, respectively.

It can be seen in Table 7 that when *NT* increases, the number of optimal solutions decreases dramatically. This is an expected situation due to the increase of the numbers of decision variables and constraints depending on the problem size.

Table 8 presents that, different *FR* values has not a significant effect on the solution status for both RDMSLB and RDMSLB-*independent* cases.

It is seen in Table 9 that, the number of optimal solutions increases while *CT* increases for both RDMSLB and RDMSLB-*independent* cases.

In the experiment, the results of RDMSLB and RDMSLB-*independent* are also compared in terms of total cost using the solutions of above mentioned 55 problems which are optimal for both cases. Experimental results indicate that, integrated balancing concept provides an improvement in total cost compared to the concept of independent balancing for all of the 55 problems. A paired samples *t*-test is also performed to compare the total costs of RDMSLB and RDMSLB-*independent*. Results of the *t*-test are given in Table 10.

**Table 7** Effects of *NT* values on the solution status

| *NT* | Solution status | RDMSLB Number of problems | RDMSLB Percentage of problems (%) | RDMSLB-*independent* Number of problems | RDMSLB-*independent* Percentage of problems (%) |
|---|---|---|---|---|---|
| 20 | Optimal | 27 | 100 | 27 | 100 |
| | Feasible | – | – | – | – |
| | Total | 27 | 100 | 27 | 100 |
| 40 | Optimal | 26 | 96.3 | 22 | 81.48 |
| | Feasible | 1 | 3.7 | 5 | 18.52 |
| | Total | 27 | 100 | 27 | 100 |
| 60 | Optimal | 10 | 37.04 | 6 | 22.22 |
| | Feasible | 17 | 62.96 | 21 | 77.78 |
| | Total | 27 | 100 | 27 | 100 |
| 80 | Optimal | – | – | – | – |
| | Feasible | 27 | 100 | 27 | 100 |
| | Total | 27 | 100 | 27 | 100 |
| Total | Optimal | 63 | 58.33 | 55 | 50.93 |
| | Feasible | 45 | 41.67 | 53 | 49.07 |
| | Total | 108 | 100 | 108 | 100 |

**Table 8** Effects of *FR* values on the solution status

| *FR* | Solution status | RDMSLB Number of problems | RDMSLB Percentage of problems (%) | RDMSLB-*independent* Number of problems | RDMSLB-*independent* Percentage of problems (%) |
|---|---|---|---|---|---|
| 0.25 | Optimal | 21 | 58.33 | 20 | 55.56 |
| | Feasible | 15 | 41.67 | 16 | 44.44 |
| | Total | 36 | 100 | 36 | 100 |
| 0.50 | Optimal | 22 | 61.11 | 18 | 50 |
| | Feasible | 14 | 38.89 | 18 | 50 |
| | Total | 36 | 100 | 36 | 100 |
| 0.75 | Optimal | 20 | 55.56 | 17 | 47.22 |
| | Feasible | 16 | 44.44 | 19 | 52.78 |
| | Total | 36 | 100 | 36 | 100 |
| Total | Optimal | 63 | 58.33 | 55 | 50.93 |
| | Feasible | 45 | 41.67 | 53 | 49.07 |
| | Total | 108 | 100 | 108 | 100 |

The results of paired samples *t*-test given in Table 10 indicates that, the difference between the total cost values of RDMSLB and RDMSLB-*independent* are significant at the level of 0.01 and mean total cost of integrated balancing approach is significantly lower than mean total cost of independent balancing approach.

In addition, by adhering to the study of Kara et al. (2011) the "percent improvement in total cost" (*PI*) is selected as the dependent variable for further

**Table 9** Effects of *CT* values on the solution status

| | | RDMSLB | | RDMSLB-*independent* | |
|---|---|---|---|---|---|
| *CT* | Solution status | Number of problems | Percentage of problems (%) | Number of problems | Percentage of problems (%) |
| 180 | Optimal | 17 | 47.22 | 13 | 36.11 |
| | Feasible | 19 | 52.78 | 23 | 63.89 |
| | Total | 36 | 100 | 36 | 100 |
| 240 | Optimal | 21 | 58.33 | 18 | 50 |
| | Feasible | 15 | 41.67 | 18 | 50 |
| | Total | 36 | 100 | 36 | 100 |
| 300 | Optimal | 25 | 69.44 | 24 | 66.67 |
| | Feasible | 11 | 30.56 | 12 | 33.33 |
| | Total | 36 | 100 | 36 | 100 |
| Total | Optimal | 63 | 58.33 | 55 | 50.93 |
| | Feasible | 45 | 41.67 | 53 | 49.07 |
| | Total | 108 | 100 | 108 | 100 |

**Table 10** Results of paired samples *t*-test

| Balancing approach | *N* | Mean | Std. deviation | *t* | Sig. of *t* |
|---|---|---|---|---|---|
| RDMSLB | 55 | 751.727 | 201.068 | 17.95 | 0.000 |
| RDMSLB-*independent* | 55 | 582.327 | 181.846 | | |

statistical analysis. *PI* is calculated by Eq. (19). Table 11 summaries the mean values of *PI* obtained in the experiment.

$$PI = [[\text{Cost}\,(\text{RDMSLB} - independent) - \text{Cost}\,(\text{RDMSLB})]\,/\text{Cost}\,(\text{RDMSLB} - independent)] \times 100 \quad (19)$$

Table 11 shows that, a total cost improvement of 23.105% is obtained when multiple straight lines are balanced in an integrated manner instead of independent balancing. It should be noted here that, the lack of $NT = 80$ in Table 11 is because none of the 80-task problems can be solved optimally. This situation can also be seen in Table 7.

An analysis of variance (ANOVA) and Duncan grouping test are also performed to analyse the results using IBM SPSS 20 statistical package. A three-way ANOVA is performed to examine the effects of three factors namely *NT*, *FR*, and *CT* on *PI* and the results are given in Table 12.

It is seen in Table 12 that, effects of *NT*, *CT* and two-way interaction of *NT* and *CT* are significant while the effects of *FR*, remaining two-way interactions and three-way interaction are not significant at the level of 0.01. These results indicate that, percent improvement in total cost is influenced by the number of tasks, cycle time and interaction of those factors. The results of Duncan groupings with the significance level of 0.01 are given in Table 13.

**Table 11** Mean values of *PI* for *NT*, *FR* and *CT*s

| *NT* | *CT* | *FR* 0.25 | 0.50 | 0.75 | Total |
|---|---|---|---|---|---|
| 20 | 180 | 16.122 | 16.122 | 23.024 | 18.423 |
| | 240 | 25.277 | 25.956 | 25.956 | 25.730 |
| | 300 | 37.285 | 37.285 | 37.285 | 37.285 |
| | Mean | 26.228 | 26.454 | 28.755 | 27.146 |
| 40 | 180 | 14.358 | 14.783 | – | 14.464 |
| | 240 | 23.094 | 23.094 | 27.972 | 24.720 |
| | 300 | 11.958 | 12.236 | 14.045 | 12.746 |
| | Mean | 16.470 | 17.253 | 21.008 | 17.957 |
| 60 | 180 | – | – | – | – |
| | 240 | – | – | – | – |
| | 300 | 18.504 | 23.510 | 29.367 | 23.794 |
| | Mean | 18.504 | 23.510 | 29.367 | 23.794 |
| Total | 180 | 15.240 | 15.787 | 23.024 | 17.205 |
| | 240 | 24.186 | 24.525 | 26.964 | 25.225 |
| | 300 | 23.092 | 24.448 | 26.590 | 24.710 |
| | Mean | 21.064 | 22.549 | 26.093 | 23.105 |

**Table 12** Three-way ANOVA results

| Source of variance | Sum of squares | *df* | Mean square | *F* | Sig. of *F* |
|---|---|---|---|---|---|
| *NT* | 837.631 | 2 | 418.816 | 12.741 | 0.000 |
| *FR* | 233.093 | 2 | 116.546 | 3.546 | 0.040 |
| *CT* | 452.714 | 2 | 226.357 | 6.886 | 0.003 |
| $NT \times FR$ | 93.357 | 4 | 23.339 | 0.710 | 0.591 |
| $NT \times CT$ | 1472.646 | 2 | 736.323 | 22.400 | 0.000 |
| $FR \times CT$ | 65.850 | 4 | 16.462 | 0.501 | 0.735 |
| $NT \times FR \times CT$ | 3.847 | 3 | 1.282 | 0.039 | 0.990 |
| Error | 1150.491 | 35 | 32.871 | | |
| Total | 34139.997 | 55 | | | |

**Table 13** Duncan groupings for *NT*, *FR* and *CT*

| Duncan grouping for *NT* | | | | Duncan grouping for *FR* | | | Duncan grouping for *CT* | | | |
|---|---|---|---|---|---|---|---|---|---|---|
| *NT* | *N* | Subset 1 | Subset 2 | *FR* | *N* | Subset 1 | *CT* | *N* | Subset 1 | Subset 2 |
| 40 | 22 | 17.957 | | 0.25 | 20 | 21.064 | 180 | 13 | 17.205 | |
| 60 | 6 | 23.794 | 23.794 | 0.50 | 18 | 22.549 | 300 | 24 | | 24.710 |
| 20 | 27 | | 27.146 | 0.75 | 17 | 26.093 | 240 | 18 | | 25.225 |
| Sig. | | 0.017 | 0.160 | | | 0.016 | | | 1.000 | 0.794 |

Duncan grouping results given in Table 13 indicates that; mean *PI* values for 20-task problems are significantly greater than that of 40-task problems. The difference of mean *PI* values for 60-task problems from that of 20-task and 40-task problems are not significant at the level of 0.01. Similarly, there is no significant difference between the mean *PI* values for different *FR* levels. Additionally, it can be seen in the Table 13 that, mean *PI* values for the cycle time levels of 300 and 240 are significantly greater than that of problems with the cycle time level of 180. The difference between the cycle time levels of 240 and 300 is not significant.

# 5 Conclusion

In this chapter, a binary integer mathematical model is proposed to adapt the resource dependent assembly line balancing approach to the problem of integrated balancing multiple straight assembly lines. The proposed model is validated on an illustrative example and an experimental study is conducted to compare the results of integrated and independent balancing approaches on the multiple straight assembly lines with resource dependent task times. The experimental results show that, when multiple straight assembly lines are balanced in an integrated manner an average total cost improvement of 23.105% can be obtained compared to the independent balances of the lines. This is a significant improvement that emphasizes the importance of balancing multiple straight lines in an integrated manner, to take the advantages of shojinka and to adhere to the main philosophy of JIT. The results also show that, above mentioned total cost improvement is influenced by the number of tasks, cycle time and interaction of those factors.

In further studies, goal programming approaches can also be applied to RDMSLB to provide flexibility for decision makers to balance their assembly lines based on their decision environments and preferred priorities. Efficient heuristics can be proposed for the RDMSLB due to NP-hard nature of the problem. Developing mathematical formulations for integrated balancing of multiple mixed model assembly lines with resource dependent task times can also be considered as a future research.

# Appendix

Precedence diagrams of the Lines 1–4 of the illustrative problem in Sect. 3 are given in Figs. 2, 3, 4 and 5.

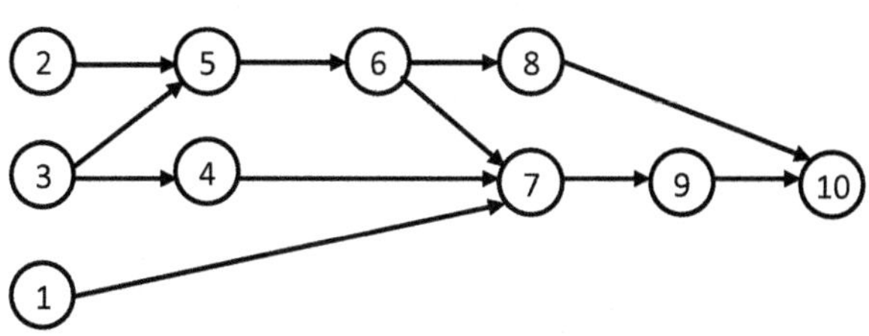


**Fig. 2** Precedence diagram of Line 1

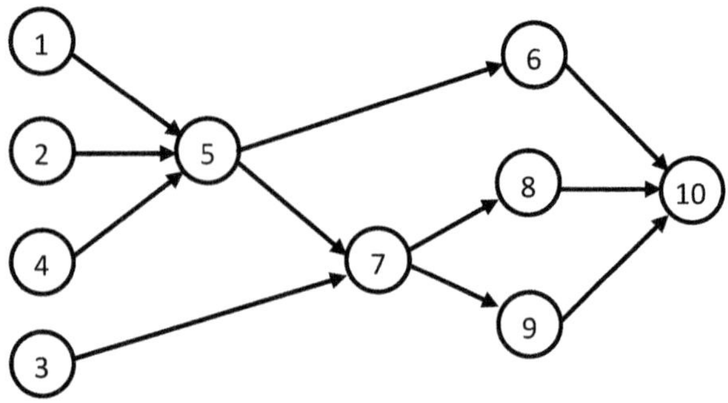


**Fig. 3** Precedence diagram of Line 2

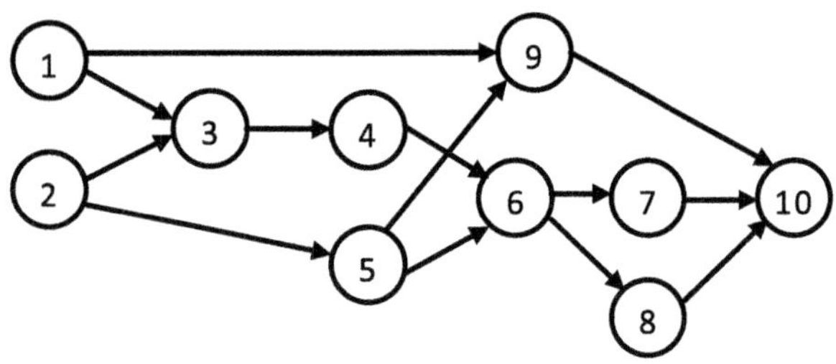


**Fig. 4** Precedence diagram of Line 3

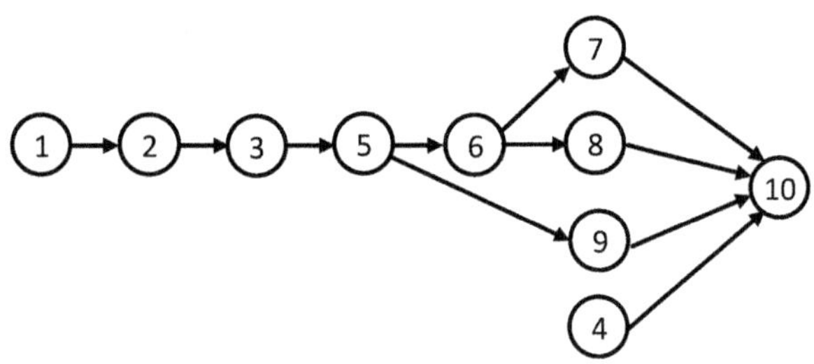


**Fig. 5** Precedence diagram of Line 4

# References


Aase, G. R., Olson, J. R., & Schniederjans, M. J. (2004). U-shaped assembly line layouts and their impact on labor productivity: An experimental study. *European Journal of Operational Research, 156*(3), 698–711.

Atasagun, Y., & Kara, Y. (2016). The multiple U line balancing problem: Recent developments and topics for future research. In *International conference on natural science and engineering (ICNASE'16)* (pp. 1471–1479). Kilis.

Battaïa, O., & Dolgui, A. (2013). A taxonomy of line balancing problems and their solutionapproaches. *International Journal of Production Economics, 142*(2), 259–277.

Baybars, I. (1986). Survey of exact algorithms for the simple assembly line balancing problem. *Management Science, 32*(8), 909–932.

Becker, C., & Scholl, A. (2006). A survey on problems and methods in generalized assembly line balancing. *European Journal of Operational Research, 168*(3), 694–715.

Bukchin, J., & Rubinovitz, J. (2002). A weighted approach for assembly line design with station paralleling and equipment selection. *IIE Transactions, 35*(1), 73–85.

Bukchin, J., & Tzur, M. (2000). Design of flexible assembly line to minimize equipment cost. *IIE Transactions (Institute of Industrial Engineers), 32*(7), 585–598.

Corominas, A., Ferrer, L., & Pastor, R. (2011). Assembly line balancing: General resource-constrained case. *International Journal of Production Research, 49*(12), 3527–3542.
Corominas, A., Pastor, R., & Plans, J. (2008). Balancing assembly line with skilled and unskilled workers. *Omega, 36*(6), 1126–1132.
Erel, E., & Sarin, S. C. (1998). A survey of the assembly line balancing procedures. *Production Planning and Control, 9*(5), 414–434.
Faaland, B. H., Klastorin, T. D., Schmitt, T. G., & Shtub, A. (1992). Assembly line balancing with resource dependent task times. *Decision Sciences, 23*(2), 343–364.
Ghosh, S., & Gagnon, R. J. (1989). A comprehensive literature review and analysis of the design, balancing and scheduling of assembly systems. *International Journal of Production Research, 27*(4), 637–670.
Gökçen, H., Ağpak, K., & Benzer, R. (2006). Balancing of parallel assembly lines. *International Journal of Production Economics, 103*(2), 600–609.
Gökçen, H., Kara, Y., & Atasagun, Y. (2010). Integrated line balancing to attain Shojinka in a multiple straight line facility. *International Journal of Computer Integrated Manufacturing, 23*(5), 402–411.
Jayaswal, S., & Agarwal, P. (2014). Balancing U-shaped assembly lines with resource dependent task times: A simulated annealing approach. *Journal of Manufacturing Systems, 33*(4), 522–534.
Kara, Y., & Atasagun, Y. (2013). Assembly line balancing with resource dependent task times: An application to parallel assembly lines. *IFAC Proceedings Volumes, 46*(9), 845–850.
Kara, Y., Özgüven, C., Yalçin, N., & Atasagun, Y. (2011). Balancing straight and U-shaped assembly lines with resource dependent task times. *International Journal of Production Research, 49*(21), 6387–6405.
Miltenburg, G. J., & Wijngaard, J. (1994). U-line line balancing problem. *Management Science, 40*(10), 1378–1388.
Monden, Y. (1993). *Toyota production system*. Norcross, GA: Engineering and Management Press.
Moon, I., Logendran, R., & Lee, J. (2009). Integrated assembly line balancing with resource restrictions. *International Journal of Production Research, 47*(19), 5525–5541.
Pinto, P. A., Dannenbring, D. G., & Khumawala, B. M. (1983). Assembly line balancing with processing alternatives: An application. *Management Science, 29*(7), 817–830.
Salveson, M. E. (1955). The assembly line balancing problem. *Journal of Industrial Engineering, 6*(3), 18–25.
Scholl, A., & Becker, C. (2006). State-of-the-art exact and heuristic solution procedures for simple assembly line balancing. *European Journal of Operational Research, 168*(3), 666–693.

Printed by Printforce, the Netherlands